What will happen to the near-Earth space environment? How can we ensure the survival of future scientific, commercial and military satellites and space stations? This book addresses the questions that must be asked as man-made debris in space around the Earth – from dust particles to rocket casings, and even radioactive materials – becomes a critical problem.

The solution lies in an international agreement for the preservation of near-Earth space. International specialists have been drawn together to address the issues, problems and policies concerned and to contribute articles to this timely volume. These cover the technical aspects and the economic, legal and international issues concerned with the future uses of space. They include the effective enforcement and monitoring of international agreements and the resolution of disputes.

Clearly written and well illustrated, this timely presentation offers the professionsal and concerned non-specialist an authoritative and comprehensive review of the problems with and solutions to space debris in the near-Earth environment.

Preservation of Near-Earth Space for Future Generations

Preservation of Near-Earth Space for Future Generations

Edited by

John A. Simpson *(University of Chicago, Department of Physics and Enrico Fermi Institute)*

CAMBRIDGE
UNIVERSITY PRESS

CAMBRIDGE UNIVERSITY PRESS
Cambridge, New York, Melbourne, Madrid, Cape Town, Singapore, São Paulo

Cambridge University Press
The Edinburgh Building, Cambridge CB2 2RU, UK

Published in the United States of America by Cambridge University Press, New York

www.cambridge.org
Information on this title: www.cambridge.org/9780521445085

First published 1994
This digitally printed first paperback version 2006

A catalogue record for this publication is available from the British Library

Library of Congress Cataloguing in Publication data

Preservation of near-earth space for future generations / edited by John A. Simpson
   p. cm.
"Based on an interdisciplinary symposium held in honor of the 100th anniversary of the
University of Chicago, 24–26 June 1992" – Pref.
Includes bibliographical references.
ISBN 0 521 44508 6
1. Space debris – Congresses. 2. Space pollution – Congresses.
3. Artificial satellites – Earth – Orbits – Congresses.
I. Simpson, John A. (John Alexander), 1916– . II. University of Chicago.
TL1499.P74   1994
363.73'0919–dc20   93-11837 CIP

ISBN-13 978-0-521-44508-5 hardback
ISBN-10 0-521-44508-6 hardback

ISBN-13 978-0-521-03675-7 paperback
ISBN-10 0-521-03675-5 paperback

Contents

Contributors

Contributors

Philip D. ANZ-MEADOR
Lockheed Engineering & Science
Company
Houston, Texas

Howard A. BAKER
Counsel, Department of Foreign Affairs
Government of Canada

William C. BARTLEY
Senior Adviser for International Affairs
Office of Science, Technology and
Health
U.S. Department of State
Washington, D.C.

Bernard BLOOM
Principal Engineer
Grumman Space Station Integration
Division
Reston, Virginia

Sergei V. CHEKALIN
Central Research Institute of Machine
Building
Russia

Stephen GOROVE
Director of Space Law and Policy
Studies
University of Mississippi Law Center
University of Mississippi

Daniel V. JACOBS
Senior International Policy Specialist
International Relations Division
NASA Headqurters
Washington, D.C.

Nicholas L. JOHNSON
Senior Scientist
Kaman Sciences Corporation
Colorado Springs, Colorado

T. Y. KAWAMURA
McDonnell Douglas Space Systems
Company
Huntington Beach, California

Donald J. KESSLER
Senior Scientist for Orbital Debris
Research
NASA, Johnson Space Center
Houston, Texas

Dr. Heiner KLINKRAD
Missions Analyst
European Space Observations Centre
(ESA/ESOC)
Darmstadt, Germany

Christopher T.W. KUNSTADTER
Senior Vice President
U.S. Aviation Underwriters Inc.
New York, New York

Dr. Winfried LANG
Professor of International Law
University of Vienna
Mission Permanente de L'Autriche
Geneva, Switzerland

Joseph P. LOFTUS, Jr.
Assistant Director
NASA, Johnson Space Center
Houston, Texas

Molly K. MACAULEY
Resources for the Future
Washington, D.C.

Jeffrey MACLURE
Foreign Affairs Officer
Office of Science, Technology and
Health
U.S. Department of State
Washington, D.C.

Jean-Louis MARCÉ
Directeur Adjoint
Centre Natl. d'Etudes Spatiales
France

Pamela L. MEREDITH
President
Space Conform
Washington, D.C.

Joel R. PRIMACK
Professor of Physics
University of California, Santa Cruz
Physics Department
Santa Cruz, California

Qi Yong Liang
Executive Director
Institute of Space Law of the Chinese
Society of Astronautics
Beijing, China

Prof. U.R. RAO
Chairman, Space Commission
Bangalore, India

Albert E. REINHARDT
Capt. USAF
U.S. Air Force Phillips Laboratory
Kirtland AFB, New Mexico

Dietrich REX
Professor
Institute for Space Flight Technology
and Nuclear Reactor Technology
Braunschweig, Germany

Robert REYNOLDS
Lockheed Engineering & Science
Company
Houston, Texas

John A. SIMPSON
Arthur H. Compton Distinguished
Service Professor of Physics, Emeritus
Enrico Fermi Institute and Department
of Physics
University of Chicago
Chicago, Illinois

Susumu TODA
Head
National Aerospace Laboratory
Tokyo, Japan

Paul F. UHLIR
Assistant Executive Director
Commission on Physical Sciences,
Mathematics, and Applications
National Research Council
Washington, D.C.

Dr. Vladimir F. UTKIN
Director
Central Research Institute of Machine
Building
Russia

H. Robert WARREN
Spacecraft Manager
RADARSAT Program Office
Canadian Space Agency
Ottawa, Ontario, Canada

Irvin J. WEBSTER
Director of Engineering Delta Programs
(Retired)
McDonnell Douglas Space Systems
Company
Huntington Beach, California

Diane P. WOOD
Harold J. & Marion F. Green Professor
and Arnold & Frieda Shure Scholar
University of Chicago Law School
Chicago, Illinois

Marie J. YELLE
Spacecraft Mechanical Engineering
RADARSAT Program Office
Canadian Space Agency
Ottawa, Ontario, Canada

PARTICIPANTS IN THE DISCUSSIONS SESSIONS

Thornton R. FISHER
Manager, Space Payloads Department
Lockheed Palo Alto Space Sciences
Laboratory, Palo Alto, California

Herbert GURSKY
Superintendent, Space Science Division
Naval Research Laboratory,
Department of the Navy,
Washington, D.C.

Klaus PAUL
Lehrstuhl fur Raumfahrttechnik
Technische Universität Munchen

Esta ROSENBERG
Office of the Secretary
Commercial Space Transportation
Office, U.S. Dept. of Commerce

Richard TREMAYNE-SMITH
Manager
British National Space Council,
London, England

Richard VONDRAK
Manager
Lockheed Palo Alto Space Sciences
Laboratory (Palo Alto, California)
Lockheed Missles and Space Company,
Inc.

Preface

This volume is based upon an interdisciplinary Symposium held in honor of the 100th Anniversary of the University of Chicago, 24–26 June 1992. The motivation for choosing as the topic the preservation of space was a personal one. The task of preserving the space around Earth for scientific, commercial or governmental spacecraft to survive without collisions with man-made debris or interference from radioactive materials, will become a challenge to all sparefaring nations as we enter the 21st Century.

An advisory group, consisting of Don Kessler, Paul Uhlir and Diane Wood, was formed in 1990 to determine the scope of the proposed Symposium and to assist in the selection of contributors. The feasibility study was supported by the Midwest Center of the American Academy of Arts and Sciences. It became clear that this was, indeed, a pending crisis which could only by attacked by all spacefaring nations working together.

The individual contributions to this volume were based upon commissioned papers made available to all the speakers before the Symposium. Edited versions of two discussion sessions were included since they raised some issues or presented points of view beyond those in the commissioned papers. Contributors from both government and private institutions participated; the views they have expressed are their own and do not necessarily represent those of their respective organizations or nations.

The Symposium was supported in all aspects by a grant from the John D. and Catherine T. MacArthur Foundation and co-sponsored by the University of Chicago and the American Academy of Arts and Sciences. The Symposium chairman and editor is indebted to Marian Rice, executive associate of the Midwest Center for the American Academy of Arts and Sciences, and her staff for their extraordinary effort in implementing the Symposium's arrangements. Viktoria Tripolskaya-Mitlyng was translator for the discussions in Russian and English. David Chaumette, Bruce McKibben and Yvette McLean provided assistance with the transcriptions of the discussions and with the collection of the papers. The Department of Physics and the Enrico Fermi Institute were hosts for the Symposium.

Finally, this volume would not have been possible without the enthusiastic support of all contributors to the Symposium.

J. A. S.

I. Introduction

1: Introduction

J. A. Simpson

In recent years there has been a growing recognition of mankind's activities affecting the future environmental health of our planet. Because of the complexity of the issues and the uncertainty of the consequences of various courses of action, however, many of these potential environmental crises for civilization unfortunately are being neglected. Though it is not yet a critical problem today, we know that the space-faring nations are introducing man-made debris (extending in size from dust particles to rocket casings) in ever increasing quantities into the space around Earth. Within the next decade or two the almost exponential increases in the amounts of these materials will present serious hazards for the survival of spacecraft, space stations and astronaughts occupying near-Earth orbits. Radiation from radioactive materials and particles will gradually close important windows for astronomical observations. In contrast with the efforts to solve some of our environmental problems which benefit some nations but not others (e.g., reduction in use of fossil fuels), in the case of the preservation of space *all* nations are beneficiaries of a solution – there will be no loser nations now or in the future, whether or not they are active in space. This factor will be important in negotiating any international agreement for the control of orbital debris. At a time when all nations perceive that preservation of space is in their own best interest, it is important for those most concerned – those nations with active space programs – to take steps toward an international agreement.

Personally, I find encouragement from the legacy of the International Geophysical Year for attempting to develop an international agreement on near-Earth space. As one of the twelve organizers for science programs (cosmic physics and space sciences) for the sixty-eight nations of the I.G.Y. (1957–1958) I found, as did others, that the preservation of our last continent, Antarctica, was high on the list of IGY's achievements. The first Antarctic Treaty was completed in 1959. Its recent renewal was a reaffirmation of the overriding concerns of nations to avoid damaging exploitation of the continent. Another more restricted example which avoids damaging exploitation is the third United Nations Convention on the Law of the Sea (1982).

With these successes in mind, I became convinced that an effective international agreement could be achieved for the near-Earth space and that we should strive for a model of an international agreement or treaty.

It is vital to recognize that any effective effort in this direction must include both civil and military participation. It also should be interdisciplinary so as to consider not only the technical aspects, but also, economic factors, legal issues, and international cooperation for future civil and military uses of space. With these requirements as guidelines, we arranged the Symposium to include all current space-faring nations. With respect to the 13-nation European Space Agency (ESA) we invited representatives covering the various points of view of the member nations. This volume contains papers by individuals from most of the nations currently active in space.

Beginning with defining the problem and projections for the future, the series of papers review the impact of space debris and radioactive material in space for future human exploration, for the space sciences (mainly x-ray and γ-ray astronomy), for commercial applications, and for military uses of space.

These reports are followed by evaluations of the techniques and practices for mitigation of and adaptation to the space environment. The diverse methods and policies of France, China, India, Russia, Japan, Canada, the European Space Agency and the National Aeronautics and Space Administration are discussed.

The solutions for the preservation of near-Earth space environment must be cost-effective over the long term and within the capabilities of the poorer space-faring nations. Reports on cost *vs*. benefits, and economic and insurance incentives were therefore presented.

The international legal issues concerned with establishing a framework for a world-wide solution to the space problem – within the rights of individual nations – have precedents, for example, in the Antarctic Treaty and more recently for the control of atmospheric pollution (e.g., the ozone layer). Lessons learned from environmental treaty making were discussed.

The regulation of orbital debris, including the effects from nuclear reactors in space, is discussed within the framework of policy choices – namely, multinational, national, or laissez faire concepts.

What should be the elements of an international agreement ? A framework treaty is debated for further consideration at later conferences. These ideas are offered within the concept of a "global commons". The Symposium recognized the importance of enforcement and monitoring the effectiveness of an international agreement and of resolving disputes.

The final discussion sessions of the Symposium contained questions and views which will be important in any future effort to establish an international agreement.

As Chairman, I sensed in closing this Symposium that several overall conclusions and questions for the future were widely accepted by the participants. They should be noted here. Even though there are some bilateral agreements to study the debris problem, we must include all space-faring nations in the initial formulation in order to achieve the objectives brought forth. The Symposium should be only the beginning of a series of interdisciplinary workshops and efforts over the next few years, with the goal of completing a model framework convention for space-faring nations within a five-year period. There was a general consensus that we should keep the United Nations informed, but not rely on the United Nations as the principal forum for successful negotiations.

The need for additional measurements of space debris was pointed out by several participants who noted that NASA and the Department of Defence were already working to develop policies and actions for minimizing debris. However, it was repeatedly emphasized that the call for more measurements and modeling should not inhibit forward movement towards an international agreement. Too often governments use a ploy of more and more study, instead of action.

If we establish a blueprint or framework convention, will governments use it as a guide? Recommendations were made to increase attendance by representatives of governments in future symposia. For example, it was suggested that within the United States we invite knowledgeable staff members of both the House and Senate for future deliberations.

The prospects for international cooperation are reviewed from different points of view, including current initiatives within the United Nations.

Least understood was how to address the question of identifying a suitable international agency capable of embracing and carrying forward an international agreement based on the principles established by the interdisciplinary symposium. The discussion of these, and related issues, is being planned for a workshop. We hope that this volume will be sufficiently provocative to stimulate further dialogue and action by those who recognize the problem.

In writing about his most recent trip to the Antarctic, Walter Sullivan (*New York Times Magazine*, 1 November 1992) noted:

> Because of international provisions against degradation of Antarctica's pristine beauty, all elements of the camp had to be removed in June when the Akademik Fedorov and the Nathaniel B. Palmer [the Russian and American ice breakers] reached the flow as it drifted toward the South Atlantic. Tin cans were stamped flat and bottles saved. So were garbage and non-burnable waste.

Can we establish equally effective international provisions against the degradation of near-Earth space?

II. Defining the Problem

2: The Earth Satellite Population: Official Growth and Constituents

Nicholas L. Johnson

Senior Scientist, Kaman Sciences Corporation, Colorado Springs, Colorado

ABSTRACT

In nearly 35 years of space activities more than 3,400 missions have reached Earth orbit or beyond. The consequences of this activity have left in excess of 7,000 satellites in near-Earth space which are trackable by terrestrial sensors. From this population only approximately 5% represent spacecraft which continue to provide useful services. A small fraction (<1%) of cataloged satellites possess radioactive materials which raise special long-term issues. The actual number of man-made objects orbiting the Earth, including debris as small as 1 cm in diameter, is estimated to be several times the official count. The spatial density of the near-Earth environment is highly non-uniform with distinct regions of elevated satellite concentrations in both low Earth orbit and in the geostationary ring. The greatest influences on the growth of the Earth satellite population are launch and space operations, unplanned satellite fragmentations, and solar activity.

HISTORICAL LAUNCH ACTIVITY

As of 1 April 1992, a total of 3,415 space missions to Earth orbit or beyond had been registered since the flight of Sputnik 1 on 4 October 1957. This figure does not include the numerous launch attempts which succumbed prior to reaching orbital velocity. However, contrary to popular opinion, no significant increase in space launch activity has occurred since the mid-1960's. By 1965 the world launch rate exceeded 100 missions per year, a level maintained (between 101 and 129) until 1991 when only 88 space missions successfully attained Earth orbit (Figure 1).

To date approximately 22,000 individual satellites have been identified and cataloged by the United States. A large number of additional satellites have also been detected and tracked but have not been cataloged due to their short orbital lives. For the past 15 years an average of 800 new satellites belonging to all space-faring nations have been added to the official American registry annually. On the other hand, the actual Earth satellite population, as well as its growth rate, has been much less dramatic (Figure 2). The lower curve in Figure 2 has been adjusted to reflect the true population status at the end

of each year, thereby correcting for cataloging delays. As an example, debris from a 1961 satellite explosion were still being cataloged nearly 30 years later.

The current population of about 7,000 objects has been increasing at a net average rate of less than 150 satellites per year for the same 15-year period (down from 250 satellites per year for the 1957–1976 period). The difference between this rate and the cataloging rate is the result of objects returning to Earth via natural orbital decay, deorbit, or retrieval and, to a lesser extent, vehicles leaving Earth orbit for missions to other members of the solar system. In addition to the official cataloged population, on average several hundred objects are being tracked by U.S. sensors while awaiting firm identification and cataloging. Hence, to the official Earth satellite population figure of 6,829 provided by U.S. Space Command as of 31 January 1992, approximately 300 satellites must be added to account for all trackable objects known to be in Earth orbit.

Space launch facilities are now operational in eight nations (China, French Guiana, India, Israel, Japan, Kazakhstan, Russia, and the United States) and on the Italian San Marco platform off the east coast of Kenya. Although some additional States have indicated a desire to launch space vehicles, e.g. Brazil, no new major facilities are expected in this decade. Historically, the Soviet Union has been responsible for 68% of all missions, followed by the United States with only 28% (Figure 3). For the past 10 years the non-U.S./non-U.S.S.R. contribution has averaged about 8 missions per year or less than 7%. The make-up of world-wide space missions in 1991 was Soviet Union – 67%, United States – 20%, European Space Agency – 9%, Japan – 2%, and China – 1%.

An examination of the national origin of the current satellite population reveals a different partitioning with approximately equal numbers for the United States and the Soviet Union and an aggregate percentage for the other space-faring nations slightly larger than the respective launch percentage (Figure 4). This difference arises from variations in technology and mission needs and from the occurrence of unplanned satellite fragmentations. The majority (about three-fourths) of Earth satellites reside below 2,000 km and represent a total mass on the order of 2,000 metric tons.

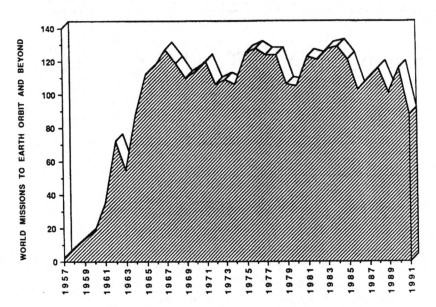

Figure 1: World space launch rate, 1957–1991.

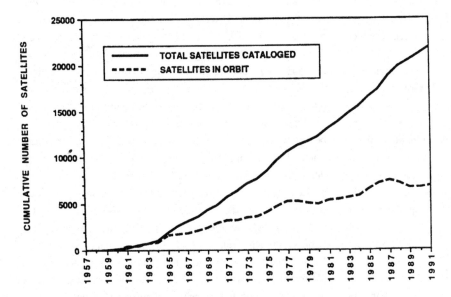

Figure 2: Cataloged satellite population history.

With the recent significant reduction in space launches from the Soviet Union (from 95 in 1987 to 59 in 1991) and based upon a projection of activities for the Commonwealth of Independent States, the global space launch rate for the remainder of this decade should remain below 100 missions per year. However, as shown in the following sections of this paper, the space launch rate may not be the most accurate measure for predicting the state of the Earth satellite population at the turn of the century.

INTERNATIONAL REGISTRATION OF SATELLITES

Within four years of the launch of the first artificial Earth satellite, the United Nations adopted Resolution 1721 (XVI) which called upon States launching objects into Earth orbit or beyond to register pertinent information on each mission with the UN. By 1974 the *Convention on Registration of Objects Launched into Outer Space* was adopted by the General Assembly and subsequently opened for signature on 14 January 1975 and entered into force on 15 September 1976.

Article IV of the Convention directs each State to provide the following information concerning each space object for which it is responsible:

"(a) Name of launching State or States;

(b) An appropriate designator of the space object or its registration number;

(c) Date and territory or location of launch;

(d) Basic orbital parameters, including:

    (i) Nodal period,

    (ii) Inclination,

    (iii) Apogee,

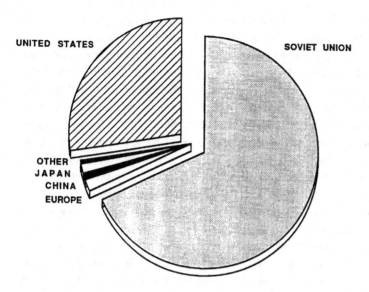

Figure 3: World space missions, 1957–1991.

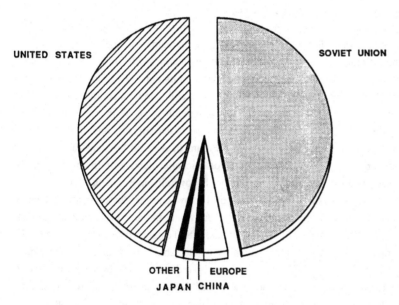

Figure 4: State of origin for cataloged satellites, 1 January 1992.

(iv) Perigee;

(e) General function of the space object."

A mandatory review of compliance with the Convention, conducted 10 years after its entry into force, revealed that most space missions were officially registered with the UN within 1–3 months after launch, although delays of a year or more had occurred. Despite the specific definition of "space object" in Article I (to include "component parts of a space object as well as its launch vehicle and parts thereof"), most States provided information only on payloads. Moreover, each State furnished information in its own format, sometimes with more or less data than cited by Article IV. In almost all cases, the description of the "general function" of the satellite was so broad as to be of little value. Interestingly, common names, e.g., Hubble Space Telescope or Mir Space Station, are neither requested nor provided.

In addition to omitting a very large number of objects in Earth orbit, the UN Registry is not a satellite catalog in the accepted sense. The current orbits of satellites are not maintained, although notice of the orbital decay of registered objects is requested. In fact, orbital parameters for some registered objects reflect transfer orbits which are used only temporarily. No attempt is made to determine the location and orbital parameters of any object at a specific time or to verify the identity of objects cataloged by State members. The latter is of significant importance in ascribing State responsibility in the event that a space object causes damage to citizens or property of another State. In summary, therefore, the UN Registry serves little practical or legal purpose.

The source of virtually all open-literature data characterizing the near-Earth environment is the United States Space Command's Space Surveillance Center (SSC), which manages and processes information collected by the world-wide Space Surveillance Network (SSN). Presently, this capable system is primarily comprised of 18 radar (9 phased-array, 8 mechanical trackers, and 1 electronic fence) and four electro-optical facilities located in North America, Greenland, Europe, and Asia, as well as in the Atlantic, Pacific, and Indian Oceans. However, only a few of these sensors are dedicated to the space surveillance mission. Most facilities serve in a collateral or contributing role while performing other important functions.

The mission of the SSC and the SSN is to detect, track, catalog, and identify all man-made objects in space. In practice, objects on the order of 10 cm or larger can normally be tracked at low altitudes, but the system sensitivity decreases to about 1 meter at geostationary altitudes. On average, 30,000–50,000 observations made by SSN sensors are transmitted daily to the Cheyenne Mountain AFB complex, within which the SSC is situated.

The SSC processes the incoming observations, updating the orbital parameters of previously known objects and examining uncorrelated observations which might represent new satellites. New objects, e.g., recently launched or released from a resident satellite, are officially cataloged only after reliable tracking has been established and an identity can be assigned. Some objects have been routinely tracked for months or years before being cataloged.

Two primary satellite databases are maintained by the SSC: the Satellite File and the *Satellite Catalog*. The Satellite File contains the most current, complete orbital parameters for each Earth satellite still in orbit, while the *Satellite Catalog* is an historical record with limited orbital data of all satellites cataloged since 1957. The *Satellite Catalog* also provides information on the type of satellite, country of origin, launch date, launch site, date of reentry (if applicable), and average radar cross-section. Data from the Satellite File and the *Satellite Catalog* are furnished to the Goddard Space Flight Center in Maryland for world-wide distribution in the form of *Two-Line Element Sets* and the *Satellite Situation Report*, respectively. Major international satellite databases, such as Great Britain's *Table of Earth Satellites* prepared by the Royal Aerospace Establishment and the European Space Agency's *DISCOS* (Database and Information System Characterizing Objects in Space) rely on these Goddard publications for almost all of their satellite orbital parameter inputs.

A frequently used tabular summary of the *Satellite Catalog* or the *Satellite Situation Report* is the "Space Objects Box Score" (Box Score for short), which can appear in a variety of formats. The Box Score normally lists by country of origin the total number of satellites which are still in orbit and which have decayed. For example, the Box Score was used to prepare Figure 4.

CHARACTERIZATION OF THE EARTH SATELLITE POPULATION

In addition to a description of a satellite by its country of origin

and its orbital parameters, a complete characterization of the Earth satellite population requires an understanding of the general nature of each object. The three basic categories of satellites are payloads, rocket bodies, and debris. However, to assess more fully the constituents of the Earth satellite population and the role they play in the evolution of the near-Earth environment, each satellite may be classified as one of five types: active payload, inactive payload, rocket body, operational debris, or fragmentation debris (see Glossary). Figure 5 illustrates the associated make-up of the Earth satellite population in 1991.

For thirty years the largest component, often accounting for one-half the entire population, has been fragmentation debris. When the first satellite breakup occurred in June, 1961, the total Earth satellite population instantaneously increased by more than 400%. During the ensuing 30 years, 108 satellite breakups have been detected, representing one-third of all satellites cataloged. Today, fragmentation debris, virtually all in low Earth orbits, makeup approximately 40% of the full cataloged satellite population.

A variety of causes are responsible for this phenomenon. The five general categories of breakups and their relative importance are deliberate fragmentations (40%), propulsion-related breakups (32%), battery malfunctions (4%), command system failure (1%), and accidental collision with another object (1%). The reasons for the remaining 22% of the known satellite breakups have not yet been determined, and thus they are labeled "unknown" (Figure 6). Not all breakups are equal in magnitude (number of trackable debris) or in the long-term hazard they pose for the overall environment. For example, since 1964 a total of 13 low altitude Soviet spacecraft have been destroyed in orbit when failures prevented controlled reentries. Although nearly 850 cataloged debris (and thousands more debris which were detectable but not cataloged) were produced, all were in short-lived orbits and none remain in orbit today. In fact, more than half of all satellite breakup debris clouds have been reduced to no more than one fragment.

A significant satellite breakup is defined as one which produces at least 10 cataloged debris which in turn remain in orbit for at least 10 years. Under this definition, which attempts to distinguish between those events which have a transitory or marginal effect on the environment and those which represent a longer term hazard to other spacecraft, 31 significant satellite breakups have occurred to date, involving 13 payloads and 18 rocket bodies (Figure 7). Perhaps more importantly, just 10 satellite breakups now account for approximately 25% of the entire cataloged satellite population. Also noteworthy is the fact that nine of the ten were rocket bodies which had performed as designed but were left in orbit after completing the assigned payload delivery task.

Another issue associated with the breakup of a satellite is the region over which the debris are spread. Some events are more energetic – both in the number of debris created and in the velocities imparted – than others. In the case of the Nimbus 6 rocket body which fragmented on 1 May 1991 after nearly

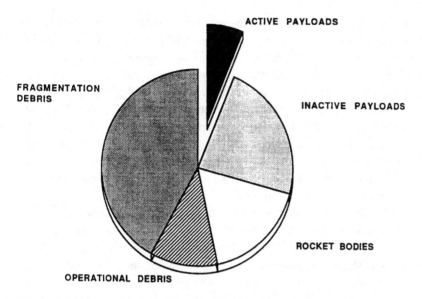

Figure 5: Make-up of cataloged Earth satellite population, 1991.

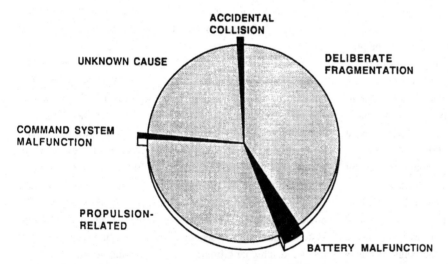

Figure 6: Relative proportions of satellite breakups by cause.

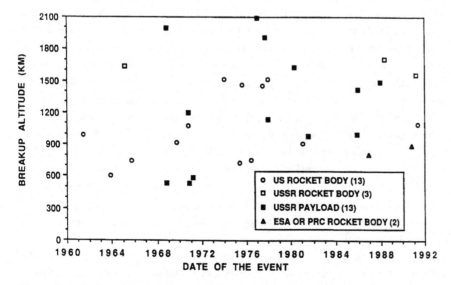

Figure 7: Altitude and date of significant satellite breakups.

16 years in orbit, almost 400 debris were initially tracked with perigees as low as 300 km and apogees as high as nearly 4,000 km (Figure 8). More than 80% of the entire Earth satellite population now penetrates the expanse of this single debris cloud. Debris from less energetic events are also important since they will pass through many altitudes as they decay.

In addition to the sometimes hundreds of trackable debris produced in a satellite fragmentation, thousands or tens of thousands of smaller particles are created, and these unseen projectiles possess sufficient energy to destroy other satellites should they collide. As the total number of Earth satellites increases, so, too, does the probability of accidental collisions. A major satellite breakup in 1981 in the most densely populated region of near-Earth space is now assessed to have been caused by an impact of a small non-tracked satellite, possibly debris from a previous fragmentation. Hence, debris can produce new fragmentations which in turn create large numbers of additional debris in a potentially cascading sequence. Ultimately, the creation of new debris in this manner can offset the rate of satellite loss via natural decay, leading to uncontrolled population increases.

Besides satellite breakups, fragmentation debris can be produced in what are generally referred to as "anomalous events". These incidents normally involve a small number of debris which are released with low velocities relative to the parent satellite. Many debris decay very rapidly, even from great heights. For example, eighteen of nineteen debris released from five spacecraft near altitudes of 1,000 – 1,100 km reentered the atmosphere within 12 years although the parent satellites may remain in orbit for as long as 1,000 years.

Anomalous event debris are often believed to be related to satellite deterioration under the harsh space environment and appear to represent objects with high area-to-mass ratios (as judged by their decay rates), e.g., thermal coverings. If this is the case, a much larger number of objects are probably produced than detected. The majority of objects identified with anomalous events belong to two specific types of satellites: U.S. Transit-class payloads and Soviet SL-3 rocket bodies. Also interestingly, all but a very few events have occurred during the last two periods of solar maximum and after the parent satellite had been in orbit more than 15 years. Particularly in the case of Transit-class payloads, multiple anomalous events have been associated with a single parent satellite.

Although not a separate category of satellites, those vehicles which carry radioactive materials have engendered special interest among various groups. Radioactive fuels have been employed on Earth satellites to produce electrical power on relatively few missions dating back to 1961. Today 54 cataloged Earth satellites from 42 flights represent nuclear reactors, nuclear fuel cores, or radioisotope thermoelectric generators (RTGs). All but two of these reside in low Earth orbits between 700 and 1,500 km (Figure 9) and carry an aggregate of approximately 900 kg of radioactive fuels, predominantly enriched uranium.

The principal concern regarding nuclear materials in space is the potential hazard presented during premature or uncontrolled reentries. The issue was brought to the forefront in 1978 when the Kosmos 954 spacecraft, which carried a nuclear reactor with an initial fuel loading of approximately 30 kg of enriched uranium, malfunctioned and reentered the atmosphere, impacting primarily in the Canadian tundra. A United Nations working group was formed shortly after this event to review operational practices of nuclear power supplies in Earth orbit and to develop safety measure guidelines for future missions. A comprehensive technical report was completed in 1981 and appropriate recommendations were endorsed later that year by the United Nation's Committee on the Peaceful Uses of Outer Space and the General Assembly.

The working group's recommendations covered a wide spectrum of issues, including operational altitudes, end-of-life storage practices, orbital lifetime predictions, vehicle integrity during reentry, and emergency safety features. The working group explicitly refrained from recommendations which limited the employment of nuclear power sources (NPS), insisting instead that "the basis of the decision to use NPS should be technical."

A redesign of the Soviet nuclear reactor of the Kosmos 954-type adopted several new safety systems. When Kosmos 1402 failed to move from its low operating orbit (~250 km) at the end of its mission in late 1982, the fuel core was separated from the reactor to ensure more complete burn-up of the radioactive materials during reentry. As a consequence, no hazardous materials were found on Earth after the return of either the reactor or the fuel core. A command system malfunction prevented the planned transfer of the Kosmos 1900 nuclear reactor to a safe storage orbit, but again new safety systems automatically separated the reactor and maneuvered it into a much higher orbit before reentry could occur.

A second concern involving the use and disposal of nuclear power supplies in Earth orbit is the possibility that such satellites might collide with other particles, creating numerous radioactive debris. This particular topic was the subject of discussions at the United Nations in 1990. The majority of nuclear reactor power supplies are currently stored near an altitude of 1,000 km, which also typically corresponds to the region of greatest spatial density and greatest probability of collision. This altitude was chosen to ensure that natural decay would not occur for several hundred years by which time the level of radioactivity on-board, particularly of fission products, would have substantially diminished. At the present, an accidental collision remains exceptionally unlikely. Moreover, although radioactive debris created in such a collision would be expected to decay more quickly, these debris would also be less likely to survive reentry, disintegrating high in the upper atmosphere.

A third issue revolves around the interference to scientific instruments on-board other satellites or high altitude balloons caused by satellites with operating nuclear reactors. In particu-

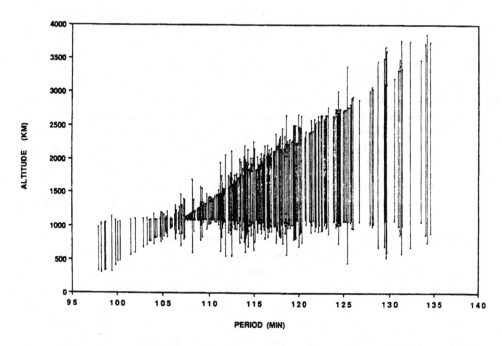

Figure 8: Altitude regimes of Nimbus 6 rocket body debris.

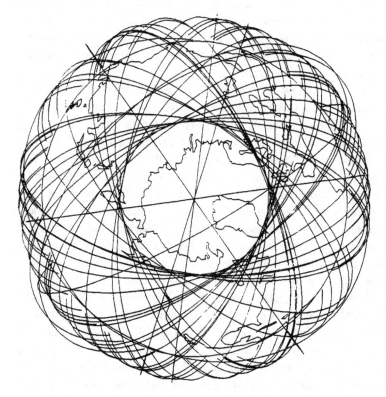

Figure 9: Orbits of low altitude satellites with radioactive materials.

lar, gamma radiation detectors are susceptible, both directly and indirectly, to insufficiently shielded nuclear reactors. No nuclear reactors have operated in Earth orbit since 1988, and reactors which have been shut-down and placed in storage orbits do not emit the offending radiation or particles. Proper design of future nuclear power supplies can prevent or minimize this interference.

POPULATION GROWTH FACTORS

As indicated previously, the average net growth rate of the Earth satellite population during the past 15 years has been approximately 150 objects per year. The future growth rate is dependent on many factors, but the most influential are (1) launch rate, (2) operational practices, (3) mission requirements, (4) satellite fragmentation rate, (5) natural decay and solar activity, and (6) deorbits.

The actual launch rate is not necessarily a good indicator of satellite growth rate; however, historically, the launch rate between 1965 and 1990 was essentially constant, and the satellite growth rate was roughly linear. Simplistically, the launch rate projection for the 1990's would suggest that the future growth rate should not exceed the historical rate. Admittedly, predicted launch rates are potentially susceptible to economic conditions and to technological advances. The latter could lead either to higher launch rates (demand of new services) or lower launch rates (improved longevity and capacity of payloads). The other growth rate factors, which are even more volatile, could easily overwhelm the importance of the launch rate.

Operational practices include the intentional release of objects during orbital deployment and mission operations. In recent years many launch vehicle and spacecraft manufacturers have moved toward minimizing such objects by redesigning the use of explosive bolts, employing bolt catchers, restraining some discarded elements with lanyards, and a variety of other techniques. Modifying launch profiles and taking advantage of natural gravitational perturbations could accelerate the decay of unwanted objects, e.g., rocket bodies, thereby, mitigating the overall satellite population growth. Although numerous debris have historically been released from manned space stations, such actions have no significant impact on the overall population since they normally reenter the Earth's atmosphere within six months.

Mission requirements, in part, refer to the issues which help to determine the selection of the payload orbit. Today, the vast majority of cataloged satellites reside in orbits with periods less than 225 minutes (the official U.S. Space Command demarcation between "low" and "high" Earth orbits). However, as Figure 10 indicates, the percentage of annual missions to high Earth orbit has increased significantly in the past 30 years. Normally, these objects are much longer lived and hence influence the long-term satellite population growth. The recent renewed interest in large constellations of small, low altitude satellites, particularly for communications, could retard or even reverse this trend while simultaneously aggravating the low altitude population density situation.

Clearly, the future rate of satellite fragmentations could be the dominant factor in the extent to which the near-Earth environment deteriorates. The historical record is not encouraging. The rate of satellite breakups has been steadily increasing: 2.2 per year during 1961–1971, 3.8 per year during 1972–1981, and 4.5 per year during 1982–1991 (Figure 11). However, the majority of these events can be prevented, either directly by reducing or eliminating deliberate fragmentations or indirectly by purging residual propellants from rocket bodies after their mission has been accomplished. If deliberate fragmentations are necessary for scientific or national security needs, careful orbit selection can greatly reduce the amount of debris left in orbit

and can accelerate the natural decay of said debris. Some spacecraft design changes, e.g., in battery construction, could further reduce the possibility of unintentional satellite breakups.

Although we have less control over the occurrence of natural collisions, prudent actions can reduce the probability of collision or the effects of a collision if one occurs. Spacecraft design and orbit selection can help in this regard. By placing future space systems in regions of lower spatial density, the threat of an accidental collision with another Earth satellite can be minimized. Figure 12 indicates the relatively chaotic motion of satellites in the rather narrow mean altitude region of 550 km to 600 km, which is typical of low Earth orbit regimes. Satellites in more elliptical orbits (not shown) also penetrate this regime during each circuit about the globe. The spatial density is worse at higher altitudes, e.g., 950–1,000 km, where the satellite concentration is four times greater.

Whereas new launches, mission operations, and satellite fragmentations constitute the primary sources for the Earth satellite population, natural decay, deorbit, and retrieval are the principal sinks for the environment. Natural decay is usually affected by atmospheric drag, gravitational perturbations, or a combination of the two. For satellites in orbits with mean altitudes below 2,000 km, atmospheric drag is the predominant factor. In highly elliptical orbits with much higher apogees, solar-lunar effects can be the principal factor determining orbital lifetime.

Above 600 km normal payloads and rocket bodies decay very slowly, typically over the course of tens, hundreds, or even thousands of years. Fortunately, the nature of operational and fragmentation debris often results in a much more rapid fall back to Earth. For example, the SPOT 1 rocket body was left at an altitude of more than 800 km, from which normal decay could be expected to take 100 years or more. However, within four and one-half years of its explosion, approximately 85% of the nearly 500 cataloged fragments had already reentered the atmosphere. Numerous debris from the more recent Nimbus 6 rocket body have completely decayed from mean altitudes above 1,100 km in less than one year.

A significant influence on this process is the Sun. Solar activity, as measured by the number of sunspots, major flares, or radio emissions, varies in a roughly sinusoidal pattern with an average period of 11 years. During periods of maximum solar activity, more energy is absorbed by the Earth's atmosphere, causing it to expand and consequently increasing atmospheric density levels at higher altitudes. The most pronounced changes in orbital lifetimes resulting from this effect can be found with satellites in altitudes between 500 and 1,000 km.

Perhaps the best known examples of this phenomenon are the Skylab and Salyut 7 space stations which fell back to Earth more rapidly than anticipated due to solar maximums in 1979–1980 and 1989–1990, respectively. A re-examination of Figure 2 reveals actual declines in the Earth satellite population during these last two periods of high solar activity. The

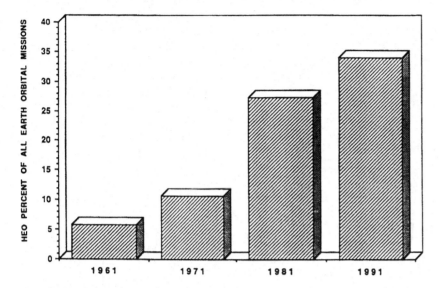

Figure 10: Increase in the use of high Earth orbits.

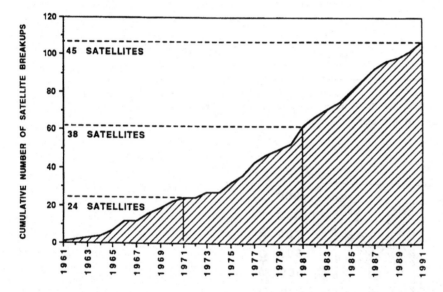

Figure 11: History of known satellite breakups.

effect would have been more pronounced had it not been for the unfortunately high rate of severe fragmentations. Thus, if future fragmentations can be curbed or controlled, periodic solar fluctuations will considerably help to purge the low altitude environment, particularly in the high density regions below 1,000 km (Figure 13). Furthermore, such increases in atmospheric density appear to have preferential effects on small debris – presumably including the very small debris which is not now trackable – due to their characteristically higher area-to-mass ratios.

Even an apparently static region may actually be in a highly dynamic state of flux. By a simple comparison of the *Satellite Catalogs* of 1 January 1991 and 1 July 1991, the satellite population of the mean altitude region between 500 km and 600 km gained two objects, one of which was newly cataloged during the period. However, a more careful investigation reveals that a

full third of the population of 1 January 1991 fell out of the regime and was replaced by satellites decaying naturally from above (Figure 14). Below 500 km, satellite lifetimes without orbital maneuvers are usually only a few years or less. Hence, if the creation of new objects can be better regulated, environmental forces can substantially aid in cleansing near-Earth space.

Similar analyses can be performed using total annual mass of spacecraft and rocket bodies placed into Earth orbit to indicate trends and sensitivities. During 1980, 24 space missions deposited a total mass of 60 metric tons into orbits between 600 and 2,000 km. At the end of the decade, this regime witnessed only 16 missions with 43 metric tons as payload longevity increased and emphasis shifted to higher orbits. On a broader scale, during 1984 a record 129 space missions were flown, placing nearly 1,300 metric tons into Earth orbits of all

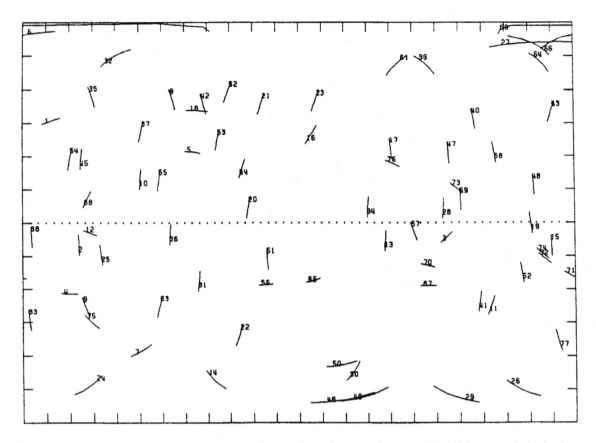

Figure 12: Relative position and motion of cataloged satellites with mean altitudes of 550–600 km. Arcs depict 2-minute tracks in March, 1992. Satellite ID number is located near start of track.

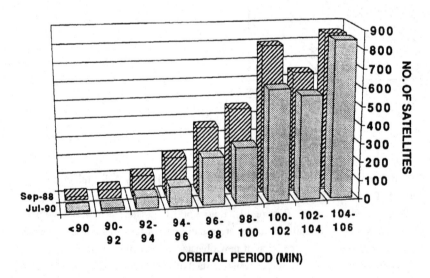

Figure 13: Decrease of cataloged satellite population during solar max.

altitudes. However, natural decay and deorbits reduced this amount to approximately 180 metric tons within five years.

Deorbiting satellites at the end of their missions is normally done only for vehicles at altitudes below 500 km due, in part, to the energy requirements. As noted above, though, the natural lifetimes of these objects are typically very short, and hence deorbiting has virtually no impact on the long-term environment. Future requirements for deorbiting (or greatly lowering the perigee of) old rocket bodies and payloads from intermediate orbits between 500 and 1,000 km could have a profound influence on the growth of the Earth satellite population. Retrieval has been used on only a few occasions to date – also at low altitudes – and likewise is insignificant to the overall population level.

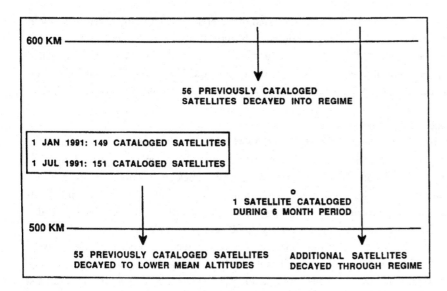

Figure 14: Dynamic state of 500–600 km mean altitude regime.

SUMMARY

During the 1980's our understanding of the near-Earth space and our awareness of its potential impact on future space operations have increased markedly. Although, as in most scientific endeavors, our knowledge remains incomplete and in some areas less precise than desired, sufficient data are available to begin the formulation of policy and regulations which can better govern the evolution of this exceedingly important and potentially fragile environment. However, without the participation of all space-faring nations and organizations, these preventive and remedial actions may be ineffective.

GLOSSARY

ANOMALOUS EVENT – unplanned separation, usually at low velocity, of one or more objects from a satellite which remains essentially intact. Anomalous events can be caused by material deterioration of items such as thermal blankets, protective shields, or solar panels. Most anomalous events are believed to be the result of harsh environmental factors, e.g., radiation, thermal cycling.

DECAY – natural loss of altitude of a satellite culminating in reentry into the Earth's atmosphere. At low altitudes the rate of decay is largely determined by atmospheric density and the satellite's area-to-mass ratio, but for satellites in highly elliptical orbits the rate of decay is usually driven by solar-lunar gravitational forces.

DEORBIT – the deliberate, forced reentry of a satellite into the Earth's atmosphere by applying an impulsive force, usually via a propulsion system. Deorbited satellites may be recovered intact, e.g., in the case of manned spacecraft or spacecraft returning scientific materials, or may be intentionally brought down over broad ocean areas as in the case of non-operational space stations.

FRAGMENTATION DEBRIS – objects which are generated when a satellite structure comes apart. Fragmentations are normally classified as satellite breakups or anomalous events (see respective glossary entries). Fragmentation debris do not include objects which are released from a spacecraft in a controlled manner, e.g., refuse ejected from a space station, or are part of the normal orbital injection process, e.g., debris created at the time of payload release from its launch vehicle. See operational debris.

GEOSTATIONARY ORBIT – nearly circular orbit with a period of approximately 1436 minutes and an inclination close to zero degrees. In such an orbit, the satellite maintains a relatively stable position with respect to a point directly above the equator. The mean altitude of geostationary satellites is approximately 35,785 km. In practice "geostationary" satellites exhibit small orbital eccentricities and slight inclinations, resulting in an apparent wobble about a fixed location.

GEOSYNCHRONOUS ORBIT – nearly circular orbit with a period of approximately 1436 minutes with any inclination. Inclined geosynchronous satellites follow a figure-eight shaped pattern, completing a full circuit once a day with the center of the figure-eight fixed directly above the equator at an altitude of 35,785 km.

HIGH EARTH ORBIT – any satellite orbit about the Earth with an orbital period equal to or in excess of 225 minutes (mean altitude greater than 5,875 km). This definition was established by NORAD and is still in use today by the U.S. Space Surveillance Network for tasking and other operational purposes.

LOW EARTH ORBIT – any satellite orbit about the Earth with an orbital period less than 225 minutes (mean altitude less than 5,875 km). This definition was established by NORAD and is still in use today by the U.S. Space Surveillance Network for tasking and other operational purposes.

OPERATIONAL DEBRIS – objects which are normally released from a spacecraft or rocket body during the course of a mission and which do not affect the integrity of the vehicles. Operational debris normally perform no useful service after release. Examples of operational debris include spacecraft-launch vehicle separation and stabilization devices, sensor covers, and temporary protective shields.

PAYLOAD – satellite which serves a specific function or mission, e.g., communications, navigation, and weather forecasting. Payloads may be active or passive and may be launched into orbit singly or in groups. When a payload can no longer fulfill its intended mission it is considered inactive or non-operational.

ROCKET BODY – stage of the launch vehicle left in Earth orbit at the end of the payload delivery sequence. Typical space missions leave only one rocket body in Earth orbit, but as many as three separate rocket bodies in different orbits are possible. Some rocket bodies may carry special devices for experimental purposes and be given names associated with the experiment. Rocket bodies are normally as large or larger than the payloads they carry and often retain residual propellants which may later be a source of energy for fragmentation.

SATELLITE – any object in orbit about the Earth or any other natural body in the solar system. The term satellite includes all individual payloads, rocket bodies, fragmentation debris, and operational debris. Orbital periods of Earth satellites may be as short as 88 minutes or as long as 1 year or more.

SATELLITE BREAKUP – disassociation, usually in a destructive manner, of an orbital payload, rocket body, or piece of debris, often with a wide range of velocities. A satellite breakup may be accidental or the result of intentional actions, e.g., due to a propulsion system malfunction or a space weapons test, respectively. Since the early 1960's debris created by satellite breakups have represented the largest single constituent of the total Earth satellite population.

SOLAR CYCLE – periodic fluctuations in the energy output of the Sun, usually measured by the number of visible sunspots, the intensity of the 10.7 cm radio flux, and the geomagnetic Ap index. In general these fluctuations exhibit a sinusoidal variation with a period of 11 years. During periods of high solar activity, the Earth's atmosphere is heated, caus-

ing it to expand and thereby increase atmospheric density at higher altitudes. In turn, this increase in atmospheric density causes satellites, particularly those in low Earth orbits below 1,000 km, to decay more quickly. Hence, the total Earth satellite population normally decreases during periods of high solar activity.

SPACE SURVEILLANCE NETWORK – collection of ground-based radar and optical sensors, operated by U.S. Space Command, which track and correlate artificial Earth satellites. When a new object is detected, tracked, and identified, it becomes a part of the official *Satellite Catalog*. Today approximately 7,000 cataloged satellites are being monitored by the SSN and several hundred more are being tracked, awaiting identification and cataloging. Most radars only track low Earth orbit satellites, while optical sensors are used almost exclusively for tracking high Earth orbit objects. A few radars can track both low and high Earth orbit satellites. The Space Surveillance Center (SSC) in Cheyenne Mountain, near Colorado Springs, manages the SSN resources and processes the incoming information. Designators assigned by the SSC are accepted by the United Nations' Committee on the Peaceful Uses of Outer Space as the international standard.

ADDITIONAL READING

Artificial Space Debris, N. L. Johnson and D. S. McKnight, Orbit Series Book, Krieger Publishing Company, 1987, 1991.

Orbital Debris, NASA Conference Publication 2360, compiled by D. J. Kessler and S.-Y. Su, National Aeronautics and Space Administration, 1985.

Orbital Debris, A Space Environmental Problem, Office of Technology Assessment, U.S. Congress, 1990.

Orbital Debris Monitor, edited by D. S. McKnight, published quarterly, 12624 Varny Place, Fairfax, Virginia, 22033.

Orbital Debris from Upper-Stage Breakup, edited by J. P. Loftus, Jr., Progress in Astronautics and Aeronautics, American Institute of Aeronautics and Astronautics, 1989. (Includes several general background papers.)

Report of the Scientific and Technical Sub-Committee on the Work of its Twenty-Seventh Session, Committee on the Peaceful Uses of Outer Space, United Nations, Report A/AC.105/456, 12 March 1990.

Report on Orbital Debris, Interagency Group (Space) for the National Security Council, February, 1989.

Space Debris a Potential Threat to Space Station and Shuttle, GAO/IMTEC-90–18, U.S. General Accounting Office report to the Committee on Space, Science, and Technology, U.S. House of Representatives, April, 1990.

Space Debris, The Report of the ESA Space Debris Working Group, ESA SP-1109, European Space Agency, November, 1988.

"United Nations Deliberations on the Use of Nuclear Power Sources in Space: 1978–1987", G. L. Bennett, J. A. Sholtis, Jr., and B. C. Rashkow, presented at the Fifth Symposium on Space Nuclear Power Systems, Albuquerque, New Mexico, 11–14 January 1988.

3: The Current and Future Environment: An Overall Assessement

Donald J. Kessler

NASA, Johnson Space Center, Houston, Texas

SUMMARY

Orbital debris is of a concern in primarily two regions of Earth orbital space: low Earth orbit and geosynchronous orbit. The hazard to spacecraft from orbital debris in low Earth orbit has already exceeded the hazard from natural meteoroids. This was predicted by models published over 10 years ago, and had been verified by measurements over the last few years. These same models also predict that certain altitudes are at, or near a "critical density," where the debris hazard will increase as a result of random collision breakups, independent of future spacecraft operational practices. Consequently, there is a need to make immediate changes in operational practices.

The current hazard in geosynchronous orbit has not likely exceeded the hazard from meteoroids. However, models and measurements of the environment in geosynchronous orbit are inadequate; therefore there is currently not an adequate long-term environment management strategy for geosynchronous orbit. A long-term strategy is required because of the increasing use of geosynchronous orbit plus the fact that objects remain in orbit essentially forever at this altitude. There is a need to understand various strategy options before making significant operational changes.

INTRODUCTION

The unlimited bounds of space could lead one to conclude that we would be incapable of causing an environmental issue in this new frontier. This may be the case for most of space; however, Earth orbital space is finite, and past spacecraft operational practices have already produced an orbital debris environment that will likely affect the design of most future spacecraft operating in near-Earth orbital space. If left unchecked, this environment could increase within the next century to the point that some operations either become too expensive or too risky. In order to effectively manage the environment, we need to understand the environment which we have already produced and the potential sources for future orbital debris. With such an understanding, we may preserve near-Earth space for future generations without significantly altering the current planned activities in space.

There are two major regions of Earth orbit where orbital debris is of concern: 1. Low Earth Orbit (LEO), usually thought of as being below 2000 km altitude. 2. Geosynchronous orbit (GEO), at an altitude of about 35,800 km. The orbital debris issues and solutions in these two regions require different approaches, so it is best to discuss them separately. Therefore this paper will be divided in two parts.

PART I – LOW EARTH ORBIT

A comparison of the hazards caused by orbital debris and natural meteoroids provides a threshold by which levels of concern can be measured. In low Earth orbit, this comparison is fairly straight forward, and provides some insight to orbital debris issues. Therefore, it is desirable to first understand the meteoroid environment.

Meteoroids

Meteoroids are part of the interplanetary environment and result from the disintegration and fragmentation of comets and asteroids which orbit the sun. Meteoroids pass through Earth orbital space, rather than orbit the Earth, with a velocity distribution averaging about 16 km/sec (Kessler, 1969). At any one instant, about 200 kg of meteoroid mass is within 2000 km of the Earth's surface. The largest fraction of this mass is in meteoroids with diameters of about 0.1 mm. A lesser fraction of the mass is between meteoroid diameters of 1 mm and 1 cm...the size interval responsible for the "falling stars" or meteors observed at night. This distribution of mass and velocity is sufficient to require shielding on some spacecraft, depending on the spacecraft size and desired reliability. Figure 1 describes the cumulative flux as a function of meteoroid diameter (Grün, et al., 1985). The average number of impacts on any surface is calculated by multiplying this flux by the product of time and average cross-sectional area of that surface exposed to the environment.

Protection against this environment requires a background in the field of hypervelocity impacts, and can become complex when unique materials and geometric properties are taken into consideration. However, as a rule of thumb, aluminum bumper shields can be constructed to have a total added aluminum thickness equal to the meteoroid diameter to be protected against; a single sheet of aluminum would have to be about 5 times more massive than an aluminum bumper, while some "multi-shock" shielding protection techniques are about half as light as an aluminum bumper.

Historically, shielding as light as thermal insulation blankets are usually sufficient for protecting vulnerable areas, such as wiring bundles, and pressurized containers, on small, unmanned spacecraft over their lifetime. This results from the fact, as can be concluded from Figure 1, that only meteoroids smaller than 1 mm in diameter are likely to hit these vulnerable areas over the spacecraft lifetime. Larger, longer duration and high reliability spacecraft, such as the planned Space Station Freedom, would require much more protection against meteoroids. Shielding weights totaling several thousand kilograms would be required to protect the vulnerable areas such as the habitation modules and fuel storage tanks. These higher weights result from the fact, as also can be concluded from Figure 1, that shielding is required to protect against meteoroids of about 0.5 cm over the hundreds of square meters of Space Station vulnerable areas in order to obtain about one chance in 10 that these areas will not be penetrated over the Space Stations planned 30 year lifetime (Christiansen, et al., 1990).

Fundamentals of orbital debris in Leo

Within the same 2000 km above the Earth are approximately 7000 man-made orbiting objects which have been cataloged by the US Space Command with a total mass of about 3,000,000 kg. Many of these objects are in near polar orbit, so that their velocities relative to one another can be as high as twice their orbital velocity, or around 15 km/sec. The average collision velocity between any particular spacecraft orbiting in near-Earth orbital space and the cataloged objects are a function of that spacecraft's inclination and ranges from about 10 km/sec for low inclination spacecraft, to about 13 km/sec for near polar orbits. Because these velocities are not too different that for meteoroids, a comparable amount of orbital debris mass in any particular size interval will produce a comparable collision probability, and the damage resulting from a collision will be similar for meteoroids and orbital debris of about the same size. Since, most of the orbiting mass is in intact spacecraft or rocket bodies...objects several meters in diameter, the hazard created by these objects is more than 4 orders of magnitude larger than the hazard from meteoroids which are several meters in diameter. However, this alone is not necessarily significant since meteoroids of this size are not a problem for spacecraft. However, when combined with other data, two issues become obvious which have both short and long term implications to the environment:

1. If only 0.01% of the orbiting mass, or less mass than in a single average spacecraft were converted into a size distribution similar to the size distribution of meteoroids, it

would create a hazard similar to the hazard from meteoroids. There have been about 100 satellite breakups due to explosions in Earth orbit...more than enough fragmentation mass to create such a hazard, but the degree of the hazard depends on the fragment size distribution resulting from these breakups. These breakups are the major concern for short term orbital debris considerations.

2. Random hypervelocity collisions will soon begin to convert the orbiting mass of satellites into a size distribution that is not too different than the meteoroid size distribution. Hypervelocity laboratory tests indicate that a hypervelocity collision between an average spacecraft and a several kilogram fragment can be expected to produce a large number of fragments in the 1 mm to 1 cm size interval.

Figure 2 gives the calculated flux (Kessler, 1981-B) of catalogued objects as a function of altitude for an orbiting spacecraft for 1987, when solar activity was low, and for 1991, when solar activity was high. Note that the high solar activity has increased the atmospheric density and reduced the flux below 600 km...this has occurred during previous high solar activity periods. By 1997, when solar activity is expected to be lower, the flux below 600 km should return to about its 1987 values. Figure 2 is averaged over inclination; the flux for spacecraft with low inclinations will be slightly lower (by about 10%) than given in the figure, while some inclinations (e.g., 80 degrees and 100 degrees), will experience twice the flux given in the figure (Kessler, et al., 1989-A). The collision rate for small spacecraft would be small against the catalogued population; however spacecraft larger than about 100 meters in diameter begin to have a significant probability of colliding with a cataloged object. It is for this reason that collision avoidance maneuvers are planned for the Space Station. Just as significantly, the total area of all cataloged objects is larger than the area of a 100 meter diameter spacecraft; consequently, since collision avoidance between all 7000 catalogued objects is impractical, there is a near certainty that two catalogued objects will collide in the relatively near future. Random collisions are the major concern for long term orbital debris considerations.

Early predictions

The earliest orbital debris studies by NASA were mostly concerned with calculating the collision probabilities between objects large enough to be catalogued by NORAD (Donahoo, 1970; Brooks, et al., 1975). All catalogued objects are larger than 10 cm in diameter. Fragmentation data from ground explosions and hypervelocity tests gathered by NASA Langley Research Center suggested that a much larger population of uncatalogued objects must exist in Earth orbit (Bess, 1975). NASA, Johnson Space Center (JSC) used the Langley data to predict a future uncatalogued population from random collisions, even if such an uncatalogued population did not currently exist (Kessler, et al., 1978). These predictions were later

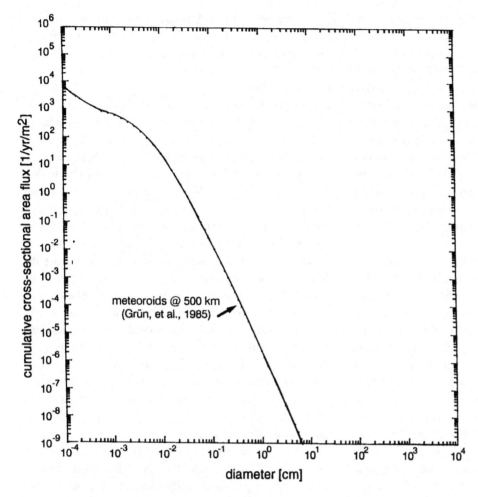

Figure 1: Meteroid environment at 500 km altitude.

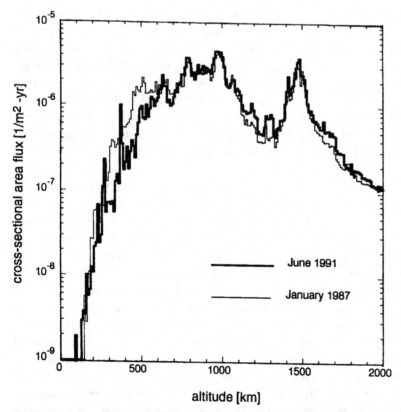

Figure 2: Average flux resulting from US Space Command cataloged population.

expanded to include an estimate of the uncatalogued population in 1978 (Kessler, 1981).

Until 1984, there were no measurements of debris in orbit, other than what could be cataloged by NORAD. Predictions of smaller debris had to be based on models which described the process of satellite breakups in terms of number and velocity of breakup fragments, and models which predicted the rate that debris was removed from the environment through atmospheric drag. To predict future environments then, as now, assumptions had to be made concerning future traffic into space and the rate that future satellites will breakup. A measure of the ability of these models to predict the future environment can be obtained by comparing these early model predictions with today's measurements.

Figure 3 shows a published prediction (Kessler, 1981), based on 1978 data, when the catalogue contained about 4500 objects. The 1978 Debris "Observed" line represents the catalogued population between 600 km and 1100 km; the "Corrected" line is a model prediction of the 1978 environment based on a comparison of the size distribution measured from explosions on the ground and the size distribution of catalogued orbiting fragments. Note that this analysis predicted that there were about twice as many objects in orbit that were larger than 10 cm than is reflected in the catalogue, and more than three times as many objects larger than 4 cm. The analysis also predicted that by 1995 there would have been 3 satellite breakups caused by random collisions, producing a population of smaller debris that produced a debris flux which exceed the meteoroid flux for sizes larger than about 1 mm. Since debris in these size ranges have now been measured, and 1995 is less than 3 years away, these predictions can be compared with recent measurements.

Measurements of uncatalogued population

The number of orbiting objects increases with decreasing size. If one were to try to catalogue all orbiting objects, eventually, the catalogue would become so large that only a statistical interpretation of the population would be meaningful. Consequently, there is a debris size where statistical measurements become more cost effective. Statistical measurements sample a fraction of the population and do not require that the each detected object be tracked so that it can be observed again.

Sampling of objects in Earth orbit is a less difficult problem than the cataloging of objects; however, the detection of uncatalogued objects requires either different sensors, or that these sensors be operated in a different mode than the sensors used to catalogue. Remote sensors are required for debris larger than about 1 mm, simply because the population of this size debris is sufficiently sparse that a large collection area is required in order to obtain a statistically meaningful sample. Below 1 mm, the population is sufficiently dense that direct impact on spacecraft will obtain a statistically meaningful sample. The least expensive remote sensors are Earth based; consequently, these

sensors have provided the best data to date. The measurements to date have been obtained using ground telescopes, ground radar, and returned spacecraft surfaces.

Ground telescopes

A 1 cm diameter metal sphere in sun light at 900 km distance would appear as a 16th magnitude star. Since telescopes larger than about 30 inches can detect stars of this magnitude, in 1983 NASA Johnson Space Center (JSC) contracted MIT Lincoln Labs to use their Experimental Test Site (ETS) to look for 1 cm debris. An advantage of the ETS was that it contained two 31 inch telescopes which could look at the same area of the sky to use parallax to determine altitude. It was felt that this feature would be essential to discriminate against the luminosity caused by much smaller meteoroids hitting the Earth's atmosphere (meteors) at about 100 km altitude. The telescopes were operated just after sunset and just before sunrise, when the debris was in sun light and the telescopes were in darkness. The telescopes were pointed vertically, and debris was observed to pass through the field of view. Nine hours of data were recorded on video tape and analyzed by Lincoln Labs. Published results (Taft, et al., 1985) concluded that the ETS detection rate was 8 times the rate expected from objects in the catalogue. However, two errors were found in the analysis, plus one of the assumptions proved to be wrong.

NASA, JSC reanalyzed the ETS data and found parallax errors which placed a larger number of the objects detected into the category of meteors. This reduced the detected orbital debris to between 2 and 5 times the catalogue rate, depending atmospheric seeing conditions. A calibration error placed the limiting magnitude of the telescopes at 13.5 for debris with the typical angular velocity of 0.5 deg/sec. Finally, independent measurements using radar, infrared wavelengths and optical wavelengths determined that the assumption that debris fragments would reflect light similar to a metal sphere was wrong. Debris fragments reflect much less light than a metal sphere...typically only about 10% of the light is reflected, although some objects reflect a larger fraction. Consequently, the limiting size measured by these telescopes was about 8 to 10 cm.

Since then, NASA has worked closely with the US Space Command to use their Ground Electro-Optical Deep Space Sensors (GEODSS), which are telescopes, similar to Lincoln Lab's ETS, except they are slightly less sensitive (limiting magnitude of about 13 at 0.5 deg/sec.), and have twice the field of view. Over a hundred hours of data have been analyzed by NASA which produced nearly a thousand orbiting objects. The US Space Command catalogue was used to predict which of the detected objects were already in the catalogue. Only about half of these objects can be identified as being catalogued objects (Henize, 1990). Consequently, these telescopic measurements have provided convincing data to NASA that at about the 10 cm threshold, the low Earth orbit catalogue is only about 50% complete. The exact limiting size measured by

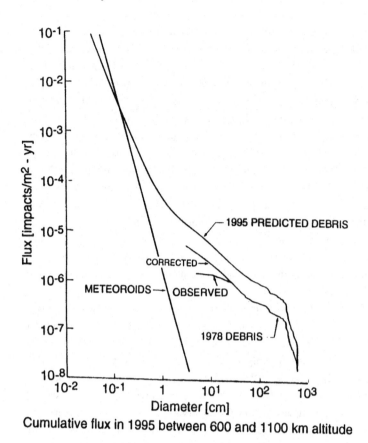

Cumulative flux in 1995 between 600 and 1100 km altitude

Figure 3: Early model predictions of orbital debris environment. (Published in 1981)

each of these two different types of sensors is a function of two different distributions relating signal return to size, and also a function of the rate of increase in the number of smaller debris with decreasing size. When these distributions are taken into account, the "average" limiting size of the telescopes can be shown to be slightly smaller than the "average" limiting size for the catalogue. The DOD has a program to better understand the size and orbits of these uncatalogued objects by tracking them with radar and ground telescopes. However, because NASA requires data on smaller debris, its program has been expanded to obtain measurements using the Haystack ground radar. This radar is now detecting a smaller population at a rate that is more than 60 times the catalogue rate.

Ground radars

The failure of ground telescopes to detect 1 cm orbiting debris forced NASA to reexamine the use of radar to detect uncatalogued objects. A major reason that US Space Command radars do not detect smaller objects is that most of their radars operate at a 70 cm wavelength; consequently, objects as small as a few centimeters in diameter are well into the Rayleigh scattering region and reflect a very small fraction of the radar signal. A shorter wavelength radar, in principle, could detect smaller objects. However, another factor in the limiting size of catalogued objects is in the method of the operation of radars to catalogue objects. In order to catalogue an object, the object must first be found within a large volume of space, then the object must be tracked by several radars over an extended period of time. This process requires a larger threshold size than a process which only required detection within a smaller volume when the object is closest to the sensor.

In 1987, NASA, JSC developed a technique of using a radar in a "beam park" mode, where the radar stares in a fixed direction (preferably vertically) and debris randomly passes through the field-of-view. In this mode, using a relatively inexpensive, high powered, moderate size X-band radar (3 cm wavelength), objects as small as 1 cm could be detected at a 500 km altitude. In 1987, interest in the hazards of orbital debris to the Space Station produced a series of events which resulted in an agreement between NASA/JSC and the US Space Command to operate the Haystack radar, located near Boston, in the beam park mode, and to develop the necessary computer programs to analyze the data. However, the Haystack radar is not optimally designed (the antenna beam width is small, consequently more time is required to obtain the necessary data), nor optimally placed (its too far north, and cannot see low inclination debris). Therefore, a Haystack auxiliary radar is being built next to the Haystack radar. In addition, another radar near the equator is to be built, although the detail of this radar has not yet been resolved. As a result of this series of events, a significant amount of ground radar data has been obtained using the Arecibo, Goldstone, and Haystack radars.

To test the concept of obtaining orbital debris data in a beam park mode, in 1989, NASA, Jet Propulsion Laboratory (JPL) used the Arecibo Observatory's high-power S-band radar, and the Goldstone Deep Space Communications Complex X-band radar to obtain orbital debris data. Neither radar was optimally configured to obtain data in this mode, although both radars were predicted to detect small debris, if it existed. In 18 hours of operation, the Arecibo experiment detected nearly 100 objects larger than an estimated 0.5 cm in diameter (Thompson, et al., 1992). The predicted number from the catalogue alone was about one. In 48 hours of observation, the Goldstone radar detected about 150 objects larger than about 0.2 cm in diameter (Goldstein, et al., 1990). The probability that at least one catalogued object would pass through the field of view during the 48 hours was about 0.13, indicating a population which is slightly more than 1000 times the catalogued population. Because little effort was made to accurately define these radars field of view, and to properly calibrate the radars, this data has fairly large uncertainties. Even so, these two experiments did demonstrate that data could be obtained is this mode of operation, and that there was a large population to be detected.

After testing the concept, NASA committed to a program of using the Haystack radar to obtain orbital debris data. The program included calibration of fragment size using a radar range and fragments from ground tests, calibration of the antenna pattern, development of a real-time Processing and Control System to process and record detections, and establishment of a data processing facility at JSC. In order to ensure that NASA was properly acquiring and analyzing the data, a peer review panel was established. The chairman was Dr. David K. Barton and included other well known experts from the radar community. The panel concluded that "the Orbital Debris Radar Measurements Project is fundamentally sound and is based on good science and engineering." They also made a number of recommendations to improve efficiency or accuracy of the data. Many of those recommendations have been implemented.

To date, over 1000 hours of data have been collected, of which over 800 hours has been analyzed, and more than 2000 objects have been detected passing through the radar beam (Stansbery, et al., 1992). Figure 4 gives the altitude of each detection when the radar beam is parked in its most sensitive position of looking vertically, compared to the detection rate expected for the catalogue alone and the rate predicted by the model given by Kessler et al., 1989. At the lowest altitudes (350 km), objects larger than 0.3 cm are detected. At the highest latitude (1400 km), objects larger than 0.6 cm are detected. The detection rate averaged over all altitudes is about 65 times the rate predicted by the catalogue alone. In the altitude band between 850 km and 1000 km, the rate is 100 times the rate predicted by the catalogue alone.

Recovered samples

Objects returned from space usually contain pits or holes from hypervelocity impacts with meteoroids or orbital debris. Outside the laboratory, these are the only two possible sources which can impact surfaces with sufficient velocity to cause melting of the surface in the impacted area. One technique to determine which of these sources caused the impact pit or hole is to use the scanning electron microscope (SEM) dispersive X-Ray analysis to determine the chemistry of material melted into the surface. This analysis has been completed for some of the pits found on Space Shuttle windows, impacts into surfaces returned from the Solar Max repair mission in 1984, surfaces on the returned Palapa satellite, and some of the Long Duration Exposure Facility (LDEF) surfaces, returned to Earth in 1991. Because LDEF was a controlled experiment, was in space for nearly 6 years, had a large surface area, and was always oriented in the same direction with respect to the orbital velocity vector, these surfaces are providing the best data to date. Analysis of LDEF surfaces is still continuing, however the data analyzed thus far exceeds the quality of the earlier data.

The largest impact crater predicted and found on LDEF was slightly larger than 5 mm in diameter, likely due to an impact by an object 1 mm in diameter The number of impact craters increased rapidly with decreasing size, with more than 3000 craters larger than 0.5 mm. The most complete chemical analysis has been conducted by Fred Hörz, the Principal Investigator for the Chemistry of Micrometeroids Experiment (Hörz, et al., 1991; Bernhard, et al., 1992). The analysis to date indicates that about 15% of the impacts in the gold surfaces, facing in the rear direction, are orbital debris. The most common orbital debris impacts are aluminum; however, copper, stainless steel, paint flecks, and silver were also found. Orbital debris impacts on rear surfaces was a surprising result because a very small set of elliptical orbital debris orbits are capable of hitting the rear surfaces (Kessler, 1992). The most probable direction for orbital debris to impact is the front and side surfaces; the surfaces facing in this direction are made of aluminum, and aluminum impacts cannot be identified. Even so, 14% of the impacts on these surfaces were identified as orbital debris; the origin of 55% of the impacts could not be identified because only aluminum was detected. If the ratio of aluminum to other orbital debris compositions found on the gold surfaces is also on the aluminum surfaces, then most of the impacts on aluminum surfaces that could not be identified would have been caused by an aluminum impact. This would increase the orbital debris impacts on the aluminum surfaces. When averaged over all orientations, the average number of orbital debris craters to meteoroid craters may be about the same, or slightly higher, as the results obtained from the Solar Max Satellite (Barrett, et al., 1988). Analysis of the LDEF surfaces is not complete.

Perhaps the most surprising result on LDEF came from the Interplanetary Dust Experiment (Mulholland, et al., 1991). This was the only experiment on LDEF which measured the time of impact. Six detectors were on orthogonal surfaces, and sensitive to impacts smaller than 1 micron. The surprising result was that most impacts could be associated with "orbital debris swarms." That is, the sensors would detect a large

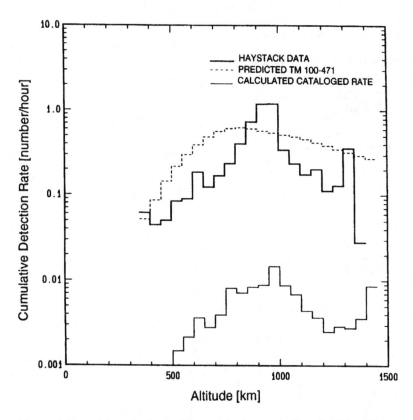

Figure 4: Rate of detection by the Haystack Radar, vertically pointed for 188 hours.

increase in flux, lasting for a few minutes, at the same points in the LDEF orbit, and these points would slowly change with time...characteristics of orbital precession rates. However, in retrospect, these results should not have been such a surprise.

The small amount of mass required to produce a large flux of less than 1 micron debris in Earth orbit, coupled with their short orbital lifetime would predict that a large number of particles could be found in orbits close to the orbit of their source. If the source is paint being removed by atomic oxygen erosion, then less than 10 grams of paint is needed to be removed from each orbiting spacecraft per year to explain these results. If the source orbit is highly elliptical, then less than 1 gram of paint need is needed. These are rates consistent with the rates expected from atomic oxygen erosion.

Other sources are possible, such as the large amount of aluminum oxide dust that each solid rocket motor expels when fired. This dust is expelled at a velocity of 3.5 km/sec, and over a range of directions, most of which would cause the dust to immediately reenter the Earth's atmosphere. Although some dust would remain in orbit, most should reenter quickly and not product swarms lasting for several months, as observed. However, this is not to say that a spent rocket stage might not slowly release sufficient dust to produce the long lasting swarms. These are possible areas of future research.

Summary environment

A summary of the best measurements to date is shown in Figure 5, compared to the natural meteoroid environment.

When compared to the 1995 model predictions shown in Figure 1, there is a general agreement (within a factor of 2) over nearly the entire size range, even though some of the measurements were made at altitudes of 600 km, or lower, where debris was predicted by some analysis to be much less than the environment between 600 and 1100 km. This could be interpreted as an indication that the environment was under predicted for sizes smaller than about 5 mm, and this may be the case. However, some of this data is indicating that elliptical orbits are important in this size range and at these lower altitudes. Elliptical orbits will produce an environment that is less altitude dependent than circular orbits. Until all of these parameters are understood, or until small debris measurements are made at higher altitudes, there will be an uncertainty in extrapolating these measurements to higher altitudes.

What should have been the easiest prediction is the catalogued population. The prediction was that would be slightly over 11000 catalogued objects in orbit by 1995. There are currently 7000, and this number might increase to 8000 by 1995, but not 11000. There are three important reasons for under predicting the catalogued population: 1. Rather than a slight increase in the amount of material launched into space, the rate of launches world wide has remained constant. 2. Two of the three highest solar activities in 200 years of record keeping occurred between 1978 and 1992. High solar activity increases the atmospheric density, causing more objects to reenter from orbit. 3. Since 1981, the US has lead an effort to minimize on-

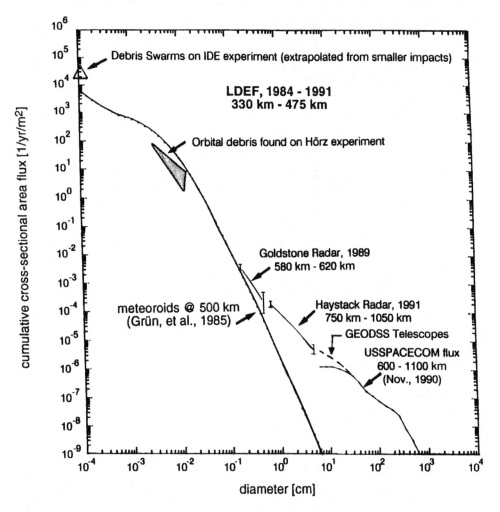

Figure 5: Meteoroid environment compared to recent measurements of orbital debris environment.

orbit explosion. As a result, fewer US, ESA and Japanese upper stages have exploded in orbit than would have occurred if this effort had not been undertaken.

For the above three reasons, the 1995 prediction of the uncatalogued population should have also been high. That is, as a result of fewer large objects in orbit, there should be fewer random collisions to generate small debris; also as a result of fewer on-orbit explosions, there should have been less small debris. The implications are that either the current breakup models are under predicting the amount of small debris which is generated and remains in orbit, or there are unmodeled sources of debris. The most likely cause is the breakup models, and that these models are under predicting the fraction of mass which goes into smaller debris. This should not be surprising since smaller fragments are more likely to be lost during ground experiments to determine breakup characteristics. The possible tendency of current breakup models to under predict should be keep in mind as similar models are again used to predict the future environment.

Future environment

The most important parameters in predicting the future orbital debris environment is the rate and consequences of satellite breakups. The major issue is the size and velocity distribution of fragments produced as a function of the amount of fragmentation energy by various fragmentation energy sources. In the past and near term, the energy sources have mostly been chemical, and we can see that this source has already produced a hazardous environment for most spacecraft. Chemical explosions can easily be controlled; so our near term environment will be a function of our efforts toward eliminating these past sources of explosions in orbit. However, in the future, the major energy source could be kinetic energy; this source is not as easily controlled. The amount of kinetic energy represented by an object as small as 1 kg, traveling 10 km/sec is not too different than the amount of chemical energy which caused past chemical explosions in orbit.

Like most chemical explosions, most of the mass of fragments from a hypervelocity collision is in the larger fragments; however, because the energy source is concentrated in a smaller amount of mass, higher temperatures are reached and melting of the impacted spacecraft occurs, which results in a small, but significant fraction of the mass being distributed in smaller fragments. These characteristics of hypervelocity collisions, combined with the increasing rate that they could occur if no changes in current practices are made within the next few years, make them important to the future orbital debris environment in low Earth orbit.

The Defense Nuclear Agency (DNA) has a program to define better breakup models (Tedneschi, et al., 1991). The analysis of hypervelocity tests by DNA are not yet complete; however, a gross characterization would be that they are in general agreement with the assumptions of earlier models; that is, at 10 km/sec, a 1 kg fragment can completely fragment a 1000 kg spacecraft, producing hundreds of 1 kg fragments, each of which could fragment another spacecraft, and also producing millions of smaller hazardous fragments, each capable of damaging an operational spacecraft. Some masses catalogued by the US Space Command are smaller than 1 kg, while there is undoubtedly some uncatalogued fragments which exceed 1 kg. The rate that catalogued objects can be expected to collide with one another is a fairly easily calculation. Based on the current population, the rate is about one collision every 20 years (Kessler, 1991A); however, to date, such an event has not been observed to occur. Estimates of the mass of uncatalogued debris predict that the current rate of satellite breakups from hypervelocity collisions is about once every 8 years (Kessler, 1991B); there is data and analysis (Johnson, et al., 1991; Johnson, 1992; McKnight, 1991) to suggest that such events have occurred. Because these rates are proportional to the square of the number density of objects in orbit, these collisional fragmentation rates can become much more frequent in the relatively near future if objects continue to accumulate at past rates.

The Earth's atmosphere will remove fragments from low Earth orbit. If fragments are removed faster than they are generated, then an equilibrium environment will be established, and this environment will not increase unless new material is added to Earth orbit. However, if collisions are producing fragments at rate faster than they can be removed by the atmosphere, then the orbital debris environment will continue to increase even without adding new material to orbit. The satellite population density which will produce collision fragments at the same rate they are removed has been defined as a "critical density." To attempt to maintain a satellite population above the critical density means that debris will increase as a result of random collisions alone. Objects nearer the Earth are removed at a faster rate. The rate that fragments are generated is not only a function of number density, but satellite size and inclination of the orbits.

Figure 6 compares the 1989 catalogue with a calculated critical density as a function of altitudes (Kessler, 1991). The figure shows that below 800 km, atmospheric drag removes fragments at a sufficiently large rate and that this region is well below the critical density line. Between 800 km and 1000 km, the current population density is above the critical density line. Above 1000 km the physical size of satellites is smaller, reducing the fragment generation rate; even so this rate exceeds the removal rate again above 1400 km. The uncertainty in this critical density line is about a factor of 3; consequently,

within this uncertainty is the possibility that a critical density has not yet been reached. However, even if it has not been reached, the population of these two altitude bands might be expected to exceed the critical density within the next few years.

Similar conclusions have been independently reached by other researchers using different modeling approaches (Eichler, et al., 1990, Talent, 1991; Farinella, et al., 1991; Lee, et al., 1990). An evolutionary model (EVOL) developed by NASA, JSC (Reynolds, et al., 1990) illustrates the same trends and is being used to evaluate the consequences of various possible operational practices. In Figure 7, this model is used to plot the 1 cm population at 400 km and 1000 km as a function of time for several operational conditions. The "case 1" curves assume "business as usual, keeping the world launch rate at the current 100 launches per year, with all objects allowed to accumulate in orbit at the end of their life, and no reductions in the rate objects explode in orbit. At the end of 100 years, the environment is a factor of 10 larger than the current environment at 1000 km altitude. At 400 km altitude, the flux varies due to the varying atmospheric density due to solar activity; even so, the flux has a general upward trend with time. The "case 2" curves assume the same conditions, except all chemical explosions are eliminated in the year 2000. Note that the rate of increase of 1 cm debris immediately begins to decrease when explosions are eliminated, especially at 400 km. However, by the year 2030, there are sufficient old rocket bodies and payloads in orbit that the satellite break up rate from random collision causes the 1 cm flux to begin to increase again. By the end of the century at 1000 km the fact that explosions were eliminated is of minor importance. Under these conditions, kinetic energy has become the most important source of energy causing satellites to break up in orbit.

Kinetic energy can be eliminated by eliminating mass in orbit. The bottom curve assumes the same conditions as the 2nd curve, except that after the year 2000, rocket bodies are required to reenter after delivering their payload, and after the year 2030, payloads are required to be removed from orbit at the end of their operational life (assumed to be 10 years). Under these conditions, the 1 cm environment continues to decline. Consequently, this model, as well as other models, is predicting that future payloads and rocket bodies must be reentered in the relatively near future in order to keep the future small debris population from increasing.

Environment management of low Earth orbit

In the short term, management of the low Earth orbit orbital debris environment is concerned with the control of explosions in orbit. Users of space are already adopting new operational procedures which are designed to accomplish this. This includes depleting the unused fuel from orbiting upper stages after payload orbit insertion and conducting military tests at low altitudes to ensure that all fragments reenter. However, it now seems clear that in the long term, these changes will make

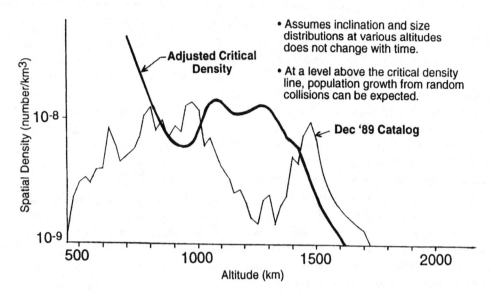

Figure 6: Critical density compared to 1989 cataloged population where the catalog is maintained.

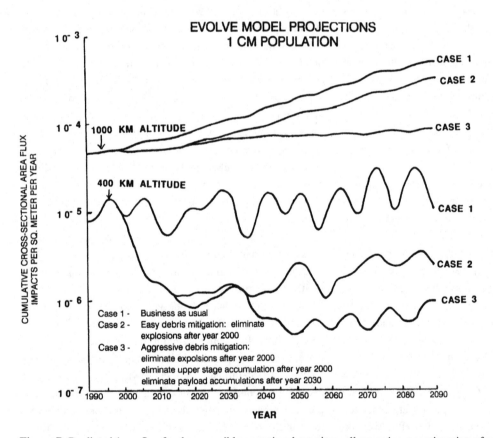

Figure 7: Predicted 1 cm flux for three possible operational practices, all assuming a continuation of current launch rates.

little difference and new, more costly changes must be made. Some of these changes should begin now.

The design process should begin now which would prevent mass from continuing to accumulate in certain regions of low Earth orbit. Various techniques have been studied which could accomplish this (Loftus, et al., 1991). These techniques include the planned reentry of rocket stages, payloads, self disposal options, retrieval, and the use of drag devices. An option

which could begin very soon is the planned reentry of certain rocket stages. With a minor amount of additional fuel and an extended battery life, many of the currently used upper stages could be left in orbits with much shorter lifetimes after delivering their payloads.

Upper stages left in transfer orbits from low Earth orbit to geosynchronous orbit can cause a significant problem to altitudes below about 500 km (Kessler, 1989). This is because at these lower altitudes, the orbital lifetime of these high energy orbits can be very long. However, the lifetime of these stages can be shortened considerably by a very small change in velocity at the apogee of the transfer orbit. The required change in velocity is so small that lunar and solar gravity can be used, if the original orbit has the proper orientation with respect to the sun (Mueller, 1985), to cause the upper stage to reenter within a year.

The reentry of payloads is more difficult because payloads do not usually have the potential propulsive capability of upper stages. Alternatives include the use drag devices (i.e., deploy a large surface area, such as a balloon, so that the atmosphere will drag the object out of orbit more quickly), or tethers. The cost of alternatives such as these and others need to be evaluated against the cost of simply increasing payload propulsion capability.

The cost of the retrieval of objects in orbit using current technology is very high, and this is a major reason for designing removal options into future rockets and payloads. However, this is not to say that future technology could not provide a dedicated retrieval system that would be inexpensive to operate. If so, such a system might then relieve some future design requirements. However, the need to not leave dead rocket stages and payloads in low Earth orbit is immediate, and should become a design consideration for all new programs.

PART II – GEOSYNCHRONOUS ORBIT

Orbital debris studies concerning geosynchronous orbit are slightly more than 10 years old (Hechler, et al. 1980). Most of the studies to date have been concerned with the larger, cataloged objects (Chobotov, 1990; Flury, 1991). No models have been developed to predict the population of small debris, nor how this population might vary with time. No measurements have been conducted to determine the orbital debris population to sizes smaller than about 1 meter in diameter.

The reasons for the lack of modeling and data are twofold: 1. The higher collision velocities in low Earth orbit cause the consequences of collisions to be more dramatic than in geosynchronous orbit. For some time there has been sufficient data to show that the environment in low Earth orbit affected the design of planned NASA missions, resulting in more resources being devoted to understanding and controlling this environment. 2. Geosynchronous orbit is farther away from Earth. While this distance has kept traffic to geosynchronous orbit smaller than to low Earth orbit, it has also made observational data more difficult to obtain; consequently, more resources may be required to obtain the necessary data than have been devoted to low Earth orbit research. Despite the lack of data, some users are unilaterally adopting a policy of moving their dead payloads to higher altitudes, "grave-yard" orbits. While this may provide some operational convenience, it may be totally inappropriate to long-term environment management.

Fundamentals of orbital characteristics in geosynchronous orbit

The orbital period of an object in geosynchronous orbit is the same as the time required for the Earth to spin one revolution, or about 23 hours and 56 minutes. If the orbit has zero eccentricity and inclination, the object will appear to be stationary (hence the orbit is sometimes referred to as geostationary orbit) over a location on the Earth's equator, and because its altitude is about 35,786 km, it can be seen by nearly an entire hemisphere. However, this orbit is sufficiently far from the Earth that the forces produced by the sun and moon will not allow the orbit to naturally maintain a zero eccentricity and inclination. Eccentricity is typically less than 0.001, keeping perigee minus apogee distances to less than 100 km. This is sufficiently small that it does not cause a ground tracking problem. However, inclination will naturally oscillate between zero and 15 degrees over a 53 year cycle. This can cause a ground tracking problem, so that a significant fuel budget is usually required for "North-South station keeping."

Without "East-West station keeping," another ground tracking problem would exist. The fact that the Earth is not a perfect sphere also causes an oscillation in the longitude of an uncontrolled satellite. This oscillation is about the nearest "stable point," located over longitudes of 75 degrees East, and 105 degrees West. If the desired position of the satellite is far away from the stable points, the satellite would oscillate nearly halfway around the Earth before returning about 3 years later. Fortunately, East-West station keeping fuel requirements are small, requiring less than 5% of the fuel budget required for North-South station keeping (Flury, 1991).

The orbital debris problem

The orbital debris problem results from the accumulation of satellites, fragments of satellites, and operational debris in orbits which pass through the paths of operational geosynchronous satellites. Any of these objects could collide with an operational spacecraft, damaging it and reducing its operational life. If a large number of objects accumulate, the hazard could significantly add to the hazards from other sources, such as collisions with natural meteoroids. Figure 8 is an expansion of Figure 2 for 1991, and compares the flux of catalogued objects in low Earth orbit with geosynchronous orbit.

At geosynchronous altitudes, there is only one natural process which will eventually eliminate a satellite from this altitude. Over extended periods of time, spacecraft and fragments of spacecraft will break up from collisions with other objects which are either in or pass through the geosynchronous region. The smallest fragments (less than about 10 microns) are affected by solar radiation, which both increases the orbital eccentricity and decreases the orbital semi-major axis, resulting in the smallest fragments being removed from orbit by hitting the Earth's atmosphere within a few months (Mueller, et al., 1985; Friesen, et al., 1992-B). If this process acted quickly to remove fragments of all sizes, then the accumulation of

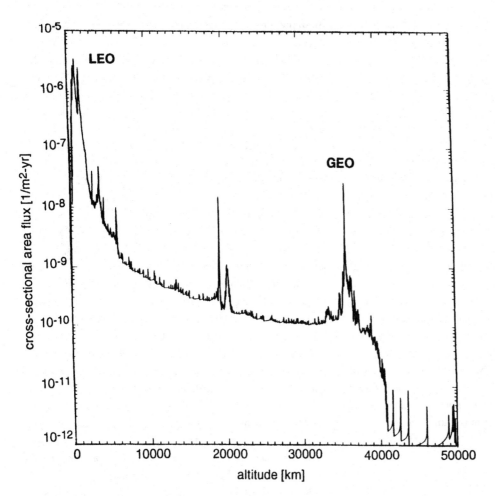

Figure 8: Average flux resulting from US Space Command cataloged population.

debris at geosynchronous altitudes problem would be minimal; however, this is not the case. If the rate of fragment generation were small, the hazard would not be significant for a long period of time. As will be developed in the following sections, collisional fragmentation rates are likely to be small; however, the rates that satellites fragment for operational or design failures are probably much larger.

Time scale for the orbital debris problem

The time required for an object to be removed from the geosynchronous altitude from solar radiation forces alone is very long. Objects smaller than about 1 cm require about 60,000 years for the orbit to decay sufficiently so that no part of the solar radiation pressure induced eccentricity causes the orbit to cross geosynchronous altitude (Friesen, et al., 1992-A). When eccentricity is induced by other forces, the time is longer. Lunar and solar gravity control the eccentricity of objects larger than 1 cm, while collisional forces are likely to control the eccentricity of collisional fragments of all sizes. Consequently, most debris is likely to cross geosynchronous orbit for periods much longer than 60,000 years, perhaps 100,000 to 1,000,000 years. An even longer period of time is required for the debris to be removed from Earth orbit, unless object is very small, less than about 10 microns. Solar radia-

tion pressure increases the eccentricity of very small debris to such a large value that the debris collides with the Earth within a few months (Mueller, et al., 1985; Friesen, et al., 1992-B).

The rate that collisional fragments are generated in geosynchronous orbit is not so clear. This rate is a function of three parameters: 1. The rate of collisions. 2. The velocity of collisions. 3. The efficiency with which the collisional energy produces fragments of various sizes and velocities. None of these parameters are well defined at this time, although a few attempts have been made to determine them.

Rate of collisions

The rate of collisions for any particular object (i.e., a measure of the hazard to that object) is proportional to the area of that object and the number of other uncontrolled objects in orbit. The rate at which any two objects collide (i.e., a measure of the rate of debris production) is proportional to the rate for a typical uncontrolled object and the number of uncontrolled objects. Consequently, debris production from collisions increases as the square of the number of uncontrolled satellites in orbit. The relationship between orbital elements and collision rates has been developed by a number of investigators.

A simple approach for obtaining the collision rate was introduced by Perek (1978). Perek calculated the collision rate by first taking the ratio of the collisional cross-sectional area of active satellites to the area of the geosynchronous ring. Inactive satellites, which would generally have an inclination of several degrees, would pass through the geosynchronous ring twice a day. The collision rate per 12 hours is then this ratio times the number of inactive satellites. A more accurate, but more complex approach was developed by Kessler (Kessler, 1981). Both approaches assume an even distribution of satellites around the geosynchronous ring. Perek's approach can be shown to be equivalent to Kessler's approach as long as orbital inclinations of uncontrolled debris are between about 0.1 degrees and 20 degrees, and orbital eccentricities are small, which is the case for most debris in geosynchronous orbit.

The simplicity of Perek's approach provides some additional insight; namely, that collision probability in geosynchronous orbit is approximately independent of orbital inclination. Consequently, this approach can be used to obtain the collision rate between any two inactive satellites if the differences in inclination are between 0.1 and 20 degrees, which is generally true for most pairs of uncontrolled satellites. If N is the number of uncontrolled satellites, then the average rate of collisions between any pair would be N(N-1)/2 times the rate of collisions between a single pair.

There are currently about 250 objects known to be in geosynchronous orbit, with average linear dimensions of about 5 meters (Royal Aircraft Establishment). Most of these objects are in orbits which confine their motion to a 100 km band, centered at the geosynchronous altitude. Assuming all of these objects become uncontrolled and are randomly distributed within this band gives an average collision rate of once every 15,000 years. This is about the same rate if one assumed an average flux of $7 \times 10^{-9}/m^2$ – yr. From Figure 8, this flux might be an appropriate average within the geosynchronous band, although it is likely that the appropriate average could be slightly higher due to the non-uniform distribution within the geosynchronous altitude band. Even so, the collision, rate is low compared to the rate in low Earth orbit; however, like low Earth orbit, it is high compared to the rate that objects are removed by natural forces. Also like low Earth orbit, this rate increases as the square of the number of objects in orbit; consequently, if the rate of accumulation of objects in geosynchronous orbit continues at its current level of 25 objects per year, then there is a 50% probability that there will be at least one collision in geosynchronous orbit in the next 140 years.

These rates are smaller than other published rates... in some cases, significantly smaller. Part of the reason is in the assumption that the satellites are randomly distributed within the geosynchronous altitude band...they are not. Some researchers have obtained significantly higher collision rates at certain longitudes (Guermonprez, 1990). However, these higher rates may not be representative of the generic hazard, but result from the desire to maintain the satellite over the same longitude. These higher rates are reduced significantly simply by

terminating station keeping. Once station keeping is terminated, the satellite begins to drift in longitude, and the distribution of satellites approaches a more uniform distribution. Researchers who assume that the satellites are simply abandoned (Hechler, 1985) obtain collision rates that are less than a factor of two different than the collision rates obtained by assuming a uniform distribution. Consequently, the long-term error in assuming a uniform distribution is probably small, although this assumption should be carefully examined.

When the size and number of satellites in geosynchronous orbit is assumed to be large, the collision rate will also be large. For example, a rate of one collision every 400 years to 600 years, and a 0.16 probability of a collision over a 20 year period was calculated by Hechler (1985), and is frequently quoted by others (ESA, 1988; Flury, 1991). This appears to be very different than the one collision every 15,000 years previously calculated. The primary reason for these large differences is in the very large size and larger number of objects in geosynchronous orbit assumed by other authors. These authors sometimes assumed 200 satellites with linear dimensions of 50 meters, and as many as 10,000 one cm orbiting fragments in geosynchronous orbit. Existing satellites in geosynchronous orbit are much smaller, and there is no hard data describing the number of small fragments in geosynchronous orbit. Even so, a comparison with the natural hazard reduces the ambiguity introduced with these assumptions.

Collision rates are proportional to area for both meteoroids and debris; consequently the relative collision rates from meteoroids and debris are the same for any assumed satellite size. The collision rate from meteoroids increases with decreasing meteoroid size; however, the damage resulting from a collision decreases with decreasing size. Consequently, a high collision rate with small debris may not be significant to the over-all hazard of the spacecraft when compared to the meteoroid hazard. A key parameter in comparing the debris hazard to the meteoroid hazard is the collisional velocity.

Collision velocity: meteoroids and satellite breakup rates

If all station keeping in geosynchronous orbit was terminated so that orbital inclinations could reach their natural long-term distribution, then the collision velocity that an average satellite would experience would range for zero to about 0.8 km/sec, and have an average of about 0.5 km/sec. The average meteoroid velocity is 16 km/sec. Therefore, for a given mass, meteoroids will collide with geosynchronous satellites at about 32 times more momentum and 1000 times more kinetic energy than a collision with another object in geosynchronous orbit. At 16 km/sec, a 0.7 kg a meteoroid would breakup the average 2000 kg spacecraft. The rate that a 0.7 kg meteoroid can be expected to collide with any one of the 250 satellites, each 5 meters in diameter, is about once every 100,000 years (Zook, 1992). Consequently, the current rate of satellite collisional breakups is probably controlled by the current number of satellites in geosynchronous orbit, rather than meteoroids; however, the time scale is very long before a satellite will break up due to either type of collision.

Satellites in geosynchronous orbit may break up more frequently for other reasons. In low Earth orbit, nearly half of the catalogued population is fragments of satellites, resulting from more than 100 explosions in low Earth orbit. Most of these explosions were due to the failure of an energy storage device, such as the tanks of upper stage which contained residual fuel, or batteries on a spacecraft. These same potential sources are equally common in geosynchronous orbit. At the rate that explosions have occurred in low Earth orbit, one should expect about 10 explosions to have occurred in geosynchronous orbit (Kessler, 1989-B)...yet, none have been officially recorded (Johnson, et al., 1991). However, there have been two reports of an observer witnessing an object exploding in geosynchronous orbit. One report was from Russia, made in February, 1992, reporting that in June, 1978, a USSR Ekran satellite was photographed as it exploded from what was believed to be a Nickel-Hydrogen battery failure (Johnson, 1992). The other was on Feb. 21, 1992, when a Titan upper stage, launched on Sept. 26, 1968, was video taped just after it appeared to explode (Bruck, 1992). However, as yet, no fragments have been catalogued from either of these events, which may not be surprising since fragments smaller than about 1 meter in diameter are difficult to detect from the ground with sufficient regularity to catalogue. Given the improbability that such events would be recorded, other, unrecorded explosions are likely to have occurred. Consequently, a satellite breakup rate due to current operational practices is likely to range between once every 1 to 10 years...a rate much higher than the highest predicted rate based on collisions. The final step in evaluating the significance of these breakup rates is in understanding the number, size, and velocity of fragments generated as a result of breakups and how the resulting debris hazard compares to the natural environment.

Consequences of breakup in geosynchronous orbit

A breakup in geosynchronous orbit has 2 possible consequences: 1. A breakup produces fragments large enough to break up another intact satellite. These fragments contribute to collisional cascading, or "a chain reaction" if an average of more than 1 large fragment per satellite breakup is generated which stays in the geosynchronous ring. If the number of large fragments is significantly larger than one, the contribution to collisional cascading will be greater. 2. A breakup produces small fragments that can collide with and damage operational spacecraft. A key question becomes how this hazard compares to the natural hazard.

Although some data is available on the number, size and velocity of fragments generated as a result of breakups, most of that data was generated under conditions very different than needed to understand the consequences of breakups in geosynchronous orbit. Missing is data from collisions at about 0.5 km/sec, and complete data on explosion fragments smaller

than 10 cm. Because satellite construction is more important at the lower collision velocities expected in geosynchronous orbit, any extrapolation of tests results leads to large uncertainties in predictions.

A "worst case" environment can be predicted by assuming that only the ratio of target mass to projectile mass, as determined by hypervelocity tests, is important in predicting the projectile mass causing catastrophic breakup at 0.5 km/sec. In this case, a 5 kg projectile could catastrophically break up the 2000 kg satellite. Ground explosions and explosions in space suggest that the largest fraction (about half) of the satellite mass goes into about this size fragment (Kessler, 1991-B), so that about 200 fragments of this size would likely be produced. The same data also suggest that these fragments would be ejected in all directions with an average velocity of about 50 meters/sec relative to the center of mass. This velocity is sufficient to spread the fragments over thousands of kilometers of altitude, so that at any one time, only about 20 of the 200 fragments would be found within the 100 km altitude band where geosynchronous satellites are located. Consequently, with this extreme assumption, collisional cascading in geosynchronous orbit is possible; but with only 20 fragments per satellite breakup to contribute to the cascading and with the first collision not expected for 140 years, the cascading would be very slow, requiring thousands of years to be noticeable.

An equally important conclusion from this extreme assumption concerns the "safe" distance to place inactive satellites outside of the geosynchronous orbit. A satellite breakup which occurred within a few thousand kilometers above or below the geosynchronous altitude would eject 5 kg fragments into orbits which passed through geosynchronous orbit. If the breakup were within a few hundred kilometers, the contribution to the hazard to geosynchronous orbit would almost be as great as if the breakup had occurred within geosynchronous orbit (Friesen, et al., 1992-B). Consequently, if the energy of future breakups in geosynchronous orbit is not too different than past breakups in low Earth orbit, the safe distance to place inactive satellites must be measured in thousands of kilometers from geosynchronous orbit in order to be effective.

Hazard resulting from explosions

Since collisions are not likely to be a significant source of debris in the near future, a more important issue might well be the consequence of past explosions in, or near, geosynchronous orbit. Two objects have been observed to explode; it is not unreasonable that 10 times this number, or 20 explosions have occurred. Assuming the same size and ejection velocity relationships as before, we should expect an average of 400 additional objects, with masses of 5 kg or larger, to be in the geosynchronous altitude band at any one time. If these fragments are capable of catastrophic breakup of any of the 250 geosynchronous satellites known to be in geosynchronous orbit, they would increase the catastrophic collision rate from once every 15,000 years to once every 8,300 years...i.e., the explosion fragments would be as important as the known satellites in geosynchronous orbit in contributing toward collisional cascading.

Explosions will also produce smaller debris which will cause a hazard to other spacecraft. However, the type of explosions which are likely to have occurred are not likely to produce a large number of small debris. For example, a low intensity explosion is predicted to produce about 1000 fragments larger than 1 gm (Kessler, 1991-B). These fragments are likely to have velocities larger than the 50 meter/sec for the 5 kg fragments, consequently, spread over a larger volume of space...however, data sources giving the expected velocity is lacking. A conservative assumption would be that the velocities are the same, implying that about 100 of the 1000 fragments would be in the geosynchronous altitude band at any one time. A total of 20 explosions would mean that 2000 fragments of 1 gm and larger are in the geosynchronous altitude band at any one time, producing a flux of 1 impact every 18 million years per square meters of spacecraft cross sectional area. The meteoroid flux for this mass is 1 impact every 3 million years per square meters of spacecraft cross sectional area, which is more than 5 times larger than the debris flux resulting from these explosions. In addition, given the low velocity of 0.5 km/sec which debris is likely to collide with spacecraft in geosynchronous orbit, a debris mass between 5 and 25 times the meteoroid mass (Christiansen, 1992) depending on spacecraft construction, is required in order to do the same damage to the spacecraft as a meteoroid. Consequently, the meteoroid flux which is likely to do the same damage as a 1 gm debris fragment is between 1 impact per square meter every 120,000 to 600,000 years, or much higher than the possible debris flux resulting from 20 past explosions.

All current satellite breakup models predict that the fraction of satellite mass which goes into smaller sizes decreases with decreasing size. On the other hand, the amount of meteoroid mass increases with decreasing size. If the debris flux of 1 gm fragments is less than the meteoroid flux, then all satellite breakup models would predict that the meteoroid flux is also larger than the debris flux for sizes smaller than 1 gm. For the orbital debris hazard in geosynchronous orbit to exceed the meteoroid hazard, many times more than the assumed 20 satellites must breakup in geosynchronous orbit.

Therefore, the ability of breakups to produce an environment in geosynchronous orbit which is more hazardous than the meteoroid environment is much less than in low Earth orbit. This should not be interpreted that one should not be concerned, but rather that there is time to properly consider the total environmental management issue in geosynchronous orbit, and to address the major sources of debris in geosynchronous orbit.

Environment management of geosynchronous orbit
The only seriously considered technique to manage orbital debris in geosynchronous orbit has been the use of a "grave-yard" orbit (Suddeth, 1985; Chobotov, 1990; Flury, 1991). Most studies show that if an intact satellite is placed in a circular orbit about 200 to 300 km away from geosynchronous orbit, it will stay there. However, if one were to do nothing but move all objects from geosynchronous orbit into such a grave-yard orbit, the same orbital debris sources of explosions and collisions would be taking place in the grave-yard orbit. As developed earlier, with only a 200 to 300 km separation distance, the orbits of fragments generated in the grave-yard orbit would still cause an increase in the hazard in geosynchronous orbits that would be reduced by less than a factor of two compared to the hazard caused by the objects fragmenting in geosynchronous orbit. Several thousand kilometers of distance is required in order to prevent a significant fraction of satellite fragments from passing through geosynchronous orbit (Friesen, et al., 1992-B). Consequently, it is important that any long term environment management include other elements.

Perhaps the most important element is to minimize the possibility of accidental explosions in, or near, geosynchronous altitude. In low Earth orbit, this has been accomplished for upper stages by eliminating excess fuel after the upper stage has delivered its payload. Other energy storage devices, such as high pressure containers and batteries should also deplete their energy source. These actions are many orders of magnitude more effective at eliminating near-term sources of debris than is the use of a grave-yard orbit.

However, in the long term, the major energy source for satellite fragmentation is kinetic energy. This energy source can only be eliminated by either eliminating the satellite mass or by minimizing the relative collision velocity between objects in the geosynchronous region. To effectively eliminate the satellite mass, the satellite must be removed from Earth orbit; this is not operationally practical since it requires a delta velocity of more than 1 km/sec. Without station keeping, the relative collision velocity of objects in geostationary orbit will increase to an average of 0.5 km/sec. There is an orbit at geosynchronous altitude where much lower velocities can effectively be accomplished for uncontrolled satellites. The orbit has been referred to as "the stable plane orbit" (Friesen, et al., 1992-A).

Use of the stable plane for environment management
At geosynchronous altitude, the precession or "wobble" of the orbital plane occurs about a plane which is inclined 7.3 degrees to the Earth's equator. It is this precession which produces orbital paths which differ by as much as 14.6 degrees, and produce collision velocities as high as 0.8 km/sec. If a satellite is orbited in this "stable plane," it would have an orbital inclination of 7.3 degrees, and a right ascension of ascending node of zero degrees, and would not have any wobble. That is, without station keeping, all objects in the stable plane orbit will always be moving in the same direction, so that if collisions occur, the relative velocity will be very small...less than 0.005 km/sec. Satellites can easily be con-

structed to avoid fragmenting at this low collision velocity. Consequently, collisions between spacecraft are not likely to produce any fragments large enough to breakup another spacecraft, and collisional cascading is not possible.

The stable plane orbit is not a geostationary orbit. That is, from the ground, a satellite in this orbit will move 7.3 degrees North and South of the equator. For ground antennas without North-South tracking, or antennas which require a high signal strength, this may not be a desirable orbit. For those ground stations with North-South tracking, it can be a highly desirable orbit, since it requires only 5% of the station keeping fuel of a geostationary orbit. Many users have already adopted the practice of not using North-South station keeping in order to extend the satellite life; however, until recently, these users were prevented from launching into the stable plane because of a ruling by the International Frequency Registration Board which limited satellite inclinations to less than 5 degrees. In March, 1991, this limitation was rescinded. Consequently, in the future, both the stable plane orbit and geostationary orbit will have users driven by economic considerations. Therefore, environment management of the geosynchronous region needs to consider both types of orbits.

From an environmental management perspective, use of both the stable plane and the geostationary orbit is preferred to using only the geostationary orbit. Use of geostationary orbit alone, without station keeping, leads to higher collision velocities than using both. From an operational perspective, if both are used, it may be desirable to require the user of one of the orbits to maintain a slight eccentricity so that the two orbital paths cannot intersect. However, this should not be necessary for satellites which do not maintain station keeping since the collision probabilities are no different than for any uncontrolled satellites at geosynchronous altitudes.

If the stable plane is used for geosynchronous operations, then the use of a near-by grave-yard orbit becomes more practical. Objects could be placed only a few hundred kilometers above the geosynchronous stable plane, and still be very near, or in, a grave-yard stable plane which is inclined slightly more than 7.3 degrees. This means that collision velocities in the grave-yard orbit would also be less than 0.005 km/sec, so that if a collision occurred, the debris would not spread to geosynchronous altitude. However, if an object is originally launched into geostationary orbit, the delta velocity required to change to a stable geosynchronous or stable grave-yard orbit is prohibitively high...nearly 400 m/sec.

A final option of the stable plane is to use the two stable points at geosynchronous altitude located over 75 degrees East and 105 degrees West as a grave-yard orbit. These two points are considered desirable operational locations because East-West station keeping is not required. However, without proper environment management, these locations would suffer the highest orbital debris flux. The tendency of objects to move toward these two points make them an even more stable grave-yard location than any other location in the stable plane or in a

higher grave-yard orbit. Collision velocities at the stable points would approach zero, and if any object had a collision velocity greater than zero, any collisions would damp out relative motion until the object came to rest at a stable point. The more mass placed in these stable points, the more stable they become. Consequently, they could represent a long term solution to management of the orbital debris in geosynchronous orbit.

Concluding remarks concerning geosynchronous orbit
An adequate environmental management strategy does not exist for orbital debris in geosynchronous orbit. The use of grave-yard orbits does not address the more serious short term sources of debris: the accidental explosions of upper stages and stored energy devices on satellites. Neither do these proposals significantly reduce the hazard caused by the long term sources of collisional fragmentation.

Current operational practices in or near geosynchronous altitudes combined with the long orbital life of debris generated as a result of these operations make an environmental management strategy desirable. Some geosynchronous operators are unilaterally performing maneuvers in the belief that they are contributing to proper environment management. With less operational expense, these operators might make a much larger contribution to environment management, once a strategy has been established.

The current hazard to spacecraft in geosynchronous orbit from orbital debris is low and is likely smaller than the hazard from natural meteoroids. However, future activities in the geosynchronous region may be on a scale much different than today's operations. We would be ill advised to preclude these operations because of poor environment management practices of today. Requiring operators to deplete excess fuel in upper stages left in geosynchronous orbit would be a much more effective management practice than requiring operators to maneuver to a grave-yard orbit.

Other options to manage orbital debris in geosynchronous orbit should be considered. Priorities based on the trade-off between operational expenses and an effective environment management strategy should be established. In order to do this, better models need to be developed. These models should be based on better data obtained from ground tests of satellite breakups, and the models should be validated with better observational data of the environment in geosynchronous orbit. Until an environmental management strategy is established which considers the cost effectiveness of all options, it is premature to establish policy adopting one option over another.

REFERENCES

Badhwar, G.D. and P.D. Anz-Meador (A). "Mass Estimation in the Breakups of Soviet Satellites," Journal of The British Interplanetary Society, Vol. 43, 1990.

Badhwar, G.D., A. Tan, and R.C. Reynolds (B). "Velocity Perturbation Distributions in the Breakup of Artificial Satellites," Journal of Spacecraft and Rockets, Vol. 27, No. 3, 1990.

Barrett, R.A., R.P. Bernhard, and D.S. McKay. "Impact Holes and Impact Flux on Returned Solar Max Louver Material," Lunar and Planetary Science XIX, March 14–18, 1988.

Bernhard, R., W. Davidson, and F. Hörz. "Preliminary Analysis of LDEF Instrument A0187–1, Location A11,R," Presented at 2nd LDEF Post-Retrieval Symposium, June, 1992.

Bess, T.D. "Mass Distribution of Orbiting Man-Made Space Debris," NASA TND D-8108, 1975.

Brooks, D.R., G.G. Gibson, and T.D. Bess. "Predicting the Probability that Earth-Orbiting Spacecraft Will Collide with Man-Made Objects in Space," Space Rescue and Safety, pp. 79–139, 1975.

Bruck, R. Personal communication, GEODSS site, Maui, 1992.

Chobotov, V.A. "Disposal of Spacecraft at End of Life in Geosynchronous Orbit," Vol. 27, No. 4, 1990.

Christiansen, E., J.R. Horn, and J.L. Crews. "Augmentation of Orbital Debris Shielding for Space Station Freedom," AIAA 90–3665, 1990.

Christiansen, E. NASA, JSC personal communication, 1992.

Donahoo, M.E. "Collision Probability of Future Manned Missions with Objects in Earth Orbit," MSC Internal Note No. 70-FM-168, October, 1970.

Eichler, P. and D. Rex. "Chain Reactions," AIAA Paper 90–1365 presented at AIAA/NASA/DOD Orbital Debris Conference, 1990.

Farinella, P. and A. Cordelli. "The Proliferation of Orbital Fragments: A Simple Mathematical Model," Science & Global Security, Vol. 2, pp. 365–378, 1991.

Friesen, L.J., A.A. Jackson, H.A. Zook, and D.J. Kessler (A). "Analysis of Orbital Perturbations Acting on Objects in Orbits near GEO," Journal of Geophysical Research, Vol. 97, No. E3, pp. 3845–3863, 1992.

Friesen, L.J., A.A. Jackson, H.A. Zook, and D.J. Kessler (B). "Results in Orbital Evolution of Objects in the Geosynchronous Region." Journal of Guidance, Control, and Dynamics, Vol. 15, No. 1, 1992.

Flury, W. "Collision Probability and Spacecraft Disposition in the Geostationary Orbit," Adv. Space Res., Vol. 11, No. 12, 1991.

Goldstein, R. and L. Randolph. "Rings of Earth Detected by Orbital Debris Radar," JPL Progress Report 42–101, May, 15 1990.

Guermonprez, V. "Coping with Geostationary Overcrowding," 5th NASA/ESA Coordination Meeting, October, 1990.

Grün, E., H.A. Zook, H. Fechtig, and R.H. Giese. "Collisional Balance of the Meteoritic Complex," Icarus, Vol. 62, pp 244, 1985.

Hechler, M. and J.C. van der Ha. "The Probability of Collisions on the Geostationary Ring," ESA Journal, Vol. 4, No. 3, 1980.

Hechler, M. "Collision Probability at Geosynchronous Altitudes," Adv. Space Res., Vol. 5, No. 2, 1985.

Henize, H. and J. Stanley. "Optical Observations of Orbital Debris," AIAA paper AIAA-90–1340 presented at AIAA/NASA/DOD Orbital Debris Conference, 1990.

Hörz, F., R.P. Bernhard, J.L. Warren, T.H. See, D.E. Brownlee, M.R. Laurance, S. Messenger, and R.B. Peterson. "Preliminary Analysis of LDEF Instrument A0187–1: The Chemistry of Micrometeoroids Experiment," NASA CP-3134, 1991.

Johnson, N.L. and D.J. Nauer. "History of On-Orbit Satellite Fragmentations," Teledyne Brown Engineering Technical Report CS91-TR-JSC-008, July, 1991.

Johnson, N.L. Reported in Aviation Week, March 9, 1992.

Kessler, D.J. "Average Relative Velocity of Sporadic Meteoroids in Interplanetary Space," AIAA Journal, Vol 7, No. 12, December, 1969.

Kessler, D.J. "Collision Frequency of Artificial Satellites: The Creation of a Debris Belt," Journal of Geophysical Research, Vol. 83, No. A6, pp. 2637–2646, June, 1978.

Kessler, D.J. (A). "Sources of Orbital Debris and the Projected Environment for Future Spacecraft," Journal of Spacecraft and Rockets, Vol. 18, No. 4, pp. 357–360, 1981.

Kessler, D.J. (B). "Derivation of the Collision Probability Between Orbiting Objects: The Lifetimes of Jupiter's Outer Moons," Icarus, Vol. 48, pp. 39–48, 1981.

Kessler, D.J., R.C. Reynolds, and P.D. Anz-Meador. "Orbital Debris Environment for Spacecraft Designed to Operate in Low Earth Orbit," NASA TM-100-471, April, 1989.

Kessler, D.J. (B). "Technical Issues Associated with Orbital Debris in GEO," Orbital Debris Monitor, Vol. 2, No. 4, 1989.

Kessler, D.J. "Collision Probability at Low Altitudes Resulting from Elliptical Orbits," Adv. Space Res., Vol. 10, No. 3–4, pp. 393–396, 1990.

Kessler, D.J. (A). "Collisional Cascading: The Limits of Population Growth in Low Earth Orbit," Adv. in Space Res., Vol 11, No. 12, pp.63–66, 1991.

Kessler, D.J. (B). "Orbital Debris Environment for Spacecraft in Low Earth Orbit," Journal of Spacecraft and Rockets, Vol. 28, No. 3, pp. 347–351, 1991.

Kessler, D.J. "Origin of Orbital Debris Impacts on Long Duration Exposure Facility's (LDEF) Trailing Surfaces," Presented at 2nd LDEF Post-Retrieval Symposium, June, 1992.

Lee, J., D. Buden, T. Albert, W. Margopolous, J. Angelo, and S. Lapin. "Technology Requirements for the Disposal of Space Nuclear Power Sources and Implications for Space Debris Management Strategies," AIAA paper AIAA-90-1368, presented at AIAA/NASA/DOD Orbital Debris Conference, 1990.

Loftus, J.P., D.J. Kessler, and P.D. Anz-Meador. "Management of the Orbital Debris Environment," Acta Astronautica, Vol. 26, No. 7, pp 477–486, 1992.

McKnight, D.S. "Examination of Possible Collisions in Space," IAF paper IAA-91-593, 1991.

Mueller, A.C. and D.J. Kessler. "The Effects of Particulates from Solid Rocket Motors Fired in Space," Adv. Space Res., Vol. 5, No. 2, 1985.

Mulholland, J.D., S.F. Singer, J.P. Oliver, J.L. Weingerg, W.J. Cooke, N.L. Montague, J.J. Wortman, P.C. Kassel, and W.H. Kinard. "IDE Spatio-Temporal Impact Fluxes and High Time-Resolution Studies of Multi-Impact Events and Long-Lived Debris Clouds," NASA CR-3134, 1991.

Perek, L. "Physical Nature and Technical Attributes of the Geostationary Orbit," UN Committee on the Peaceful Uses of Outer Space, 1978.

Reynolds, R.C. "Review of Current Activities to Model and Measure the Orbital Debris Environment in Low-Earth Orbit," Adv. Space Res., Vol. 10, No. 3, pp. 359–372, 1990.

Royal Aircraft Establishment. "The RAE Table of Earth Satellites 1983–1985," 1986.

Stansbery, E.G. and C.C. Pitts. "Radar Measurements of the Orbital Debris Environment," ESA SP-342, October, 1991.

Stansbery, E.G. "Characterization of the Orbital Debris Environment Using the Haystack Radar," JSC-32213, April 24, 1992.

Suddeth, D.H., "Debris in the Geostationary Orbit Ring: 'The Endless Shooting Gallery' – The Necessity for a Disposal Policy," NASA CR-2360, 1985.

Taft, L.G., D.E. Beatty, A.J. Yakutis and P.M.S. Randall. "Low Altitude, One Centimeter, Space Debris Search at Lincoln Laboratory's (M.I.T.) Experimental Test System", Adv. Space Res., Vol. 5, No. 2, pp 35–45, 1985.

Talent, D.L. "Analytic Model for Orbit Debris Environment Management," Journal of Spacecraft and Rockets, Vol. 29, No. 4, pp. 508–513, July-Aug., 1992.

Tedeschi, W., J. Connell, D. McKnight, F. Alladadi, A. Reinhardt, R. Hunt, and D. Hogg, "Determining the Effects of Space Debris Impacts on Spacecraft Structures," IAA paper, IAA-91-594, 1991.

Thompson, T.W., R.M. Goldstein, D.B. Campbell, E.G. Stansbery, and A.E. Potter. "Radar Detection of Centimeter-Sized Orbital Debris: Preliminary Arecibo Observations at 12.5 cm Wavelength," Geophysical Research Letters, Vol. 19, No. 3, 1992.

Zook, H.A., D.S. McKay, and R.P. Bernhard. "Results from Returned Spacecraft Surfaces," AIAA paper AAIA-90-1349, AIAA/NASA/DOD Orbital Debris Conference, 1990.

Zook, H.A. NASA, JSC personal communication, 1992.

4: The Current and Future Space Debris Environment as Assessed in Europe

Dietrich Rex

Technical University of Braunschweig, Germany

1 INTRODUCTION

Spaceflight – wasn't that meant to be the conquering of the universe by mankind ? We speak of astronauts, which means "star travellers", or even of cosmonauts, "seaman in the cosmos", and also the term "space" itself makes no distinction to "cosmos" or "universe". These words, coined at the time of the upcoming "space age" are evidence of a fundamental misconception, slipped into the minds of the enthusiastic elder generation and still alife in science fiction literature. The cosmos, the universe, is something nearly infinitively large, should one ever be concerned about overcrowding it with man-made objects ?

However, space flight as being exercised today and also in all foreseeable future, has nothing to do with the "universe". It mainly utilizes a shell around the earth with a thickness of about one third of the earth's radius, it utilizes extensively the geostationary ring about 6 earth radii away from the earth's surface and it performs a few missions in the Solar System. So the space mainly used by man is by no means infinitively large, in fact compared to the universe it is ridiculously small, though its use with its manifold technical and social progress and its perspectives justifies the term "space age".

While the elder generation was unaware of the problem, it now becomes evident that this comparatively small volume used for spaceflight can easily become overcrowded by space activities. Considering the high velocities of orbiting objects and the even higher velocities relative to each other, the risk of collisions especially in the near earth environment, but also in the geostationary ring, is no longer negligible and will become a serious obstacle for spaceflight in the future if space activities are just continued as in the past.

The risk would have been avoidable if only those objects had been brought and left in orbit which are really needed there: the number of those space objects presently in active use is less than a half percent of all man-made objects (larger than 1 cm) in the near earth environment. The bulk is space debris, drifting around on their orbits as useless and uncontrolled remainders of 34 years of spaceflight. Obviously somehow our habits have to be changed.

This perception now prevails not only among scientists and engineers, but also among many managers in the administrations and among the decision makers. It is common conviction that the geostationary ring, but also the near earth space, is a limited natural resource important for the use by all mankind and by the generations to come, and that it has to be protected.

2 SURVEY OF EUROPEAN ACTIVITIES

In 1987, the Director General of ESA set up an ESA Space Debris Working Group, consisting of experts from the member states. Preceding to this first highlight, and paving the way to it, there had been re-entry prediction campaigns in several European countries for potentially dangerous re-entry events, a study on space debris in my institute at the Technical University of Braunschweig financed by the German Ministry of Research and Technology and finally and especially important, the activities of Dr. Flury, ESOC, to channel all these activities in the framework of ESA.

The ESA Space Debris Working Group, within 18 months layed down all the available technical knowledge, the European facilities and first proposals for mitigation of the space debris problem within ESA in a report "Space Debris". The report can be compared in its contents to the US-report "Orbital Debris" issued in 1989 by an Interagency Working Group. By the European report it became clear that Europe heavily depended on US knowledge and data in the space debris field and that increased European activities should be initiated. Since then, European expertise has been enhanced and independent data have been and are being obtained by the LDEF-impact evaluation, by the evaluation of the IRAS-infrared observations, by hypervelocity impact testing and by using the German radar facility FGAN, but still the main knowledge of the orbiting debris population depends on radar and optical data from the US.

As a successor of the Space Debris Working Group, there is now the ESA Space Debris Advisory Group which reviews the ESA space debris study plan and the various studies on space debris performed in industry and institutes and considers space debris related activities within the ESA programmes. ESA has set up a Space Debris Coordination and Technical Analysis

Group for the in house activities. All ESA space debris activities are included in the mutual exchange with NASA through the "ESA-NASA-Cooerdination Meetings" taking place every half year.

The ESA council, in 1989, adopted its "Resolution on the Agency's Policy Vis-a-vis the Space Debris Issue" which establishes basic guidelines to be observed in the ESA programmes. Following this policy, many measures have already been taken in ESA programmes to avoid further debris generation and to protect hardware against impacts. The residual fuel from the Ariane 4 upper stage is now vented after the delivery of payloads into low earth orbits, so that the explosion which happened with the Ariane V16 flight in Nov. 1986 cannot occur again. Presently the same venting is not feasible for missions into the geostationary transfer orbit, but it is considered for the future.

Several European satellites in the geostationary orbit have been re-orbited at the end of their active life into orbits sufficiently higher than the geostationary ring, so that the risk of future collisions is eliminated. Also for most of the future European geostationary satellites it is the intention to proceed in the same way.

In the Columbus programme, the shielding of the pressurized European module to be attached to the International Space Station has been intensively studied. A programme of hypervelocity impact tests with various shielding geometries is under way in order to secure the modules against impacts of debris objects up to sizes of 1 (or possibly 2) centimeters.

ESA has a long term study plan to initiate and finance studies on space debris by European institutes and industry. Some such studies have already been completed, others are under way. In addition to this programme also national investigations are being performed. In the following, I can mention just a few of these activities.

As a basic source of information ESA/ESOC has developed the DISCOS data base on all trackable debris objects and supplies it to users for their own work. A second major step to obtain detailed space debris expertise in Europe is the development of an "ESA Space Debris Reference Model", the work on which is under way by ESA contract to my institute at the Technical University of Braunschweig. The model will comprise all objects larger than 0.1 mm in size and their historic and future evolution and will describe the three dimensional spatial distribution.

Institutes in several European countries are evaluating the impacts on various surfaces of the LDEF-satellite retrieved after 6 years' exposure to the space environment. This analysis is an important data source for objects smaller than a few tenth of a millimeter. In this size range, the discrimination between man-made debris particles and natural meteoroids is essential. This is tackled by both, chemical analysis and analysis of the angular surface distribution.

The IRAS infrared astronomical satellite performed its mission 1983. It was found that the astronomical observation data,

as a by-product, also contain much information on space debris objects which by chance crossed the field of view. These data have been extracted from a vast data bank and are now being analysed to determine the debris characteristics.

The feasibility of obtaining more data on space debris particles in the important millimeter and centimeter size range by optical/infrared observation either from the ground or onboard an orbiting detector is being studied by an ESA study and also by a German national study. The results seem encouraging both for ground based (low cost) and space based (higher cost) detectors, but at present no concrete mission plans have materialized due to lack of funding. In my own institute, since 6 years now funded by the German Ministry of Research and Technology (BMFT) and by the German Space Agency (DARA), all aspects of space debris in low earth orbits are investigated. This research includes analysis of the generation of debris by explosion and collision events, the time evolution of debris clouds, the deterministic orbit propagation of all known and simulated objects, scenarios for the future including the possibility of a collision chain reaction. Also the debris impact flux distribution on a reference object and its optimal shielding is analysed as well as means to remove debris objects from space. The following chapters of my paper are mainly based on this work.

This research programme of our institute, since two years, is performed in close cooperation with NASA (mainly NASA – Johnson Space Flight Center in Houston, Texas), based on an agreement between BMFT/DARA and NASA. Meetings every half year between the two groups have proven to be very efficient to understand and possibly eliminate any differences of the results obtained while maintaining the differences of the approaches used to have a cross-check. These meetings are in addition to the ESA-NASA-Coordination Meetings already mentioned.

3 SPACE DEBRIS DISTRIBUTION AND DYNAMICS

3.1 Overview

The term space debris refers to all man-made solid objects on orbits around the earth which are not or not any more used to fulfill any task. It comprises all abandoned satellites and objects related to missions (ejected covers, clamp-bands, bolts, …), rocket upper stages and the vast number of objects generated on orbit by fragmentation events. These fragmentations are unintentional explosions (high intensity and low intensity, especially of rocket upper stages), intentional military explosions and some collisions.

In this paper the debris population will be treated in three size classes

- all objects larger than 0.1 mm, about 10^{10} objects
- all objects larger than 1 cm, about 80,000 objects.
  Active satellites or space stations cannot be shielded against this class

- objects larger than 10 cm, about 7,000 objects.
  This is the population tracked by US Space Command and individually known by their "Two Line Elements" (TLE)

Fig. 1 gives a break-down scheme of this whole population. The extreme increase of the numbers towards lower sizes should be noted, which cannot be brought to scale by the circles in Fig. 1. Only about 350 satellites are actively used at present, these alone would not represent any space debris problem.

This break-down and all other properties of the debris environment presented here is the result of the present state of modeling for the untrackable objects smaller than 10 cm. Some methods of this modelling are contained in the following chapters. The objects larger than 10 cm are directly represented from the TLE measurement data and other information available.

3.2 The tracked objects

The lower size limit of this class is between 5 and 10 cm depending on the orbital altitude, due to the detection limit of the US Space Command radar facilities. As Fig. 1 indicates, also in this larger size class about 50% is fragmentation debris, 25% unused mission related objects, mainly rocket upper stages. Although during the fragmentation events the fragments receive some additional velocity, this does usually not change very much the orbital inclination, so that they also represent the inclination distribution of the parent bodies. As Fig. 2 indicates, the inclination distribution reflects the latitude of the launch sites and special purpose orbits as e.g. the sun-synchronous inclination 97...100°.

The eccentricity distribution in Fig. 3 and the perigee- and apogee altitude distributions in Fig. 4 and 5 show that apart from some special classes, namely the geostationary-transfer and the Molnija-type orbits, the rest of the LEO population has very low eccentricities.

This will be also reflected by the angular distributions of impacts on an orbiting object in section 3.4. Fig. 6 shows the historical increase of this population class since the beginning of spaceflight in 1958. It is not unreasonable to extend this general trend into the future. Fig. 6 also shows that the result of our modeling of fragmentation processes as described briefly in the following chapters is in good conformity with the measured data.

3.3 The 1 to 10 centimeter class

With respect to the risk of impacts on active satellites or space stations, this is the most important class. These objects are too large to render them ineffective by shielding, but they are too small to be detected from ground by the normal radar stations. So they are unknown objects which can only be modelled by a statistical simulation of their generation processes and the orbital behaviour over time. Also, their number is by about a factor of 10 higher than the number of tracked objects. The bulk of this size class originates from explosion events in space.

Fig. 7 lists the time history of all explosion events in the past. Each of these events has generated a debris cloud, which in number, size distribution and orbits of the fragments depends on the intensity and other characteristics of the explosion. In order to model the whole fragment population, each explosion event is simulated, the size and orbital distributions of the fragments are obtained at the time of the event and then propagated up to present. The fragment size distribution for each event is fitted to the size distribution of the tracked fragments of that same event, as is indicated by Fig. 8. The general shape, i.e. the mathematical function for the curve in Fig. 8 is derived from explosion experiments on the ground, and the mass distribution of the tracked fragments is derived in a statistical manner from the radar cross-section measured.

While this simulation may be fairly uncertain for the single event, on the average over more than 100 events the uncertainties level out.

Fig. 9 and 10 show the behaviour of a typical debris cloud in the first hours after the explosion event, and in the following months and years. Shortly after the event the debris distributes in a ring in the orbital plane of the parent object, and later it distributes in a shell around the earth due to differential perturbing forces. During that later time also objects are eliminated from the debris cloud due to atmospheric drag, which reduces the debris flux in the cloud as indicated by Fig. 11. This later effect of course depends on the orbital altitude and would nearly by absent above 1000 km.

3.4 Distribution and behaviour of all objects larger than 1 centimeter

By summing up all fragments from the more than 100 known fragmentation events, adding them to the tracked objects of the US SpaceCom Catalog and making some corrections to account for additional optical measurements, a model of the whole debris population of objects larger than 1 cm is obtained. Our computer model of these objects contains 85,000 individual objects, i.e. orbits, sizes, time history of them are described in a *deterministic* way, although the majority of them have been generated initially by a *statistical* simulation of the explosion event. The spatial distribution of the density of objects per km^3 as described by this model is shown in the next figures. Fig. 12 shows the present altitude distribution with characteristic peaks at 800, 1000 and 1500 km. Fortunately, present manned missions between 300 and 500 km are in a region where the atmospheric drag, the so called self-cleaning effect of the atmosphere, reduces the density considerably compared to the peaks.

The peaks at 1000 km, especially that at 1500 km, will not be reduced by drag in any reasonable time, so objects brought to these altitudes just accumulate there. So it needs no sophisticated computer simulation to predict that overcrowding there is merely a matter of time, unless spaceflight activities are generally changed so as to retrieve or de-orbit all payloads deployed in these altitudes. The beneficial properties of

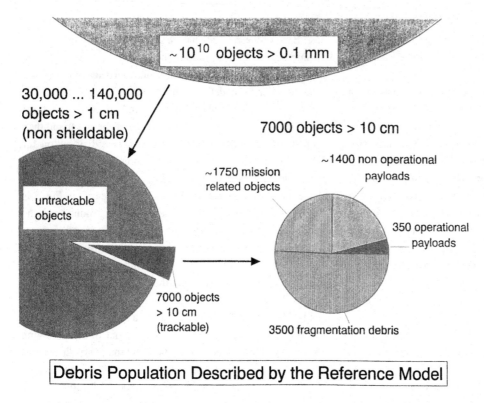

Figure 1: The composition of the orbital debris in earth orbits larger than 0.1 millimeter.

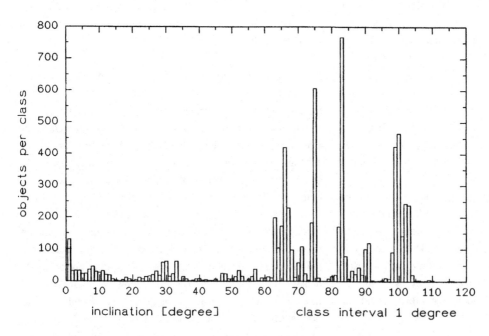

Figure 2: The inclination distribution of the trackable debris population (larger than ~10 cm).

the atmospheric self-cleaning effect is dramatically demonstrated by Fig. 13 which indicates the vast debris peak which would be there at altitudes around 200 km without the atmospheric drag.

Fig. 14 is one example of the scenarios obtained for the future development, here with a more or less direct extrapolation of present days' activities. Again, the sensitivity of higher altitudes against pollution is obvious. This scenario does not include interactive collisions as a source for new debris, which is treated separately in section 3.6.

In the following, some examples of the more detailed information which can be obtained from such a debris model are presented. The two diagrams in Fig. 15 show the debris impacts representative for a space station at 450 km altitude and 28.5° inclination. The broad velocity distribution up to about 14 km/s is important for shielding design. What is most

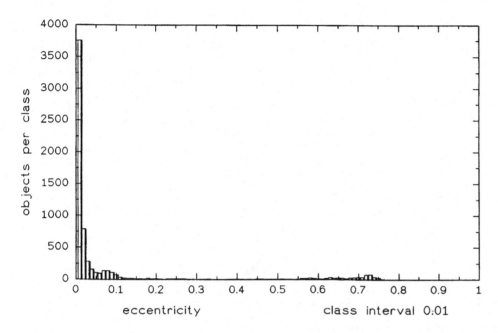

Figure 3: The eccentricity distribution of trackable debris population.

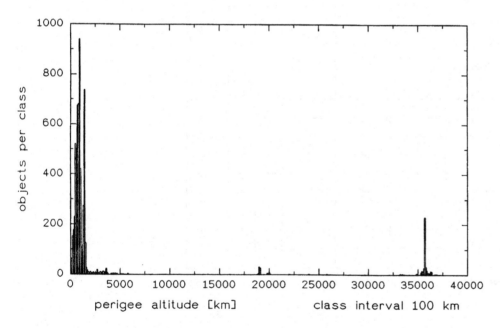

Figure 4: The distribution of perigee altitude of the trackable debris population.

remarkable, is the distinct directional distribution of the impacts visible in Fig. 15.

Nearly all impacting objects are approaching the space station in the horizontal plane, and within that plane distinct angles to both sides of the direction of flight predominate. This peculiar behaviour is strongly related to the inclination- and eccentricity-distributions shown in Figures 2 and 3. These directional properties of the debris flux could be – but are not really at present – utilized to optimize the shielding design for space stations. The angular and velocity distributions of debris impacts on an orbiting object depend on the orbital inclination of that object as shown in Figures 16 and 17. The configuration

of satellites with respect to debris-sensitive surfaces (tanks, optical apertures etc.) can be designed using this knowledge.

3.5 The class of particles smaller than 1 centimeter

It is expected that manned objects can be shielded against particles of this size range. But for the design of that shielding and also for the risk assessment of unshielded objects (e.g. satellites) it is essential to have a model of this size class providing its altitude distribution and the angular distribution of any impacts.

However, the origin of that flux, especially in the millimeter and submillimeter size range, is still not fully understood. So,

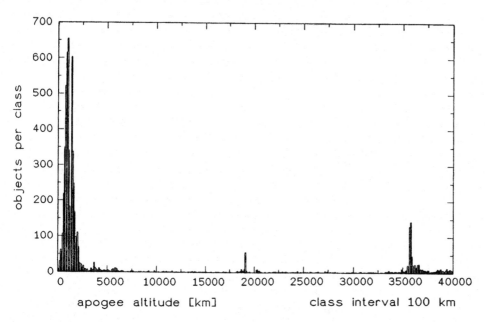

Figure 5: The distribution of apogee altitude of the trackable debris population.

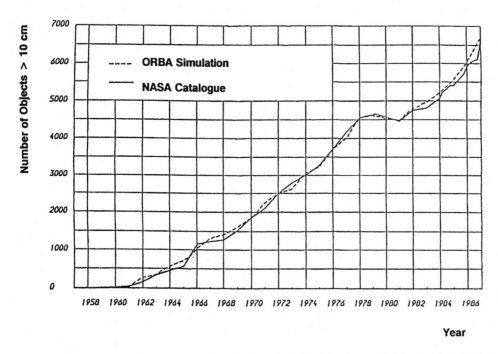

Figure 6: The increase of the number of objects since the beginning of space flight ——— from TLE data – – – – debris from modelling.

in the following, only some first and preliminary approaches to obtain a model for that size range can be sketched. From the evaluation of impacts on surfaces retrieved from space it is evident, that there must be a huge flux of small particles in the submillimeter range, see Fig. 18. The normal low intensity explosion (e.g. most chemical reactions of residual fuel in rocket upper stages) does not create so many very small particles. Only high intensity explosions (i.e. detonations) and collisions (e.g. impacts of smaller debris particles on abandoned satellites) would provide sufficient energy for a particle spectrum with a large portion of small particles, as shown in Fig. 19.

So for a model of this size range one has to assume such high energy events. Such high energy events would provide much higher velocity increments to the fragments (in the km/s range rather than in the 100 m/s range as in low energy explosions), which means a much broader eccentricity- and inclination-distribution of the fragments compared to their parent objects. This in turn influences the angular distribution of impacts on the various surfaces of a satellite. So a proof of the model can be made by calculating the angular debris impact distribution for the LDEF satellite and comparing it to the measured sum of man-made debris impacts and meteoroids, Fig. 20. Although some agreement can be stated, the statistical

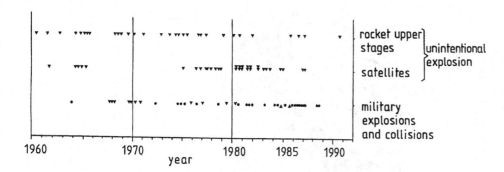

Figure 7: Fragmentation events in earth orbits since 1960.

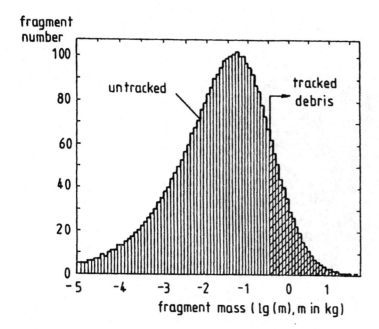

Figure 8: Mass spectrum of a typical explosion event (Ariane V16) matched to the distribution of the tracked fragments.

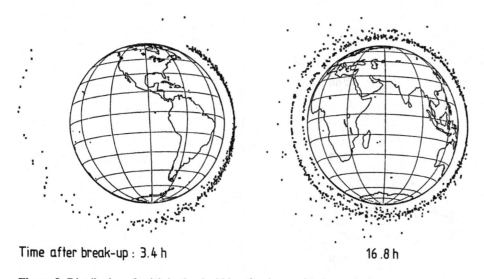

Figure 9: Distribution of a debris cloud within a few hours after the explosion event.

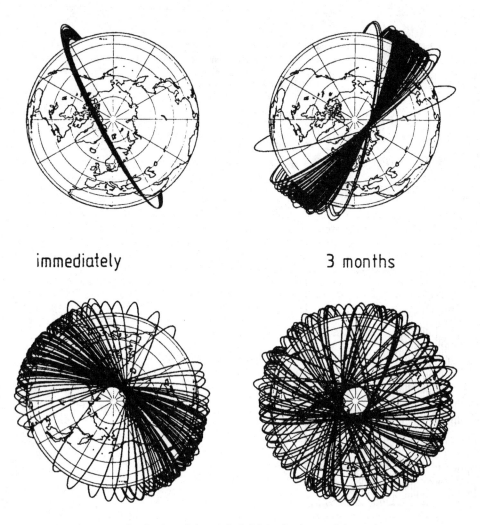

immediately   3 months

Figure 10: Long term distribution of the original debris cloud around the earth.

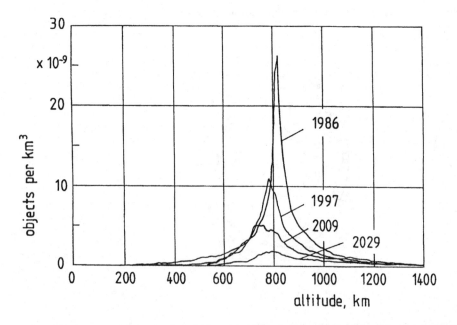

Figure 11: Decrease of the density within a debris cloud due to elimination of fragments by atmosphere drag.

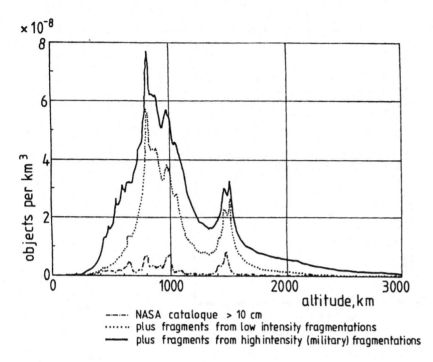

Figure 12: Altitude distribution of the debris density of all objects larger than 1 cm.

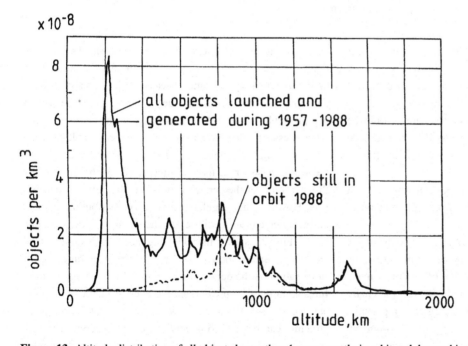

Figure 13: Altitude distribution of all objects larger than 1 cm presently in orbit and decayed in the past.

material is not yet sufficient for a final confirmation of a new small particle model.

3.6 Collision chain reactions
As mentioned before, at present the bulk of the orbital debris has been generated by explosions of orbiting objects. Only a few, perhaps 2 or 3 collisions may have contributed, although the number of collisions already occured in the past is not known with certainty. With increasing numbers of objects in orbit, the number of collisions between any of these rises with

the square of the numbers, so that the average time between two collisions, which is presently believed to be around 2 to 5 years, may considerably become shorter. Then eventually, collisions may become a major source of new debris. This, obviously, is a very unfavorable development which should be avoided by all means by preventive measures.

Even more unfavorable is a certain possibility within the collision evolution which is called the collision chain reaction. When two larger objects collide they disintegrate completely. The resulting broad fragment size spectrum may contain sev-

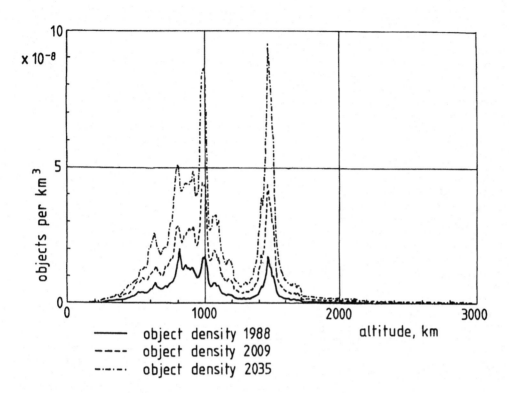

Figure 14: Simulation of the debris distribution in space for the future.

eral larger fragments, which in turn are sufficient to trigger another complete disintegration and so forth, so that an avalanche of debris can be created. Of course the time interval between two such events would be large initially, but nevertheless a self-sustained and uncontrolled situation could be reached in this way, which would be able to fill certain altitude regions around the earth to such an extent, that no further space flight would be possible there.

Closer analysis of this phenomenon shows that the present LEO environment may not be very far from the initiation of such a process.

Fig. 21 shows which sizes of debris objects would mainly be involved in a collision chain reaction process. Since these are mainly the large objects, the knowledge of which is much more certain than that of the small debris objects, the chain-reaction can fairly well be assessed. Fig. 22 shows a potential increase of the object numbers in the low earth environment resulting from a simulation of the process.

In order to prevent such a process, it would be necessary to limit the number of large objects in the altitude regions of highest density. This must be achieved early enough. Once the threshold is exceeded, then only active removal of objects from orbit would be a remedy (see section 4.3).

4 COUNTERMEASURES AGAINST SPACE DEBRIS

4.1 Shielding

Manned spacecraft planned for a longer residence in space have to be shielded against space debris impacts in order to ensure the necessary safety level for astronauts. This for the first time becomes a necessity for the pressurized modules of the international space station because of its size and mission duration. Without discussing the shielding hardware design here, it can be stated that the knowledge of the debris environment is essential for the mass optimization of double or triple wall shieldings. A simple not refined shielding would cover the whole pressurized module irrespective of the impact flux distribution over the surface and its specific velocity distribution. However, the angular and velocity distribution on each surface element are decisive for the shielding design at that place. Fig. 23 shows the shielding mass of a double wall shielding required for various levels of optimization using the information on the debris flux available. In larger space structures as in the four pressurized modules of the planned space station, the self-shielding between the various parts of the structure can be used to largely reduce the amount of shielding. Fig. 24 shows the flux on the various surfaces of the pressurized module array.

4.2 Collision warning, collision avoidance orbital manoeuvres

By optical, IR or radar sensors onboard a space station a certain portion of uncatalogued objects due to make an impact on the space station later could be potentially detected early enough to allow for a collision avoidance orbital manoeuvre.

It has to be investigated whether the portion of pre-detected collision objects would be large enough to justify the effort and whether the number of false alarms could be limited to a reasonable level. Impacting objects could be identified by the

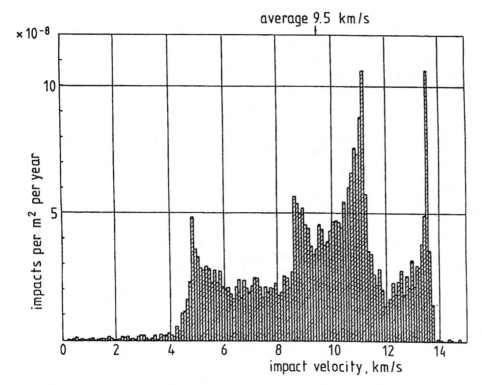

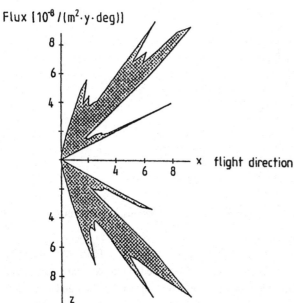

Figure 15: above: velocity distribution of debris impacting on a space station at h = 450 km and inclination 28.5°
below: angular distribution of space debris impacting on a space station in the horizontal plane. Angles out of the horizontal plane are nearly zero.

fly-by repetition rate some orbits (i.e. several hours) prior to the impact, if an object recognition is possible due to time interval and approach angle, see Fig. 25. Key issues are the sensitivity of the detector (i.e. the range for objects down to centimeter size) and the precision of the orbit prediction both for the debris object and the space station. The collision avoidance manoeuvre of the space station could be made part of the normal orbit raising scheme and so would not require any extra propellant. The manoeuvre would in any case result in an along-track avoidance as shown in Fig. 26.

The methods discussed and analysed in this connection are also in close relation to debris detection missions to support the debris model, as mentioned in section 2.

4.3 Active removal of debris from space
Many ideas to remove space debris from space are discussed in the literature. Here we will not consider those mostly questionable schemes to clear orbital regions from small particles. As has been discussed in chapter 3.6, the larger objects are those which could trigger a collision chain reaction.

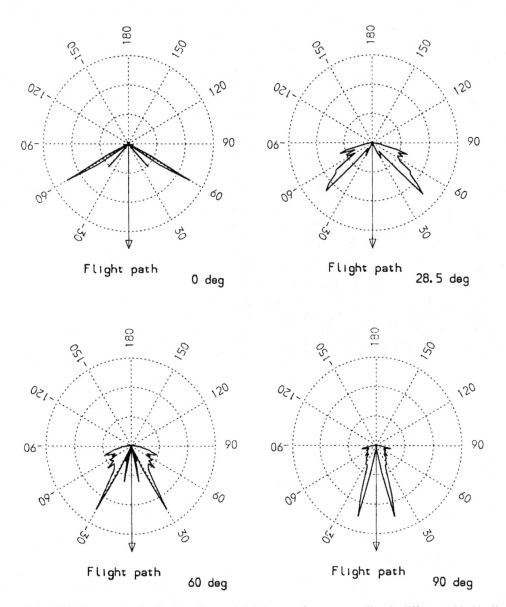

Figure 16: The angular distribution of space debris impacts for target satellites in different orbital inclinations, h = 500 km.

Therefore this threat could be counteracted by removing a considerable number of large objects from space, if it turns out that now already or later the critical density for such avalanche of debris generation has been reached. In any case, this would be a costly operation, and since no single nation would be responsible for it, it only could be performed in an multinational mission. All plans to use a garbage collecting vehicle which then either retrieves the debris to the ground or makes it re-enter by thrusting are highly uneconomic as can easily be shown.

The new space tether technology could be a way to solve the problem. An unmanned debris remover vehicle called "TERESA" in our study, operated in a multinational mission, would be launched into an inclination which contains many large debris objects. In fact, the debris is clustered in a few inclination classes as can be seen in Fig. 2, so that without costly inclination changes the remover vehicle could perform rendez-vous manoeuvres with many

debris objects one after the other. The TERESA vehicle as shown in Fig. 27 is equipped with an extendable and retrievable tether of typically 1 mm in diameter and 100 km in length (typical mass 200 kg). So, after the first space debris object has been gripped, it can be roped down to a lower altitude.

There it is detached from the tether and re-enters to the atmosphere (with impact into one of the large oceans) due to orbital mechanics. Orbital mechanics would also have the effect of bringing the remover vehicle to a higher elliptical orbit by that manoeuvre, but this can be counter-acted either by conventional thrust or by electromagnetic braking with the tether-wire in order to bring TERESA to the vicinity of the next object to be removed. Fig. 28 shows a typical manoeuvre-sequence of TERESA. So with very little expenditure of propellant a number of typically 100 objects of several 100 or 1000 kg each could be removed from orbit within some years. TERESA has been studied in some depth and it

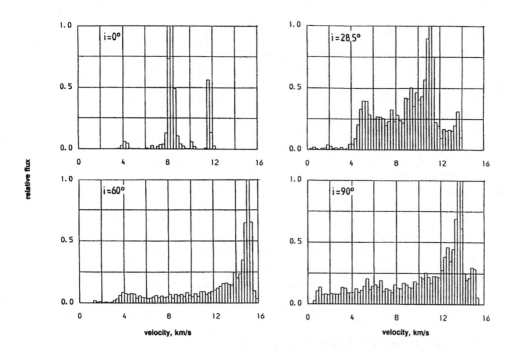

Figure 17: Velocity distributions of space debris impacts for target satellites in different inclinations (as in Fig. 16), h = 500 km.

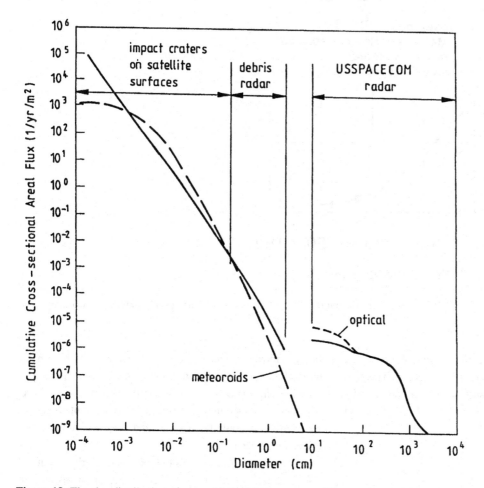

Figure 18: The size distribution of space debris objects meas-ured from various sources.

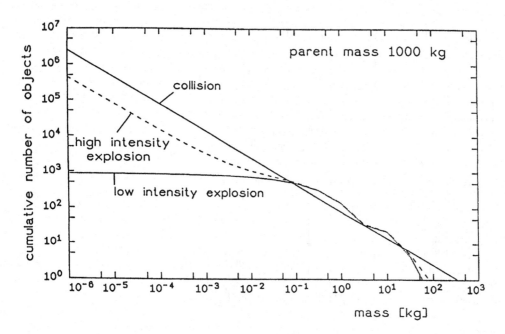

Figure 19: The size spectra (cumulative) of fragments from 3 different fragmentation events.

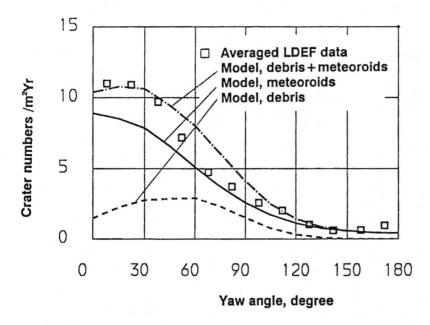

Figure 20: Attempt to interpret the measured impact crater distribution on LDEF by the sum of meteoroids and debris including elliptical orbits.

seems to be feasible and also superior to other concepts, but still would imply the cost of a major space mission. Therefore all efforts should be taken that such debris removal mission will not become necessary. This calls for early debris avoidance in all missions.

5 DEBRIS AVOIDANCE

It is the conclusion from all space debris studies, that space flight has a long term future only if all unnecessary orbiting objects related to space missions are avoided. This has technical and cost effects which finally must be borne by all space

faring nations. Obviously this will only be possible on the basis of an international agreement to avoid unequal competition. It seems that if such debris avoidance schemes are introduced early enough in rocket design and mission planning, the cost burden will only be marginal and does not hamper space utilization in general.

Of course, not all measures to limit space debris are equally effective, and the technical and reliability problems and the cost involved in the various technical solutions may be very different. This is an important field of technology where much work in detail has to be done and it should be done soon. The following is a preliminary compilation of steps for debris

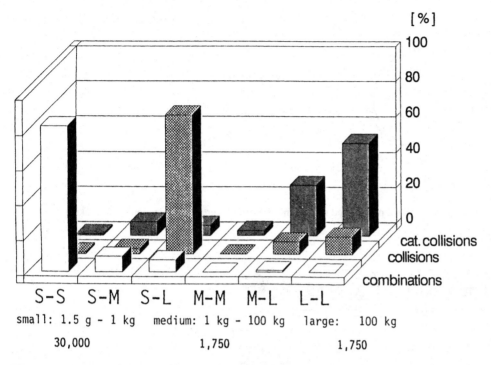

Figure 21: Debris size classes preferable involved in certain collision events.

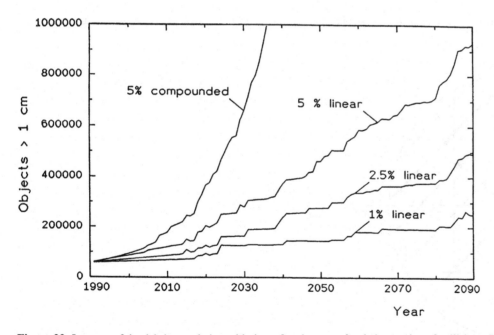

Figure 22: Increase of the debris population with time after the start of a chain reaction of collisions. Example of a simulation with various yearly increase rates of the basic population (1% to 5% per year linear and 5% compounded).

avoidance with an attempt to list them in the order of feasibility and cost effectiveness. In fact, some of them are already used in practice.

- Avoidance of the release of unnecessary objects related to a mission, e.g. ejection of solid rocket motors, of flanges, covers over experiments etc., bolts and other equipment. Most of this can be hinged or otherwise fixed to the main body.
- Avoidance of all kinds of explosions on orbit. This is a

large field corresponding to the many explosions happened in space with partly unknown reasons. The main objective is to avoid rocket upper stage explosions due to residual fuel. This can easily be achieved by venting the fuel after the payload has been delivered to orbit. This provision alone would eliminate the bulk of small sized debris generation in the future. However, since many rocket upper stages with residual fuel are still in orbit and, as experience shows, explode even after many years,

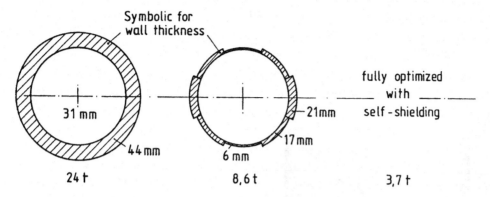

Figure 23: Mass of a double wall shielding system for a pressurized manned module, for different optimization levels making use of the detailed knowledge of the impacting flux.

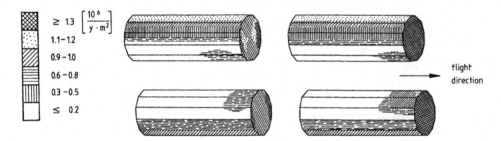

Figure 24: The debris flux distribution on a configuration of four pressurized modules taking account of the angular flux distribution and the self-shielding among the modules.

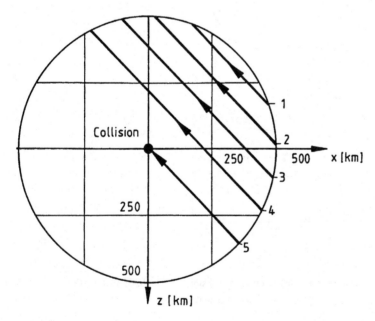

Figure 25: Passage of a debris object through the field of view in the horizontal plane around an orbiting target object.

the fuel venting now will not be an immediate remedy against all future explosions.

Other types of explosions (e.g. battery overpressure) must be studied and remedies be found.

Intentional explosions for military purposes should be stopped or at least performed in very low altitudes, so that no long living debris is generated.

- Deorbiting of rocket upper stages after delivery of the payloads. With each payload deployed on an earth orbit the rocket upper stage is left on a similar orbit. This does not apply for shuttle launches in very low orbits, but for higher orbits usually a intermediate stage is used which also remains in orbit. The elimination of these rocket upper stages would cut the number of large objects in

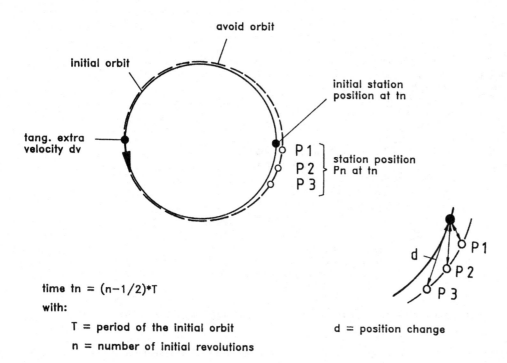

time tn = (n−1/2)*T
with:

    T = period of the initial orbit
    n = number of initial revolutions

d = position change

Figure 26: Example for a collision avoidance manoeuvre.

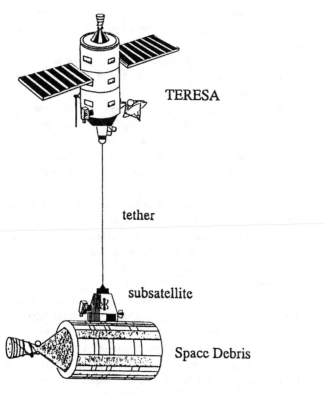

Figure 27: The TERESA concept for a space debris remover vehicle using a tether.

the orbital lifetime by enhanced atmospheric drag. This approach is questionable since the target area for collision is enlarged, which counteracts the shorter lifetime. But it would be a simple and low cost device.

  – or by active retarding thrust using the attitude control thrusters or additional small solid rocket thrusters. The amount of propellant required is low (e.g. 20 kg for 1000 kg rocket body mass), but the manoeuvre is complicated and the attitude control system must still be operable.

It seems that this latter procedure may become the normal way of eliminating rocket upper stages from higher circular orbits and from GTO in the future. It will probably only be possible to implement this procedure when a new generation of rockets is designed.

- De-orbiting of payloads (satellites, platforms, manned space stations) after their active use is terminated. The procedure is similar as described for rocket upper stages (drag enhancement or active thrusting), but the devices used for this manoeuvre are required to be operable at the end of the mission, i.e. often several years after the launch. But nevertheless this procedure should be seriously envisaged for the preservation of higher orbits in the region above 800 km altitude, which otherwise is bound to become overcrowded.

- Re-orbiting of satellites in the geostationary orbit after their active use is terminated. Since de-orbiting, i.e. the re-entry to the earth's atmosphere, is too costly for GEO, for all foreseeable future only re-orbiting, i.e. lifting the satellite about 300 km beyond GEO, can be envisaged. This would prevent collisions in the geostationary ring and so is

space to nearly one half, although their orbital lifetime tends to be lower than that of the payloads due to their smaller mass to area ratio. They could be de-orbited, i.e. re-entered into the atmosphere, by

  – either drag enhancement, i.e. the deployment of a balloon or other large area/lightweight device to shorten

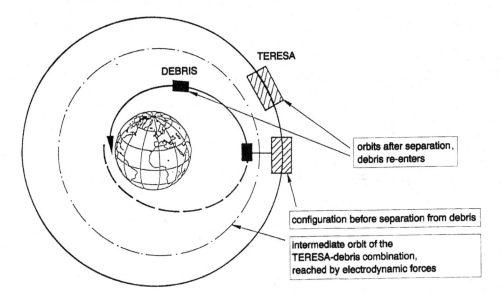

TERESA

DEBRIS

orbits after separation,
debris re-enters

configuration before separation from debris

intermediate orbit of the
TERESA-debris combination,
reached by electrodynamic forces

Figure 28: Sequence of orbits for a debris removal by TERESA, repeated in a similar manner for each debris object.

an essential measure to preserve this very important environment. In fact, this has already been performed with many GEO satellites, since it is easily achieved with the east-west-station keeping thrusters. However it is not fully understood whether the method is really a long term solution. After some decades a so called "graveyard orbit" higher than GEO is filled with objects, drifting around in an uncontrolled manner, and the debris of possible collisions among them may not be harmless to the geostationary orbit itself. It is urgent, that this very important question is studied further and the resulting decisions are taken.

At the end of this overview of the space debris problems as seen in Europe, I would like to put it this way: In utilizing space, an awareness of our responsibility is needed in a broad and international scope. For LEO it must become a generally accepted procedure that a mission consists of "deploy and de-orbit" of an object, and these two are equally essential. Without that awareness we are very likely to loose an environment which is so promising for the development of mankind.

LITERATURE

Klinkrad, H.; Jehn, R., "The Space-Debris Environment of the Earth", ESA Journal, Vol. 16, No. 1, pp. 1–11, 1992.

de Jonge, A.R.W.; Wesselius, P.R.; van Hees, R.M.; Viersen, B., "Detecting Orbital Debris with the IRAS Satellite", ESA Journal, Vol. 16, No. 1, pp. 13–21, 1992.

Space Debris, The Report of the ESA Space Debris Working Group, ESA SP-1109, Nov. 1988.

US Interagency Group (Space), "Report on Orbital Debris for the National Security Council", Washinton D.C., February 1989.

McDonnell, J.A.M.; Sullivan, K.; Stevenson, T.J. and Niblett, D.H., "Particulate detection in the near earth space environment aboard the Long Duration Expose Facility LDEF: Cosmic or terrestrial ?", Proceedings of IAU coll., No. 126, "Origin and evolution of interplanetary dust", Kluwer Acad. Pub. Co., 1991.

Kessler, D.J., "Orbital Debris Environment for Spacecraft in Low Earth Orbit", AIAA 90-1353, AIAA/NASA/DOD Orbital Debris Conference: Technical Issues & Future Design, Baltimore, MD, April 16–19, 1990.

Kessler, D.J.; Cours-Palais, B.G., "Collision Frequency of Artificial Satellites: Creation of a Debris Belt", Progress in Astronautics and Aeronautics, Vol. 71, pp. 707–736, Martin Summerfield Series Editor, 1978.

Loftus, J.P.; Kessler, D.J.; Anz-Meador, P.D., "Management of the Orbital Environment", 42nd Congress of the IAF, October 5–11, 1991, Montreal, Canada, paper IAA-91-590.

Flury, W., "Activities on Space Debris in Europe", 42nd Congress of the IAF, October 5–11, 1991, Montreal, Canada, paper IAA-91-589.

Zhang, J.; Rex, D.; Sdunnus, H., "Orbital debris model in the submillimeter region and its validation with impact data", Proceedings of "Workshop on Hypervelocity Impacts in Space", University of Kent at Canterbury, U.K., 1–5 July 1991.

Bendisch, J.; Rex D., "Collision Avoidance Analysis", AIAA Orbital Debris Conference, Baltimore, 1990, paper AIAA-90-1338.

Eichler, Peter und Rex, Dietrich, "Chain Reaction of Debris Generation by Collisions in Space – A Final Threat to Spaceflight?", 40. IAF Congress, paper IAA 89–628, Malaga, Spain, 1989 published in: Acta Astronautica, Vol. 22, pp. 381–387, 1990
also in: Heath, Gloria W. (Editor) "Space Safety and Rescue 1988–1989, Sciences and Technologie Series, Vol. 77, pp. 329–343, AAS Publications Office, San Diego, California, ISBN 0-87703-327-7.

Rex, Dietrich; Eichler, Peter und Zhang, Jingchang, "A Review of Orbital Debris Modeling in Europe", AIAA/NASA/DoD Orbital Debris Conference, paper AIAA 90–1354, Baltimore, MD, USA, 1990.

Eichler, Peter und Rex, Dietrich, "Debris Chain Reactions", AIAA/NASA/DoD Orbital Debris Conference, paper AIAA 90–1365, Baltimore, MD, USA, 1990.

Eichler, Peter und Bade, Anette, "Removal of Debris from Orbit", AIAA/NASA/DoD Orbital Debris Conference, paper AIAA 90–1366, Baltimore, MD, USA, 1990.

Eichler, Peter und Rex, Dietrich, "Kollisionen durch Raumfahrtmüll auf Erdumlaufbahnen – langfristige Entwicklung und mögliche Gegenmaßnahmen", 39. Raumfahrtkongreß der Hermann-Oberth-Gesellschaft, 28.6–1.7.1990, Garmisch-Partenkirchen.

Rex, Dietrich; Bendisch, Jörg; Eichler, Peter und Zhang, Jingchang, "Protecting and Manoeuvring of Spacecraft in Space Debris Environment", 28. COSPAR, paper MB.2.2.1, The Hague, Netherlands, 1990 published in: Advances in Space Research, Vol. 11, No. 12, pp 53–62, 1991.

Eichler, Peter und Bade, Anette, "Strategy for the Economical Removal of Numerous Larger Debris Objects from Earth Orbits", 41. IAF Kongreß, paper IAA 90–567, Dresden, FRG, 1990 to be published in: Heath, Gloria W. (Editor) "Space Safety and Rescue 1990–1991, Sciences and Technologie Series, AAS Publications Office, San Diego, California.

Eichler, Peter und Rex, Dietrich, "The Risk of Collisions between Manned Space Vehicles and Orbital Debris -Analysis and Basic Conclusions", Zeitschrift für Flugwissenschaft und Weltraumforschung (ZFW), Vol. 14, No. 3, pp. 145–154, 1990.

Bade, Anette und Eichler, Peter, "Möglichkeiten zur Debris-Entsorgung mit Hilfe von Seilsystemen", 3. Raumfahrt-Kolloquium an der Fachhochschule Aachen, 13. Dezember 1990, paper 91–038 published: Proceedings 3. Raumfahrt-Kolloquium an der Fachhochschule Aachen, DGLR-Bericht 91-03, ISBN 3-922010-59-8.

Eichler, Peter und Bade, Anette, "TERESA – Entsorgung von Weltraummüll", Raumfahrt Journal 2/91, pp 45–47

Eichler Peter; Bendisch, Jörg und Zhang, Jingchang, "Closing the Data Gap of Space Debris: Ground Based or Space Based Sensors", 4th European Aerospace Conference, paper 27.03, Paris, May 13–16, 1991 published: Proceedings 4th EAC, ESA SP-342, Oct. 1991.

Eichler, Peter, "Gefährdung von Satelliten und Raumstationen durch Kollisionen mit Weltraummüll", DGLR-Symposium: Leichtbaustrukturen unter kurzzeitiger Beanspruchung, Bremen, 6.–7. Juni 1991 published: DGLR-Bericht 91-04 Leichtbaustrukturen unter kurzzeitiger Beanspruchung (Impact, Crash), ISBN 3-922010-63-6.

Bade, Anette und Eichler, Peter, "Der Einsatz von Seilsystemen zur Entsorgung von Weltraummüll", DGLR-Jahrestagung 1991, paper 91–179, 10.–13. September 1991, Berlin published: Jahrbuch 1991 I, DGLR-Jahrestagung 1991, Berlin, pp 561–570, GW ISSN 0070-4083.

Eichler, Peter, "Analysis of the Necessity and the Effectiveness of Countermeasures to Prevent a Chain Reaction of Collisions", 42. IAF Congress, paper IAA 91–592, 5.–11. October 1991, Montreal, Kanada published: Proceedings 42. IAF Congress
also in: Acta Astronautica, Special Edition for World Space Congress, September 1992
also in: Heath, Gloria W. (Editor) "Space Safety and Rescue 1990–1991, Sciences and Technologie Series, AAS Publications Office, San Diego, California.

Bade, Anette und Eichler, Peter, "The Removal of Large Space Debris Objects with the Help of Space Tethers", to be published in: Zeitschrift für Flugwissenschaft und Weltraumforschung (ZFW).

Eichler, Peter; Sdunnus, Holger und Zhang, Jingchang, "Reliability of Space Debris Modeling and the Impact on Current and Future Space Flight Activities", World Space Congress 1992, 28. COSPAR Plenary Meeting, paper B.8–M.42, 28.8–5.9.1992, Washington D.C., USA.

5: Human Survivability Issues in the Low Earth Orbit Space Debris Environment

Bernard Bloom
Grumman Space Station Integration Division

INTRODUCTION

Descriptions of the magnitude of the low earth orbit (LEO) meteoroid and debris populations are provided by others at this conference [1], [2], [3]. These populations suggest the existence of a significant orbital debris flux which may endanger the survival and functional success of large, long duration vehicles. LEO debris populations place two interrelated constraints on designers and operators of manned vehicles intended to operate in, or transit, this region. Firstly, survival of manned vehicles and their flight crews requires careful consideration of numerous, and often indirect, failure modes and associated risk mitigation procedures. Secondly, catastrophic failure of a manned vehicle by hypervelocity impact may lead to measurable increases in the debris population, which then place additional risk and cost on other LEO users. For both reasons, the risk to survival of vehicles to be used in LEO for human exploration presents significant new challenges for vehicle managers. This paper presents an overview of the critical issues now being recognized and addressed by manned vehicle designers. It describes some of the elements of the process being used to assess and mitigate the true risks imposed by the low earth orbit space debris environment.

An essential feature of human survival problem in the LEO orbital debris environment is that many of the issues are best presented in probabilistic form. As is well known by conference attendees, the "natural" orbital debris environment is often represented as a plot of cumulative object flux vs size, as is illustrated in Figure 1. A cursory review of the ordinate indicates that even for large structures (~1000 m²), the nominal probability of encountering a debris particle capable of penetrating well protected structures (as a figure of merit, one capable of resisting 1 cm sized impacts) is currently on the order of 1% per year. Because the small sized LEO populations are thought to be growing, the mean interval between 1 cm impacts is decreasing. For a 1000 m² exposed area, this interval is now calculated to be approximately 50 years using nominal +1 cm flux values. Uncertainty in knowledge of the magnitude and composition of the debris environment adds to the probabilistic nature of the debris risk. Current uncertainty in the +1 cm populations is on

the order of a factor of two. While recent measurement programs are reducing these uncertainties, such data are only now beginning to become available for vehicle designers.

Vehicle and crew response to penetrating impacts also has important probabilistic aspects. For example, whether a penetration will result in a catastrophe depends on the specific impact location, the operational condition of the vehicle (e.g., its flight attitude and altitude), the whereabouts of its crew, and similar factors.

Therefore, even for long duration LEO missions, the risk to loss of mission from debris impact is difficult to predict. Nevertheless, the risk is manageable. The probabilistic nature of the debris problem challenges human explorers to focus on the degree of risk one is willing to take, the level of uncertainty one is willing to bear, and the costs associated with managing risk to any particular level. Such questions are often difficult for program planners who are responsible for convincing democratic publics to accept the inherent riskiness of costly human space exploration.

THE THREE DEBRIS THREAT RANGES

For the purpose of managing risk in the LEO environment, the cumulative flux curve shown in Figure 1 is divided in three regions as shown in Table 1. The shieldable regime is defined as that part of the size spectrum against which additional passive protection in the form of thin, multiple wall structures is feasible. The upper edge of this regime is in the 1–2 cm debris object diameter range, with the boundary determined by the degree of protection required, the ballistic limit capabilities of advanced shield technology, and the marginal costs of launching such shielding compared to alternate means of increasing survivability. The large object regime is defined as that portion of the size spectrum at which orbital debris is observable and trackable. Current ground-based capabilities of the United States Space Command network of sensors and computational facilities places the lower edge of the large object regime at approximately 10 cm debris diameter. Objects in this regime may be actively avoided by appropriate warning and maneuver protocols. Objects between ~ 1 cm and ~10 cm are described

Nominal 400 km Altitude Collision Probability (1) (1000 sq m: 2000-2010)	Orbital Debris Regime				
	Shieldable		Mid-Range	Avoidable	
	5 mm	10 mm	1-10 cm	> 10 cm	>20 cm
Total/Critical Surface Area(2)					
1	78%	14%	12%	2%	1.8%
2				4	3.6
3				6	5.4
4				8	7.4
Consequence of Collision	Potential Loss of Crew or Vehicle ──────────────►			Probable Loss of Crew or Vehicle	

(1) Assumes 1991 Kessler model environment and 2% growth/year in sub 1 cm particle populations. Does not account for uncertainites in knowledge of debris flux.

(2) Assumption is that only a fraction of a manned vehicle's exposed surface is "critical" if penetrated by small objects but that as the size of peneterator increases the fraction of vehicle catastrophically vulnerable increases. For SSF this ratio is ~ 4

Table 1: Description of orbital debris risk regimes.

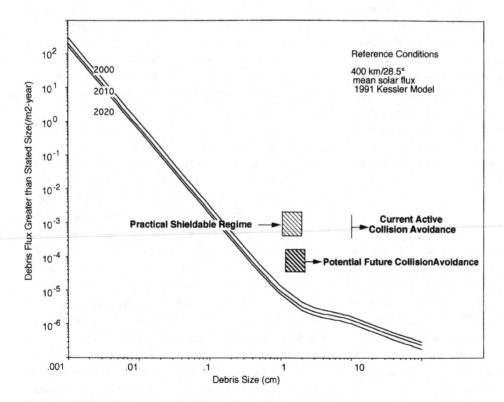

Figure 1: Orbital debris cumulative flux vs size.

as being in the mid-range regime. Although it is currently impractical to protect LEO satellites by passive shielding methods, it is also not currently feasible to avoid mid-range collisions by active avoidance owing to the detection/tracking limits of ground based radars.

Recent developments in the application of high frequency ground based radars to the debris tracking problem may enable future vehicle managers to actively avoid objects as small as 2 cm. Should such technology be brought into existence, the shieldable and avoidable regimes could be overlapped in such a manner that mid-range regime risk would be eliminated.

Two attributes of these regimes, the probability and consequences of impact, are shown in Table 1. The probability of experiencing shieldable regime impacts is a strong function of the particle size a protection system is designed to defeat. The populations of sub-1 cm debris are currently modeled as decreasing as diameter, $d^{-2.5}$. Within the 5–10 mm regime, significant protection is affordable by shielding. Once the ground-trackable regime is reached, the rate of reduction diminishes. Risk in the large object regime will depend on the probability of actually observing and tracking these objects near the lower edge of the regime. The second attribute is that as impactor size increases, the conditional catastrophic likelihood of a collision increases. Satellite breakup studies are indicating that certain catastrophic failure modes may not even have thresholds until objects are of mid-ranged size. Current design practice is to assume that all impacts with trackable objects will be catastrophic. A collorary of the latter two observations is that in the ground-trackable regime, the surface area to be considered in vulnerability calculations should be the entire projected area of the vehicle, not just the area of the shieldable regime "critical components."

It also follows from the foregoing that a comprehensive approach to vehicle protection will involve protection and risk analysis for each of the three threat regimes. Surfaces whose small object penetration could lead to catastrophic failure need to be shielded. Trackable objects must be avoided. If the residual risk from mid-range objects exceeds vehicle-level survivability requirements, means must be provided for the detection, tracking, and mitigation of impact risk in this regime. Mid-range threat mitigation presents the added challenge that it may have to be performed on-orbit. Mid-range protection engagement scenarios involve response times, from first detection to mitigation, on the order of tens of seconds. These short durations derive from the rapid closing speeds (~ 10 km/s at 500 km) involved and the difficulties of predicting conjunctions several revolutions before collision.

RISK CATEGORIES

Broadly put, orbital debris presents two types of risk to manned vehicles. These may be classified as the risk that certain mission-related functions will be lost or degraded as a result of orbital debris impact. Examples include long term erosion of thermal coatings, photovoltaic power degradation, loss of active thermal coolant, degraded power storage capability, and loss of avionics capability. Functional degradations may occur as a result of single impacts or by the cumulative impacts of small particles. Such functional losses can be accommodated by redundancy, designing to end-of-mission needs, and by planning on-orbit replacement or repair of injured components. For long duration manned missions such planning is an extension of the usual vehicle design process.

The second risk class is termed catastrophic risk, the risk from any impact which may endanger the physical survivability of the manned vehicle or its crew. Catastrophic risk is associated with single impacts which may cause immediate loss of crew or vehicle or may render a vehicle unsavable, even

though the inevitable loss may not occur within minutes of the impact. Thus, examples of catastrophic risk include both crew loss as a result of immediate, rapid depressurization of a penetrated crew cabin and the uncontrolled deorbiting of a vehicle subsequent to loss of control of the vehicle's guidance and navigational capabilities. In addition, catastrophic failures may result from the secondary effects of an impact (such as the creation of secondary debris), even though the direct effect of the primary impact was otherwise inconsequential. The timescale for catastrophic events may range from seconds to months after the initial debris impact. Catastrophic failure modes from orbital debris are vehicle-specific; a partial list of potential catastrophic failure events being studied by the Space Station Freedom program is shown in Table 2 for the components shown in Figures 2 and 3.

CATASTROPHIC RISK PROBABILITY

The likelihood of catastrophic failure from orbital debris impact is determined by the product of three factors. The first is the critical flux, F_c, defined as the the flux of objects which are large and fast enough (relative to the target component) to be able to induce a catastrophic effect. F_c is derived from a component's critical debris size, d_c, which is component-specific and the size distribution of flux in the orbiting debris population. In general, the goal of protection system designers is to decrease a component's penetration probability by raising the critical value needed for penetration. Protection by shielding enables the designer to take advantage of the fact that cumulative debris flux appears to diminish as $d^{-2.5}$, as is indicated in Figure 1 for the sub-1 cm regime.

The second factor is the effective exposed area of the vehicle's critical components, A_e. Effective area depends both on the physical size of the critical component and the vehicle's attitude relative to the incident debris flux. This stems from the debris flux directionality and is made more complicated because a vehicle's orientation to its velocity vector may itself be continually changing. The third factor affecting risk is the duration, T_e, that each critical component of a manned vehicle is exposed to the LEO debris flux. As long duration, manned vehicles may be assembled and modified over a period of time on orbit, exposure time is not a constant for all critical components. Crew member exposure is different matter still from vehicle exposure because a manned vehicle may only be inhabited a portion of the time and individual crew members each have their own specific flight durations.

Orbital debris risk to manned vehicles in LEO is growing because each of these factors has been increasing since the beginning of human space flight. Orbital debris fluxes have been growing. The National Aeronautics and Space Administration (NASA) estimates that since 1988 the 1 cm or greater flux has been increasing by at least 2% per year, a rate which is being used as a planning value for current Space Station Freedom design purposes [4]. Manned vehicle sizes are

Hazard After Penetration	Crew Modules					Stored Energy Components				
	Living Quarters	Work Spaces	Airlocks	Descent Vehicle	EVA	Propulsion	Life Support	Thermal Control	Power Storage	GN&C
Crew module depressurizaztion	X	X	X	X						
Fragmentation Impacts	X	X	X	X						
Uncontrolled crack propagation	X	X	X	X						
Suit penetration: injury to crew member					X					
Dynamic Response to Low & High Pressure Venting - Induced Loads exceed structural integrity	X	X	X	X		X	X	X	X	
- Induced torques exceed vehicle control authority (loss of ability to maintain orbit integrity)	X	X		X		X	X	X	X	
Burst of High Pressure Tanks (after penetration ordelayed rupture after non-penetrating vessel impact)						X	X	X	X	
Explosive detonation or deflagration of propellants						X	X			
Mechanical rupture and fragmentation of rotating H/W										X
Impacts onto Single Point Failure Component Surfaces						X		X		X

Table 2: Potential debris-induced catastrophic failure modes.

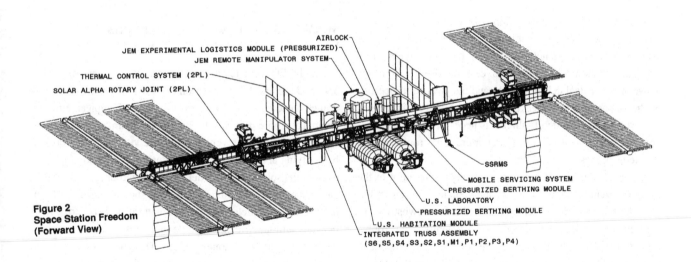

**Figure 2
Space Station Freedom
(Forward View)**

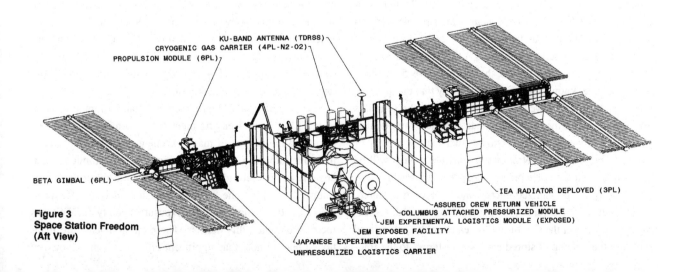

**Figure 3
Space Station Freedom
(Aft View)**

increasing and such vehicles have begun to operate at higher altitudes (at which critical fluxes are higher). In particular, exposure times of manned vehicles to the LEO flux will be significantly increased within the decade with the construction of Space Station Freedom.

One measure of debris risk which quantifies these three factors is the probability of no penetration, PNP, of a critical component, expressible via the Poisson relationship as:

$$PNP = \exp [-T_e \Sigma\Sigma (F_{cij} A_{eij})] \tag{1}$$

in which the summation indices, i and j, respectively refer to the ith critical component of a vehicle and the jth surface of that component. Figure 4 uses PNP to illustrate relative orbital debris risk to manned spaceflight. In this figure, PNP refers to the penetration of the shields that are either integral with the component or placed in front of the hardware being protected. Risk is defined as 100 x (1-PNP). The essential point of the figure is that as the critical area-exposure time product increases, risk will rise unless it is offset by decreasing the critical flux on exposed surfaces. Critical flux is understood to be that flux which will penetrate essential hardware; it is that fraction of the LEO flux which exceeds the ballistic limit properties of the impacted surface.

In order to maintain penetration risk constant as area-exposure time products increase, it is necessary to increase the critical size of protection shielding. An active form of hypervelocity penetration research seeks to develop the relationships between the critical debris size, d_c, and the physical properties of the impactor and target. Knowledge of the ballistic limit properties of orbital debris shields is necessary (see later section) but insufficient, though, to determine catastrophic risk to manned vehicles.

A critical component penetration is only the initiator of the chain of events leading to loss of the vehicle or its crew. For some failure scenarios the chain may be broken downstream of the penetration, thereby mitigating the potential catastrophic effects of the penetration. Therefore for a given component, the probability of no catastrophic failure, PNCF, will generally be higher than the associated PNP. Designing manned vehicles to maximize survivability (the capability of a a vehicle to sustain an impact without experiencing catastrophic failure) requires a quantitative understanding of each of the event chains (more accurately, event "trees") that may end in catastrophe. A typical event tree is shown in Figure 5. At each node of the tree, knowledge of the probability that one particular branch will be followed is required to assess the vehicle's survivability given the initiating impact. In the example, it is crucial to know whether a hydrazine fuel tank will vent or physically burst upon penetration and whether a burst will be driven by the release of chemical energy (e.g., hydrazine detonation) or by the pressurized energy under which it is stored. The magnitude of the venting or burst effects depend, respectively, on details of the size of holes created in the pressurized tanks and the amount of stored energy transferred to the frag-

ments from a burst vessel. Whether such energy releases induce catastrophic failure is dependant, as well, on their magnitudes.

A generalized statement of the preceding paragraph is shown in equations (2) and (3).

$$PNCF_i = [1 - (1-PNP_i) * P_{ic/p}] \tag{2}$$

$$PNCF vehicle = \pi PNCF_i \tag{3}$$

The subscript i refers to the ith critical component and the term pic/p refers to the probability that a catastrophic event will ensue if the ith component is penetrated. This formalism is intended to show how designers may combine the risks from impact on quite different components. As suggested by Figure 5 and Table 1, evaluating such expressions (i.e., obtaining insight into the values of $P_{ic/p}$ requires knowledge of diverse effects in applied mechanics, satellite attitude control, and human factors. The cost of obtaining such detailed knowledge rises geometrically with the size and complexity of a manned vehicle. Therefore, a more conservative approach, the one currently being employed, is to assume that penetration of a passive protection system is itself catastrophic.

SURVIVABILITY ISSUES AND DEBRIS ENVIRONMENTAL KNOWLEDGE

Manned LEO vehicles operate in the altitude band between 350–500 km and it is this region of near-Earth space that is of immediate concern to vehicle designers. A typical altitude planning profile for Space Station Freedom is depicted in Figure 6. It shows the vehicle's altitude varying over a 60 km band, flying at the higher altitudes during solar maximum portions of the solar cycle in order to reduce aerodrag on the vehicle. Superposed on this profile is a shorter zigzag pattern reflecting the vehicle's successive natural descents to a rendezvous altitude with the Shuttle and subsequent reboosts to begin the next 3–6 month pattern. Designers need to have accurate characterizations of the debris flux in this region for all phases of the vehicle's mission. For long duration flight manned vehicles such information necessarily requires measurements at and above these altitudes to account for the significant fractions of debris now in higher LEO orbits that will pass through the manned vehicle LEO region in the next decades through the effects of drag and the collisional and explosive breakups of existing satellites.

Several other space environmental threats to human survival in the LEO natural environment are reasonably well understood and the subject of many excellent reviews. Much study has gone into the effects of ionizing radiation, atomic oxygen erosion, ultraviolet radiation, and the natural meteoroid environment. [5], [6], [7] The recent return and analysis of experiments on the Long Duration Exposure Facility (LDEF) has recently enriched this field [8]. These environments are mentioned here to note that significant effort has been devoted to

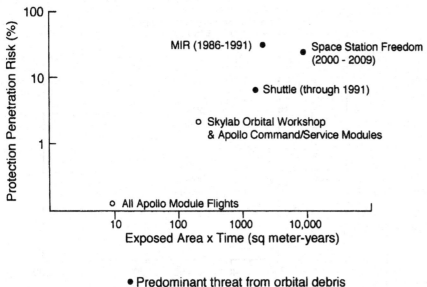

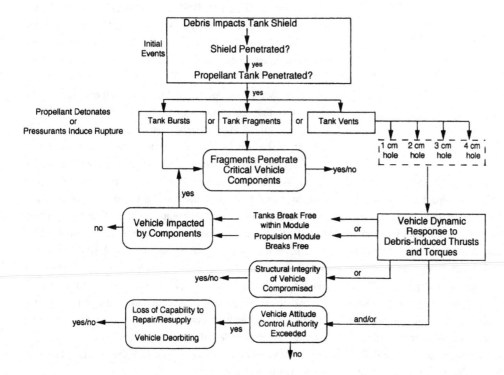

Figure 4: Meteoroid/orbital debris penetration risk for human spacecraft.

Figure 5: Event tree for propellant module.

understanding the risk these phenomena create for LEO satellites. For instance, knowledge of the dependence of penetration ionizing radiation (mainly energetic geomagnetically trapped protons) with altitude within LEO is necessary both for the protection of crew and electronic parts within manned vehicles. Such models [9] are widely used by vehicle designers.

For manned mission planners similar efforts will be needed to characterize the behavior of orbital debris flux. Designers are now faced with significant uncertainty concerning the number and properties of objects in the size regimes capable of inducing catastrophic failures. Until recent months, flux models in the critical 5–15 mm band were based on log-log linear interpolation between observations of craters and holes in hardware retrieved from LEO and catalogue radar observations in the > 10 cm regime. Yet, knowledge of the absolute values of penetrating debris flux is essential for proper risk assessment. As is indicated by equations (1) – (3), the probability of catastrophic failure is most directly coupled to the absolute magnitude of debris flux $\geq d_c$. Current flux uncertainties at 1 cm are still on the order of a factor of 2.

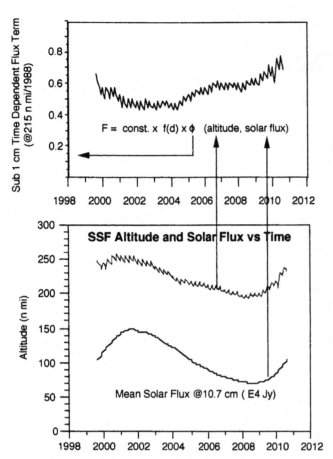

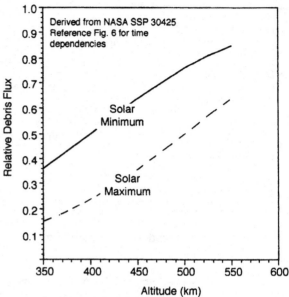

Figure 7: Altitude sensitivity of debris flux.

Figure 6: a) Relative debris flux vs. time. b) Variation of debris flux and vehicle altitude with solar cycle.

Recently carefully calibrated radar observation campaigns have been expanded [11] by NASA to permit a statistically useful sampling of LEO 1 cm flux. Observation times are still limited. At the Haystack radar, objects > 1 cm are observed at rates of 0.1 detection/hour in the altitude band centered at 450 km. With observing times ~ 300 hours per year available for observing at this size and altitude, designers will be working with only a few dozen detections during the next several years. Relative variations with altitude, inclination, and solar flux are also critical for mission planning purposes. Figure 7 illustrates the altitude dependence of orbital flux based on the current baseline Space Station Freedom model. Implicit in this figure is that a ± 50 km change in mean value of a vehicle's altitude profile would lead to a + 50% / − 28–38% change in debris flux. Because such sensitivities may prove to be attractive to program planners for their risk reduction potential, it will be necessary to increase confidence in the validity of such models. Similarly, the variation of debris flux with angle of incidence is another key design parameter. Figure 8 shows the directionality of debris incident on a plate normal to the RAM direction at 28.5° inclination. Such directionality information is currently being used by shield designers to optimize protection on those surfaces which will experience the greatest impact risks. For instance, shield

design tools such as BUMPER or ESABASE Debris Model directly incorporate the debris object inclination distributions upon which Figure 8 is based. This distribution is in turn based on measurements which will need periodic updating and fore-casting.

One further aspect of the debris environment that needs mention is that the composition, shape, and effective mass density of critical sized debris objects is poorly understood. Currently, the shield designer must make an assumption regarding the characteristics of an impactor in order to size a design. Within the Space Station Freedom program, the often used assumption is to treat debris of a given mass as having a diameter equivalent to an aluminum sphere. Depending upon the ballistic limit equations used to size a shield, this type of assumption may lead to unfortunate results. A simplified version of one penetrator equation [10] is shown in equation (4), in which t_w is the thickness, in g/cm^2, of an aluminum rear wall of a two-wall (Whipple) shield needed to defeat penetration by a particle of diameter, d, normal velocity component, V_n, wall spacing S, and particle bulk density of ρ_p. The constant k contains information on the mechanical and physical properties of the projectile and rear wall.

$$t_w = k \, d^{1.5} \, (\rho_p)^{1/2} \, V_n / S^{1/2} \qquad (4)$$

Note that the assumption of an aluminum penetrator would result in a 70% plate undersizing if the impactor turned out to be a remnant of a steel booster casing. Weight penalties for a "100% steel" assumption are probably not acceptable nor war-ranted, given the large amount of aluminum launched into low earth orbit. Undersizing protection systems is also undesirable. Therefore, measurements of the actual distribution of fragment densities in LEO or assessments based on known materials-launched-into-space is needed.

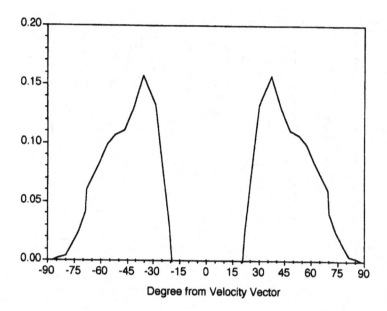

Figure 8: Debris incident angle distribution.

SHIELDABLE REGIME PROTECTION

Shielding critical components is the first line of defense against the debris environment. Figure 9 indicates that a one-to-two order of magnitude risk reduction is possible provided that shields can be designed to effective ballistic limits near the 1 cm size. Therefore, considerable attention has been placed on the design of light weight protection for spacecraft structures. This section is not intended to be a review of shielding technology but rather to point to some of the issues that face practical application of the approach to survivability management.

During early manned spaceflight, aluminum dual-wall "Whipple" shields were deployed to protect against the natural meteoroid environment. Shields for this purpose were designed to defeat low density objects impacting with higher relative velocities (~ 20 km/s). The LEO shieldable regime debris population consists of higher density objects impacting at lower relative velocities (a mean of ~10 km/s at 28.5°) and first line debris protection must take this fact into account. One implication of this point is that the full effectiveness of the dual-wall shield does not get realized if insufficient kinetic energy is available to completely fragment the impactor on the outer wall. Therefore early equations used to size dual-wall shields [11] are not considered fully applicable to the present orbital debris problem.

Figure 10 is a recent estimate of a 45° angle of incidence ballistic limit curve family for an aluminum system. As is indicated by the notes on the figure, the details of such curves depend on a large number of projectile and target factors. Testing and analysis programs are therefore needed to define the ballistic response to objects in the debris population. But also as noted on the figure, the ability of terrestrial laboratories to simulate the impact velocities in LEO is currently limited to

the sub-7.5 km/s light gas gun regime for practical purposes. Yet as is suggested in Figure 8, the larger portion of LEO impacts occur in the +7.5 km/s regime. This circumstance highlights the need for work in advancing the state of hydrodynamic simulation of impacts in this regime to supplement data acquirable through direct testing.

Such knowledge is valuable for the obvious reason that debris shielding adds costly launch weight and does not improve other vehicle functions. The effectiveness of dual-wall shielding is suggested by Figure 9. However, the marginal benefit of the technique decreases at approximately 1 cm (aluminum spherical projectiles), an effect which is indicated by Figure 11. This figure shows the risk-weight curve at the margin for a triple wall (2 bumper) shield of a large cylindrical structure. Note that significant improvements in shield efficiency are more readily achievable by increasing the standoff distance (bumper-rear wall spacing) than by adding bumper weight. Extending standoff distance reduces the impulse per unit area delivered to the rear wall of such designs and is a practical means for increasing the critical particle size of orbital debris shields.

The technique has its limits. Standoff lengths do require the availability of unused launch volume, mechanisms to deploy extendable shields after launch, or the assembly of large standoff shields in space by robotic or EVA means. As with weight, free volume within launch vehicles is often not available. The other two methods just referenced present vehicle designers with potential unfavorable cost implications.

Therefore, efforts to reduce the unit weight of existing shields are also needed.

One such effort takes advantage of the multishock principle and is well described elsewhere. [12], [13] Combined with the use of lighter weight ceramic fabric bumpers, weight reductions by as much as 50% compared to the dual-wall aluminum

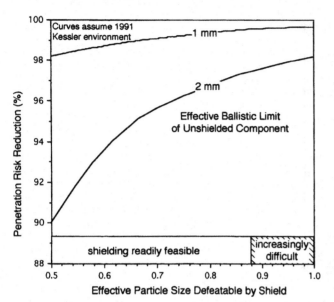

Figure 9: Penetration risk reduction afforded by shielding.

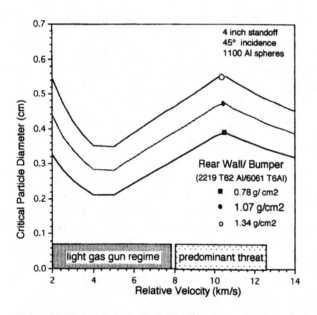

Figure 10: Example of ballistic limit curves for aluminium whipple shields.

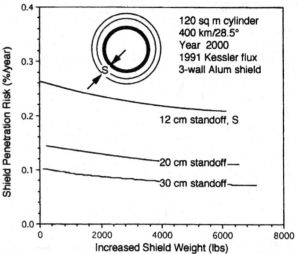

Figure 11: Weight limitations of passive shielding.

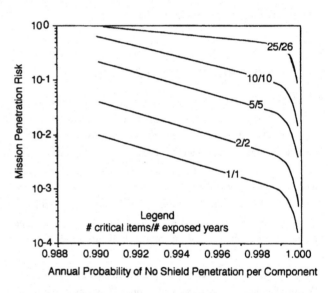

Figure 12: Relationship between component and vehicle-level shield penetration risk.

shield may be possible. Such technology may prove particularly useful for the progressive augmentation of the shields initially deployed on manned vehicles. Augmentation on orbit might best be performed robotically by spooling out a length of such fabric from a central cannister, a concept for which the new fabrics appear suited.

Protection-through-shielding is limited by the size and duration of large manned vehicles. For small vehicles exposed for short durations (e.g., one or two critical components for three years) usual dual-wall shields can be used successfully to limit system catastrophic risk. However as illustrated in Figure 12, as the size and mission duration of a manned vehicle increases, the design requirements on passive shielding to achieve speci-

fied mission-level catastrophic risk objectives reach the point of diminishing returns. The figure indicates that the PNP required at the component level to maintain a constant level of penetration risk is driven sharply to higher values as mission durations and vehicle complexity is increased. Practical passive shielding limitations (in the form of weight, volume, and materials) are presently being approached at annual shield PNP values of .999 for components on the order of the cylinder modeled in Figure 11. with correspondingly higher PNPs for smaller components following the relationship expressed as equation (1). One measure of such limits is that to achieve such probabilities with current technology requires the vehicle designer to allocate on the order of 5–10% of total vehicle weight to passive shielding. Even for such allocations, the lifetime-expectation of at least one shield penetration for large, long duration vehicles may well exceed a 10–15% shield penetration risk.

SURVIVING SHIELD PENETRATION

Passive shielding is only one component of debris risk management. Survivability of manned vehicles in the LEO debris environment requires careful consideration of the potential adverse effects of, and means of mitigating, debris-induced penetrations of essential vehicle hardware. The survivability analyst needs to consider the consequences of penetration as well as the probability of its occurrence. As indicated by the large scope even for the partial listing of failure types in Table 2, the task is varied and challenging. For instance, information needed for quantitative probabilistic risk assessment is quite specific, requires the application of specialized tools, and presents a very large multidimensional analysis matrix.

One approach to this problem has been developed by the Space Station Freedom program. A comprehensive analysis and test plan, which follows the logic of Figure 13, is being implemented to determine the major contributors to debris-induced catastrophic risk and the cost-effective means of managing it. The plan separates the consequences of impact by particles just above the ballistic limit in the shieldable regime from the large object avoidable regime. The latter is treated as a problem in risk management in which the probability of collision avoidance is traded against the number of maneuvers needed to assure collision avoidance, the capabilities of the ground-based sensor network to detect and track large objects, and the costs of increasing either capability.

Shieldable regime assessments, in turn, are performed by component type (manned module, high energy storage components, single point failures) and the potential failure mode after a penetration. This approach is taken for three reasons:

- Loss of crew within the manned modules requires consideration of interrelated effects specific to humans. As pointed out by Williamsen [14], cabin penetration may cause loss by hypoxia (rapid depressurization, followed by loss of useful consciousness, and the inability to effect a safe egress), impact by primary or secondary debris fragments, and the effects of blinding light flash, deafening noise (among other such effects)
- The dynamic response of the vehicle is expected to vary considerably with the forces and torques generated by the low pressure manned modules and higher pressure components. As is suggested by Figure 14, these drivers vary over three orders of magnitude.
- The Space Station Freedom shield protection design happens to be differentiated by the fact that the habitable module shields (single and double bumper Whipple design) use the pressure wall of the modules as the rear wall of the protection system whereas the other shielded components are protected by shields placed in front of, and non-integral with, the protected components.

Features of crew survivability assessment
Penetration of a crew cabin results in depressurization of the penetrated module and at the same time presents a potential realtime injury to crew members resulting from the primary and secondary fragments deriving from the interaction of the debris cloud and equipment racks in the interior of the penetrated module. Factors affecting crew ability to cope with the event include whether they are awake or asleep, the specific module in which they are located, and the specifics of their escape route relative to their own location. These factors are probabilistic functions of crew timelines, impact locations, and internal module conditions immediately after the impact. Only a certain time is available to effect a safe egress for the crew. This time is determined by the size of the penetration, which in turn is determined by the size, impact velocity, impact obliquity, and ballistic limit properties of the impacted surface.

Models of crew survivability must take into account the probabilistic nature of the problem just described. One model being developed by Goodwin and Rapids [15], uses a Monte Carlo approach to assess the probability of crew loss given a penetration. Its essential features are shown in Figure 15 in which the parenthetical numbers refer to input variables listed in Table 3. Model outputs are the times necessary to safely egress a penetrated module. By performing runs over the configuration space assuming specific module-to-module hatch protocols the model produces a spectrum of minimum egress times. If such results are superimposed on the available time, as is done in Figure 16, these results can be used to determine the likelihood that a penetrating impact could produce a catastrophic outcome. It also can be used to estimate the number of crew members affected by the event. This model is now being refined for application to the hatch protocol and other risk mitigation concepts for Space Station Freedom.

The power of such an technique is illustrated by another probabilistic approach to this problem described by Williamsen. He obtains a spread of values of $P_{c/p}$ (reference eq (2) of 0.13 to 0.23, depending on the protocols for maintaining hatches between compartments open or closed. These values may be interpreted as representing a range from 1 in 7 to 1 in 4 chance that a penetration, were one to occur, would lead to loss of crew. One should recall that the initial probability of experiencing a penetration is itself a probability deriving from the environmental flux and component shield design. By combining a practical shield design (PNP ~ .9995 per year) with such internal values, it becomes possible to limit catastrophic risks (from those specific effects described above) to ~ 0.01% per year per module. Such a risk level is equivalent to a catastrophic event occurring one the order of once in 1000 years for a ten module vehicle.

Features of stored energy component assessments
Pressurized tanks present a different set of survivability problems. Consider a pressure vessel located behind a penetrated shield. If a debris penetration occurs, what are the potential competing events which may lead to a catastrophic outcome for the vehicle or its crew? The vessel may be penetrated (but

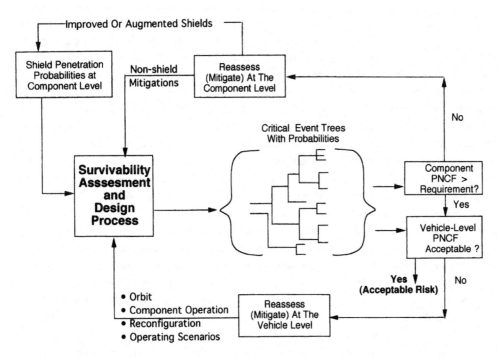

Figure 13: Survivability assessment process.

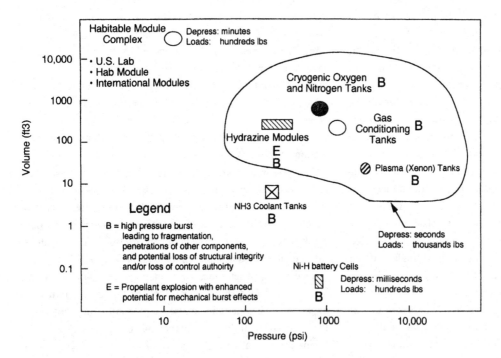

Figure 14: Space station freedom pressure vessels and potential catastrophic failure modes.

not necessarily) by the debris cloud emerging behind the shield's rear wall. If penetrated, the gas or high pressure fluid may be released in the form of a sonically vented jet, which may excite the vehicle's normal structural modes. Such vents may last for seconds to minutes, depending on the size of the vessel. If large enough, such vents may damage the structural integrity of the vehicle.

At the same time, the vehicle's attitude control system will attempt to respond to the torques arising from such vents.

Controllability of the vehicle may be compromised. Servicing a manned vehicle requires the ability to dock or berth with the servicing vehicle. Therefore complete loss of control presents a potential for catastrophic loss of the vehicle. Further, as LEO vehicles are sensitive to drag forces (as indicated by Figure 6) serious control problems have to be overcome within months in order to return the vehicle to a safe altitude.

A common feature of the venting problem is that vehicle response depends upon the size and nature of of the holes gen-

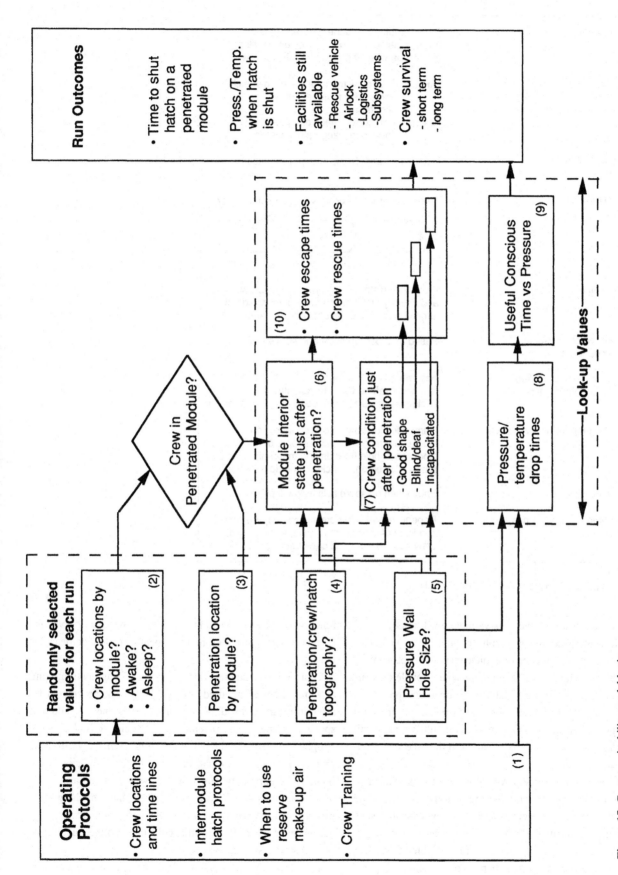

Figure 15: Crew survivability model logic.

1. Survival Protocol	Crew combine to rescue all, then close all hatches instricken module; they do not close all hatches when the pressure falls too low
2. Crew location vs. times	On-duty: One crew member handling vehicle, three crew members at work in laboratories Asleep: Two crew in each of 2 modules with interconnecting hatches open
3. Penetration locations by modules	Determined from Bumper code calculations
4. Impact/crw/hatch topography	Three topographies occur with equal frequency 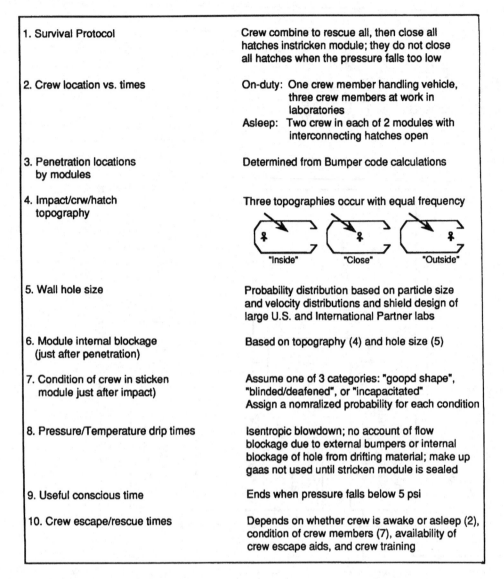
5. Wall hole size	Probability distribution based on particle size and velocity distributions and shield design of large U.S. and International Partner labs
6. Module internal blockage (just after penetration)	Based on topography (4) and hole size (5)
7. Condition of crew in sticken module just after impact)	Assume one of 3 categories: "goopd shape", "blinded/deafened", or "incapacitated" Assign a nomralized probability for each condition
8. Pressure/Temperature drip times	Isentropic blowdown; no account of flow blockage due to external bumpers or internal blockage of hole from drifting material; make up gaas not used until stricken module is sealed
9. Useful conscious time	Ends when pressure falls below 5 psi
10. Crew escape/rescue times	Depends on whether crew is awake or asleep (2), condition of crew members (7), availability of crew escape aids, and crew training

Table 3: Factors used in crew survivability analysis.

erated. Techniques to predict this variable must be developed to permit confidence in the results of analysis. The threat matrix is large and prediction of vehicle response is probabilistic. A vehicle with n pressurized components capable of venting in 3 orthogonal directions, at various operating pressures, and able to excite m normal modes at N points of structural interest, presents at least 3xnxmxN potential ultimate strength analyses *per hole size*. Serious attention to the hole size problem is required.

But the penetrated pressure vessel may not vent. Rather, it may fragment during impact, releasing such fragments into the local environment. Fragmentation models indicate that such low energy fragmentations can create dozens of fragments which are both large enough and of sufficient velocity relative to the vehicle to penetrate adjacent parts of the vehicle. Figures 17 and 18 presents the results of the simulation of the burst of a 68 kg /1 m^3 hydrazine filled pressure vessel. Plotted in the upper curve are the number of penetration-capable fragments generated from hypothetical explosions of the propellant as a function of the energy yield of the event. The lower curve represents two versions of current fragment spread velocities after the explosion.

Figures 17–18 suggest that hundreds of fragments moving at a minimum of hundreds of meters/sec can be created by such an event. Depending upon the event's energy yield and specific geometry, the potential for a vehicle-crippling catastrophe is evident. In the particular simulation shown in these figures, perhaps important features of the explosion were ignored for simplicity. For instance, would the adjacent fuel tanks also explode? What happens when the outrushing fragments reach the inner surfaces of the debris shields: is more debris generated by the resulting collisions; do the shields mitigate the effects of the explosion; etc?

The preceding discussions of catastrophic event modes is meant to be illustrative of the scope of the survivability assessment problem. One conclusion that is encouraged is that many tools will have to be developed or adapted to confront a complex problem. Application of the satellite breakup modeling

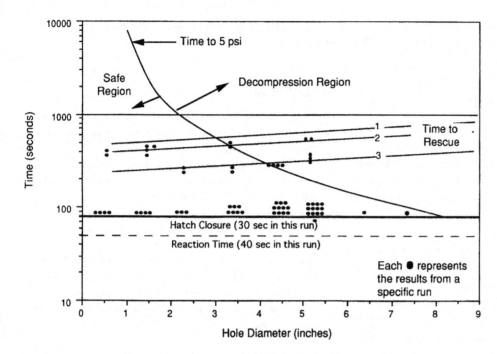

Figure 16: Typical results from crew survivability model run.

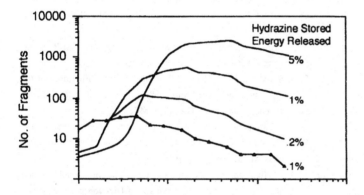

Figure 17: Fragment number vs velocity.

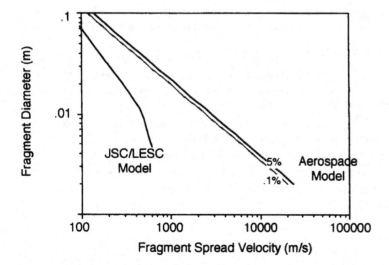

Figure 18: Fragment size vs velocity.

tools now being applied to the environment forecasting problem discussed by others at this symposium will have to be applied to the manned vehicle survivability problem as well.

EFFECTS ON THE ENVIRONMENT

Another feature of the energetic burst event is that the new fragments also contribute to the total LEO debris population. How serious this possibility may be is being addressed by others at this symposium. For the energetic burst event described above, thousands of fragments > 10 cm diameter are injected into the LEO environment. Many of these fragments will have ballistic coefficients and orbital elements such that they will only create a transient threat to LEO satellites (including itself). A smaller fraction of the fragments will be injected into higher orbits with correspondingly longer lifetimes. One estimate of a serious breakup event is that it would contribute 10 fragments large enough to be tracked and to stay in orbit for 10 years. As manned vehicles may present relatively large debris collision cross sections, adoption of vehicle design requirements may have to take into account their subsequent effect on LEO debris populations. This subject is beyond the intended scope of this paper but the reader is referred to reference 16 for further discussion.

SUMMARY

Humans will be leaving the Earth's surface and working for extended periods is low earth orbit to explore the surface, to learn how to function in deeper space, and to perform work properly suited to low earth orbit. LEO will be the environment through which humans will enter and return from extraterrestrial journeys. Learning to function in this environment, with its increasing debris populations, is a challenging task. People have only recently been learning how to work continuously in this environment.

Part of this learning involves the survivability of large long duration vehicles and their crews. This paper raises some of the issues now being addressed by vehicle designers and managers. These can be summarized as follows:

Survivability is an inherently probabilistic feature of the problem. Debris is now being modeled as a flux of random particles distributed over a range of sizes and relative velocities. Current forecasts place the mean risk of a significant impact at a magnitude of 1% per year for large manned vehicles. The response of a vehicle to such impacts is dependent on how well shielded are the vehicle's critical surfaces and how well the subsystem operating protocols have been worked out to enhance survivability. Doing so involves the application of knowledge gained from many disciplines, including ironically,

studies of satellite breakup mechanics intended to forecast the next generation's debris environment.

REFERENCES

1. N. Johnson, "Sources of Orbital Debris", Preservation of Near-Earth Space for Future Generations, Symposium at the University of Chicago, June 24, 1992.

2. D. Kessler, "Overall Assessment of the Orbital Debris Environment", Preservation of Near-Earth Space for Future Generations, Symposium at the University of Chicago, June 24, 1992.

3. D. Rex, "Overall Assessment of the Orbital Debris Environment", Preservation of Near-Earth Space for Future Generations, Symposium at the University of Chicago, June 24, 1992.

4. "Space Station Program Natural Environment Definition for Design", Section 8 (Meteoroid and Orbital Debris Environment), SSP 30425, Space Station Freedom Program Office, NASA, June, 1991.

5. H.K.A. Kan, "Space Environment Effects on Spacecraft Surface Materials", *Radiation Effects in Optical Materials*, SPIE, Vol. 541, p. 164.

6. H. Zook, D. McKay, and R.P. Bernhard, , "Results from Returned Spacecraft Surfaces", AIAA/NASA/DoD Orbital Debris conference, Baltimore, MD, April 16–19, 1990.

7. E.G. Stassinopoulos and J. P. Raymond, "The Space Radiation Environment for Electronics", Proceedings of the IEEE, Vol 76, No. 11, November, 1988.

8. A. Levine, "LDEF-69 Months in Space:First Post-Retrieval Symposium", Kissimmee, Fl, June 2–8, 1991.

9. D.M. Sawyer and J.I. Vette " AP8 Trapped Proton Environment for Solar Maximum and Solar Minimum", Rep NDSSC,76–06, National Space Science Data Center, Greenbelt, MD, 1976.

10. E. Christiansen, "Performance Equations for Advanced Orbital Debris Shields", AIAA Space Programs and Technologies Conference, Huntsville, AL, March 24–27, 1992.

11. J.P.D. Wilkinson, "A Penetration Criterion for Double-Walled Structures Subject to Meteoroid Impact", AIAA Journal, Volume 7, No. 10, October 1969.

12. B.G. Cours-Palais and J.L. Crews, "A Multishock Concept for Spacecraft Shielding", International Journal of Impact Engineering, Vol/ 10, p.135, 1990.

13. G. Edeen, "Space Station Freedom Multi-Shock Shield Study", Johnson Space Center Structures and Mechanics Division, NASA, March 1991.

14. J.E. Williamsen "Orbital Debris Risk Analysis and Survivability Enhancement for Freedom Station Manned Modules", AIAA Space Programs and Technologies Conference, Huntsville, AL, March 24–27, 1992.

15. J. Goodwin and R. Rapids, Briefing to Space Station Freedom Program Office, June 4, 1992

16. D. McKnight, "Orbital Debris Standards: Are They Necessary or Possible?", AIAA Space Programs and Technologies Conference, Huntsville, AL, March 24–27, 1992.

6: Protecting the Space Environment for Astronomy

Joel R. Primack

Physics Department, University of California, Santa Cruz, CA 95064

A common misconception of space is illustrated by the *Star Wars* movie: an explosion occurs and the screen is filled with debris, but a moment later it clears. The debris is gone without a trace. In fact, debris and charged particles injected into near-Earth space are trapped by Earth's gravity and by the geomagnetic field, and they become hazards to spacecraft until they are removed by interaction with Earth's upper atmosphere.

Eventually debris particles will collide with enough air molecules to slow them down, go into decaying orbits, and reenter the atmosphere. When the sun flares up in its eleven year cycle, it heats the upper atmosphere and makes it expand so that debris and spacecraft in lower orbits are subjected to increased drag. But the higher the original orbit, the less air there is to collide with. Above about 800 km, the atmosphere is so thin that the lifetime of orbital debris may be many decades. Above 1000 km, debris may orbit for centuries.

Imagine near-Earth space as the hillsides of a steep valley. A lake lies at the bottom, washing debris off only the lowest part of the hillsides regularly. The lake is our atmosphere. Not only is near-Earth space filling with debris, the atmosphere itself is polluted with stratospheric chlorine from every launch using solid rocket fuel, which also dumps tons of particles into the atmosphere and near-Earth space.[1]

Space is the most fragile environment that exists because it has the least ability to repair itself. Thus near-Earth space is at great risk from human activities and it is in great need of protection by scientists and humanity at large.[2] Scientists should be especially concerned, both because we place many crucial scientific instruments in near-Earth space, and also because we are in a unique position to foresee the problems human activities are causing and to propose measures to mitigate or avoid them.

In this paper, I will discuss protection of the near-Earth space environment primarily from the viewpoint of an astronomer. I will first explain why near-Earth space is such a desirable location for astronomical instruments, and briefly describe some of the most important examples. I will then consider the main environmental threats associated with human activities in space – space debris and the consequences of using offensive weapons in space, and "light pollution" caused

by orbital activities including radio interference and charged particles from space reactors – and discuss various proposals to deal with them. Finally I will try to put the problem in an astronomical time perspective.

ASTRONOMY FROM NEAR-EARTH SPACE

Scientific instruments have been placed in orbit to get a view upward unimpeded by the opacity, heat, and distortion of the atmosphere, and a view downward from which to survey vast areas of the Earth. This symposium is concerned with near-Earth space. I will take this to be the Low-Earth Orbit (LEO) region, from the lowest practical orbits, about 300 km altitude, up to about 2000 km. Most of the important astronomical satellites have been placed in LEO, between about 450 and 900 km.

On the best nights at the best ground-based observatories such as Mauna Kea in Hawaii, objects closer together than about 0.5 seconds of arc cannot be resolved.* Even when the sky is completely clear, images are distorted by small fluctuations in the atmosphere. This prevents astronomers from seeing the structure of distant galaxies, for example. After its correcting optics are installed, the Hubble Space Telescope should have about a factor of ten better resolution than can be achieved from the ground.+ This will permit detailed studies of galaxies roughly ten times farther away; the number of galaxies that will become available for such study correspondingly expands by a factor of a thousand. One of the first projects will be to observe Cephid variable stars in many galaxies, which may at last let us accurately measure the expansion rate of the universe and thus, indirectly, the time since the Big Bang.

The atmosphere is transparent to visible light, and to some other infrared and radio bands of electromagnetic radiation. But it is opaque to wavelengths shorter than visible light. Thus we cannot observe much infrared light from the ground, nor any ultraviolet light, X-rays, or gamma rays. Some of the most important satellite astronomical observations have been made in these bands.

The Cosmic Background Explorer (COBE) satellite has observed the fluctuations in the first light of the universe – the heat radiation that was emitted as the hot primordial plasma first cooled and became transparent. This happened only about 200,000 years after the origin, long before the first stars formed. This radiation has been redshifted to radio wavelengths by the expansion of the universe. Although such radiation can readily penetrate the atmosphere, COBE was able to see tiny fluctuations in the temperature in different directions that had eluded more sensitive instruments on the ground and on balloons, because COBE could survey the entire sky without interruption or background radiation from the hot atmosphere.

The temperature fluctuations COBE detected are relics of ancient differences in the density of the primordial universe from place to place. These initial conditions are what led over billions of years to the formation of galaxies and larger-scale structures in the universe, according to popular but – before COBE – unconfirmed theories such as Cold Dark Matter.[3] The fluctuations can also help clarify the origin of the universe: how the Big Bang happened, whether its initial stages involved cosmic inflation, and perhaps even the nature of the meta-universe ("eternal inflation"?) from which the Big Bang may have arisen.[4]

In the seventeenth century, Newton's separation of physics into differential equations and force laws that are universal, and initial conditions that depend on special circumstances, provided a paradigm that still guides the field, even though the universal laws themselves have been revised several times. Darwinian evolution plays a similar role in biology, connecting the structures of organisms and of natural communities with the underlying molecular genetics. Geology just became a theoretical subject a quarter century ago with the confirmation of the plate tectonics paradigm. The new COBE data and other space observations should give astrophysicists at last a solid foundation on which to construct an overarching theory of the origin and evolution of the universe, an achievement that is also bound to have deep implications for the development of human culture.

Geography of space

Proximity to Earth is both the main advantage and the main disadvantage of LEO for astronomical instruments. Proximity to Earth is a disadvantage for astronomy because the nearby Earth hides half the sky. Moreover, a LEO satellite's movement from night to day and back during each 90-minute orbit necessitates complex spacecraft telemetry, pointing, and power systems. It is an advantage since it permits relatively low-cost deployment of large satellites. The nearby atmosphere also keeps the lower LEO altitudes relatively free of space debris and geomagnetically trapped particles.#

The available launch vehicles have allowed us to place in LEO two giant astronomical observatories that are as large and heavy as a school bus: Hubble Space Telescope (11,500 kg, in a 600 km orbit), and Compton Gamma Ray Observatory (16,000 kg, in a 450 km orbit).

The smaller COBE (2,270 kg) and the now-defunct Infrared Astronomy Satellite (IRAS) are in 900 km polar orbits, which is also considered LEO. Because infrared light penetrates the dust of the Milky Way galaxy far better than visible light does, the IRAS catalog of infrared galaxies is the basis of a redshift survey which has provided the only "all-sky" maps available of the distribution of nearby galaxies. These maps are helping to determine whether the invisible "dark" matter that apparently makes up 95% of the mass of the universe is distributed like the visible galaxies, and whether there is enough matter altogether to reverse the expansion of the universe eventually.[5]

Only smaller instruments have been placed in higher orbits. We can launch a school bus to 600 km, or a school bus-sized rocket to 600 km which in turn launches a spacecraft the size of a microwave oven or a refrigerator to 36,000 km (geosynchronous orbit). International Ultraviolet Explorer (IUE, 462 kg) is the only astronomical satellite that has been successfully placed in an approximately geosynchronous orbit (GEO). The European astrometry satellite Hipparchos was intended for a GEO orbit but is stranded (although nevertheless working) in a highly elliptic (600 - 36,000 km) orbit.[6]

SPACE DEBRIS AND WEAPONS AS THREATS TO SPACE ASTRONOMY

At any moment, only about 200 kg of meteoroid mass is within 2000 km of the Earth's surface. Within this same altitude range there is roughly 3,000,000 kg of orbiting debris introduced by human activities. Most of this mass is ~3000 spent rocket stages and inactive payloads. Approximately 40,000 kg of debris is in some 4000 additional objects several cm in size or larger currently being tracked by radar, most of which resulted from more than 90 satellite fragmentations. The main threat to satellites in LEO is from the 1000 kg of 1 cm or smaller debris particles, especially the approximately 300 kg of debris smaller than 1 mm.[7] A marble-size chunk of debris striking a spacecraft at the typical relative orbital speed of 10 km/s (36,000 km/hr) is as potent as a large safe falling from the roof of a ten-story building. The much more numerous BB-size fragments of debris have the same destructive energy as a bowling ball moving at 100 km/hr. An average small satellite in an 800 km orbit now has about a one percent chance per year of failure due to collision with a BB-size piece of debris.[8] The danger to a large satellite such as Hubble Space Telescope is even greater. And the amount of small debris is doubling roughly every decade.

Random collisions between man-made objects in LEO are still relatively rare, but the density of such objects may already be sufficiently great at 900–1000 km and 1500–1700 km that a chain reaction or cascade of collisions can be sustained.[9] Further growth of the debris population will increase the threat at even lower orbital altitudes. The resulting debris environment will obviously be very hostile to satellites in LEO. I will consider at the end of this paper various measures that should be taken to prevent this from happening.

Astronomers have just in the past decade begun to enjoy the privilege of using powerful observatories in space, and our understanding of the universe has already been wonderfully enriched. While the debris threat can be evaded by lofting such satellites to very high orbits, this would add great cost or result in much less scientific return for the same investment. Such high orbits could be scientifically justified only for a few astronomical satellites for which a near-Earth location is quite undesirable, such as the long-proposed (but not yet funded) NASA Great Observatory known as the Space Infrared Telescope Facility (SIRTF).

Offensive weapons in space pose perhaps the nastiest threat to satellites in LEO. Fortunately, offensive weapons have not yet been introduced into space except for a few tests (such as a Soviet space mine explosion which generated hundreds of pieces of trackable debris). For example, kinetic kill vehicles such as the U.S. Strategic Defense Initiative's proposed thousands of "Brilliant Pebbles" are sure to generate great quantities of space debris just during their initial deployment, and far more if they are ever used. Any kind of space warfare will put all satellites at risk. Perhaps worst of all would be the deliberate injection into LEO of large numbers of particles as a cheap but effective antisatellite measure, or the explosion of nuclear weapons in space (prohibited by the Outer Space Treaty, but routinely considered by military planners), which would indiscriminately destroy unprotected satellites by electromagnetic pulse (EMP) or nuclear radiation.[10] It would be wise to take long-range steps during the current period of relaxation of international tensions to discourage the development and prevent the deployment of any weapons capable of attacking spacecraft.

LIGHT POLLUTION IN/FROM ORBIT

Optical

"Light pollution" from space increasingly threatens both ground-based and space-based astronomy. The atmosphere is transparent only in the optical/infrared and the radio windows. The main problem in the optical is bright satellites, which have increased in number by about a factor of 3 since 1970, while the sensitivity of astronomical detectors has increased by more than a factor of 10.[11] Wide field (Schmidt) telescope exposures within two hours of sunrise or sunset are now contaminated by an average of five satellite tracks. Among the really silly bright satellites that have been proposed are cremated human remains in highly reflective cannisters in polar orbit (Celeste Corp.), a ring of a hundred 6m-diameter aluminized spheres (Eiffel Tower 100th Anniversary), and a 1800 m^2 reflective sail (Art Satellite).[12] Reflection even from small satellites and space debris can damage instruments on astronomical satellites such as Hubble Space Telescope.[13]

Radio

The main space environment problem for radio astronomy is unfiltered sideband radio emissions from networks of satellites. Each kind of atom and molecule in galaxies radiates at a characteristic frequency. The Soviet GLONASS navigation satellite system has since 1982 increasingly interfered with radio astronomy in the important radio frequency band at 1612 MHz used for study of hydroxl (OH).[14] Two or three of the GLONASS satellites are now always above the horizon. Their sideband flux densities are so much higher than those from astronomical sources that even far sidelobes of radio telescopes pick up background signals.

There was a similar problem from the six U.S. Global Positioning System satellites launched before 1986 interfering with observations of the 1667 MHz OH band; the more recent GPS satellites have filtered this sideband out.[15] Motorola has recently been persuaded by radio astronomers to avoid interference at 1612 MHz from its proposed 77-satellite "Iridium" cellular phone remote communications network.[16] Radio astronomers pushed for increased protection at the International Telecommunications Union (ITU) World Administrative Radio Conference (WARC) in February 1992. But it appears to be inevitable that most of the radio window will increasingly fog over. As bad as the problem is for radio astronomy, it may be even worse for the ambitious new program to search for radio signals from extraterrestrial intelligence.[17]

GAMMA RAY ASTRONOMY AND SPACE REACTORS

Gamma-ray astronomy can only be done from satellites or high balloons. The many unshielded nuclear reactors that have operated in LEO have produced unwelcome background radiation that has interfered with observations by both satellite and balloon-borne gamma-ray detectors.[18] In addition to direct gamma radiation, which is only a problem when the source is visible to the balloon or satellite, the electrons and especially the positrons (anti-electrons) that are copiously emitted by such reactors are then trapped by the earth's magnetic field. They become temporary radiation belts.

The U.S. orbited one small space reactor in 1964. The USSR orbited more than 30 reactor-powered military Radar Ocean Reconnaissance Satellites (RORSATs) in 1965–1988 in low 250 km orbits, plus two higher-altitude "Topaz" reactors in 1987–8 in 800 km orbits.@ The geomagnetically trapped electrons and positrons produced at a rate of about 10^{13} per second by the Soviet space reactors have caused serious interference with gamma-ray astronomy. Those from the higher altitude reactors caused the worst problems, since they were removed more slowly by collisions with molecules of the thinner upper atmosphere. The U.S. recently purchased a Russian "Topaz" reactor, and has for a nearly decade been developing (for SDI directed energy weapons or unspecified NASA uses) a space reactor called SP-100 that is a hundred times more powerful.[†]

I have been a leader of a joint group of the Federation of American Scientists (FAS) and the Committee of Soviet Scientists for Global Security. In May 1988 we called for a ban on nuclear power in Earth orbit.[19] Our arms control arguments for the ban were as follows:

- It would restrain the weaponization of space. Reactors are probably necessary power sources for directed energy weapons, but as far as I can tell they are not essential for *any* other orbital application.
- It would be verifiable because orbiting reactors emit easily detectable infrared, gamma rays, and particles.
- It would have been a good deal for both the U.S. and (the former) USSR, trading off U.S. SDI reactor-powered satellites against Soviet RORSATs, which are the main target of proposed U.S. antisatellite (ASAT) weapons.

Our environmental/nonproliferation arguments for such a ban were that:

- Reentry of two of the RORSAT space reactors (in 1978 over northern Canada, and in 1983 over the ocean) has already caused radioactive contamination of the Earth's surface and atmosphere. Future larger reactors such as SP-100 would contain much more radioactivity. Break-up on reentry could cause devastating contamination, but engineering them to re-enter intact could be equally dangerous by allowing recovery of enough high-enriched ^{235}U to make many nuclear weapons.
- Used reactors have been routinely kicked up to ~950 km "disposal orbits," exacerbating the space debris problem, which is especially serious at that altitude.[20]
- Interference with gamma-ray astronomy could become increasingly severe as instrumental sensitivity and reactor power both increase.

I want to emphasize that our group has not opposed nuclear power for lunar, planetary, or deep-space applications.[21] Indeed, an FAS report by Steven Aftergood on the Radioisotope Thermal Generator (RTG) aboard the Galileo planetary exploration spacecraft was submitted by NASA to help persuade a judge to allow the launch.

Aftergood recently revealed the secret SDI "Timberwind" project, to develop and test nuclear thermal rockets.[22] The hydrogen propellant would be heated by a particle-bed reactor, to achieve specific impulse $I_{sp} = v/g \approx 10^3$ s, a factor of ~ 3 better than chemical fuels. The military uses of such rockets remain obscure, despite the fact that perhaps two hundred million dollars have been spent on this project in the past few years. A recent report to the U.S. National Space Council[23] stated that "the only prudent propulsion system for Mars transit is the nuclear thermal rocket." This proposal needs further study. It is not clear under what assumptions nuclear thermal propulsion is favored, nor where such devices could be safely

tested, and it would be foolish to use them until their reliability is established. The problems include disposal of the reactor even if a nuclear rocket launch or boost succeeds, radioactivity in the exhaust, and accidents. Even if there are no accidents, long-duration tests of powerful unshielded reactors in LEO will interfere with gamma-ray astronomy.

The continuing interest in some agencies of the U.S. government in the use of nuclear reactors for power or propulsion in near-Earth space has already had another unfortunate side effect. After more than a decade of discussion, the scientific and technical panel of the U.N. Committee on Peaceful Uses of Outer Space had tentatively agreed, with U.S. concurrence, on detailed draft guidelines on the use of nuclear power. The draft was not as strong as I would prefer, but it would nevertheless have been quite useful. Its spirit is indicated by draft principle 3: "In order to minimize the quantity of radioactive material in space and the risks involved, the use of nuclear power sources (NPS) in outer space shall be restricted to those space missions which cannot be operated by non-nuclear energy sources in a reasonable way." But in early 1991, U.S. Defense Department complained to the State Department that the draft guidelines could pose obstacles to a nuclear-powered ballistic missile defense system, and NASA officials said they could constrain interplanetary human missions. So at the U.N. panel meeting in February 1991, the U.S. delegation proposed major changes that would drastically weaken the proposed guidelines.[24]

In December 1992, the High Energy Astrophysics Division of the American Astronomical Society issued a protest against plans of the Strategic Defense Initiative Office to launch a Russian "Topaz II" reactor into a 1500 km orbit, which would cause very serious interference with the Compton Gamma Ray Observatory (especially the BATSE detector) and with several other sensitive gamma ray satellites. I hope that the incoming Clinton Administration will adopt strong new rules to prevent such pollution of the near-Earth space environment.

ACTIONS TO PROTECT THE SPACE ENVIRONMENT FOR ASTRONOMY

Scientists can foresee problems of which others are unaware. Our dual role in helping to avert a space "tragedy of the commons"[25] is to increase the understanding of relevant basic science, and to define and advocate needed policies. My list of such policies is as follows:

- Avoid fragmentation of satellites from accidents or anti-satellite weapons tests, the main cause of space debris. Prohibit explosions of any kind in space.
- Do not introduce attack weapons into space.
- Ban nuclear reactors in orbit.
- Minimize light pollution from orbit.
- Develop launch vehicles that do not deplete ozone.
- Design boost and deployment systems for satellites that minimize the production of space debris. Require all

satellites in LEO to carry a mechanism, such as rockets or inflatable devices to increase drag, that will cause them to reenter when their useful life is over. Scientists designing space research satellites should take the lead in engineering such devices, thereby establishing a model for the rest of the space community.

A code of acceptable behavior should be drafted stating that that nothing should be launched into space unless it fulfills certain basic requirements to protect the space environment. Unnecessary launches of objects that will produce or themselves become long-lived space debris are far worse than ocean dumping, and should be prohibited by international agreement.

LONG-TERM VIEW

The space age is only 35 years old, yet we humans may already have placed so many artificial objects in the near-Earth environment that random collisions between them can produce a cascading number of debris fragments that will threaten and eventually prevent scientific and other uses of low earth orbit.[9] Such a debris belt would have other unfortunate consequences: for example, fragmentation of this debris by further collisions can eventually produce enough dust to cause a lingering twilight as it is illuminated by sunlight, a new and particularly unpleasant sort of light pollution.[26] It will without doubt be necessary for all space agencies to take active steps to prevent the buildup of debris, and it is an encouraging first step that NASA and ESA have succeeded in eliminating the Delta and Ariane upper stage explosions that were a major source of orbital debris. But much more effort will be needed, and it may even be necessary to deploy special spacecraft to remove some of the space debris at the altitudes where the critical density for a cascade have already been reached. Designing such devices will be a useful exercise, not least because it will help to impress on public officials the cost of space debris.

National political leaders usually take a short-range view, hardly ever stretching past the next change of government; astronomers measure time in millions and billions of years. We must help to educate the general public to think with at least an intermediate perspective of centuries and millenia about the environmental degradation that our increasingly powerful technology is causing on and near our beautiful but fragile planet – the only one like it that we know in the entire universe.

Acknowledgements. I am grateful to Steven Aftergood and my other FAS/CSS colleagues for an enjoyable and productive collaboration on space nuclear power issues, and to Donald Kessler and John Michener for sending me useful unpublished material. I thank Nancy Abrams and Rachel Somerville for editorial assistance.

Note added. After this paper was prepared and presented, I learned of a related volume, titled *The Vanishing Universe*, edited by Derek McNally (Cambridge University Press, in press). In particular, the article by David Malin, "The Disappearance of the Night Sky," is relevant to the section above on Light Pollution in/from Orbit.

ENDNOTES

*   A second of arc is 1/3600 of a degree.

\+   With "adaptive optics" techniques that are now being developed to compensate many times per second for atmospheric fluctuations, very high resolution astronomical images can be obtained with ground-based telescopes – but only over a small patch of the field of view, a few arc seconds across. Space Telescope images should be sharp from corner to corner.

\#   The Van Allen belt radiation becomes too intense for many astronomical instruments to tolerate by about 1000 km altitude above the equator and at lower altitudes near the Earth's magnetic poles and over the South Atlantic Anomaly. The structure of the radiation belts is complex and moreover variable during the 11-year solar cycle; the upper electron belt extends to beyond GEO.

@   These Soviet space reactors each produced approximately 100 kw of thermal power from roughly 30 kg of high-enriched ^{235}U fuel.

†   SP-100 is designed to produce ~2.5 million watts of thermal energy with ~200 kg ^{235}U.

REFERENCES

1. Steven Aftergood, "Poisoned Plumes," *New Scientist* (7 September 1991) 34. M. J. Prather et al., "The Space Shuttle's Impact on the Stratosphere," *J. Geophys. Res.* **95** (D11) 18,583 (Oct. 20, 1990) calculate that a launch schedule of nine Shuttle plus six Titan IV launches per year would add about 0.25% to the current stratospheric source of chlorine.

2. G. B. Field, M. J. Rees, and D. N. Spergel, "Is the space environment at risk?" *Nature*, **336**, 725 (1988).

3. G. R. Blumenthal, S. M. Faber, J. R. Primack, and M. J. Rees, "Formation of Galaxies and Large-Scale Structure with Cold Dark Matter," *Nature* **311**, 517–525 (1984). Reprinted in *Inflationary Cosmology*, L.F. Abbott and S.-Y. Pi, eds. (World Scientific, Singapore, 1986) pp. 316–324; in *The Early Universe: Reprints*, E.W. Kolb and M.S. Turner, eds. (Addison-Wesley, 1988), pp. 617–625; and in *Particle Physics and Cosmology: Dark Matter*, Mark Srednicki, ed. (North-Holland, Amsterdam, 1990), pp. 96–104.

4. Andre Linde, *Particle Physics and Inflationary Cosmology* (Gordon & Breach, New York, 1991).

5. E.g., M. A. Strauss, M. Davis, A. Yahil and J. Huchra, *Astrophys. J.*, **361**, 49 (1990); A. Dekel, E. Bertschinger, A. Yahil, M. A. Strauss, M. Davis, and J. Huchra, in preparation (1992).

6. See, e.g., Y. Kondo, ed., *Observatories in Earth Orbit and Beyond* (Kluwer Academic Publishers, Dordrecht, Holland, 1990).

7. D. J. Kessler, R. C. Reynolds, and P. D. Anz-Meador, "Orbital Debris Environment for Spacecraft in Low Earth Orbit," NASA TM 100–471 (April 1988).

8. These nice examples are from J. M. Ryan, "Tossed in Space: Orbital

Debris Endangers Instruments and Astronauts," *The Sciences* (July/August 1990) 14, which is the best-written brief introduction to the subject that I have read. Other standard references include N. L. Johnson and D. S. McKnight, *Artificial Space Debris* (Orbit Books, Malabar, Florida, 1987), and *Space Debris* (European Space Agency, Paris, 1988).

9. D. J. Kessler and B. G. Cour-Palais, "Collision Frequency of Artificial Satellites: The Creation of a Debris Belt," *J. Geophys. Res.* **83** (A6) 2637 (June 1, 1978); D. J. Kessler, "Collission Probability at Low Altitudes Resulting from Elliptical Orbits," *Adv. Space. Res.* **10** (3)393 (1990); D. J. Kessler, "Collisional Cascading: The Limits of Population Growth in Low Earth Orbit," in *Space Dust and Debris, Adv. Space. Res.* **11** (12)63 (1991).

10. See, e.g., J. R. Wertz and W. J. Larson, eds., *Space Mission Analysis and Design* (Kluwer Academic Publishers, Dordrecht, Holland, 1991), esp. § 8.2.

11. S. van den Bergh, "Effects of Space Debris and Satellite Interference on Astronomy," in *Proc. Int'l Colloq. on Environmental Aspects of Activities in Outer Space*, K.-H. Boeckstiegel, ed. (Heymanns-Koln, 1990) p. 71.

12. P. Murdin and R. F. Malina, in *Light Pollution, Radio Interference, and Space Debris*, D. L. Crawford, ed. (Astronomical Society of the Pacific, San Francisco, 1991) pp. 139, 145.

13. M. M. Shara and M. D. Johnson, *Publ. Astron. Soc. Pacific*, **98**, 814 (1986).

14. J. A. Galt, "Interference with Astronomical Observations of OH Masers from the Soviet Union's GLONASS Satellites," in *Light Pollution, Radio Interference, and Space Debris*, D. L. Crawford, ed. (Astronomical Society of the Pacific, San Francisco, 1991) p. 213. R. J. Cohen, "The threat to radio astronomy from radio pollution," *Space Policy* **5** (May 1989) 91. M. Mitchell Waldrop, "Taking Back the Night," *Science*, **241**, 1288 (1988).

15. J. A. Galt, "Contamination from Satellites," *Nature*, **345**, 483 (1990).

16. R. Stone, "Radioastronomers Seek a Clear Line to to the Stars," *Science*, **251**, 1316 (1991).

17. J. Tarter, "As Bad as Things Are for RadioAstronomy in the Protected Bands – They Are Worse for SETI," in *Light Pollution, Radio Interference, and Space Debris*, D. L. Crawford, ed. (Astronomical Society of the Pacific, San Francisco, 1991) p. 273.

18. See Joel R. Primack, "Gamma-Ray Observations of Orbiting Nuclear Reactors," *Science*, **244**, 407, E1244 (1989), and additional papers in this issue of *Science*}. Also F. Makino, et al., "Artificial Radiation Observed with Ginga Satellite," in *Proc. Workshop on Space Debris*, May 11, 1989 (Institute of Space and Astronautical Science, Kanagawa, Japan, 1990) p. 65.

19. J. R. Primack, et al., "Space Reactor Arms Control: Overview," *Science and Global Security*, **1**}, 59 (1989), and subsequent papers in this issue.

20. See also P. D. Anz-Meador and A. E. Potter, Jr., "Radioactive Satellites: Intact Reentry and Breakup by Debris Impact," in in *Space Dust and Debris, Adv. Space. Res.*, **11** (12)37 (1991).

21. S. Aftergood, D. W. Hafemeister, J. R. Primack, O. F. Prilutsky, and S. N. Rodionov, "Nuclear Power in Space," *Scientific American*, **264** (6) (June 1991) 42. Steven Aftergood, "Nuclear Safety Issues Associated with NASA's Galileo Mission to Jupiter" (Federation of American Scientists, Washington, D.C., Sept. 1989).

22. J. R. Asker, "Particle Bed Reactor Central to SDI Nuclear Rocket Project," *Aviation Week & Space Techn.* (April 8, 1981) 18.

23. *America at the Threshold: Report of the Synthesis Group on America's Space Exploration Initiative*, (U. S. Government Printing Office, Washington, D.C., 1991).

24. A. Lawler, "U.S. Backpedals on Nuclear Rules," *Space News* (March 25–31, 1991) pp. 1, 20.

25. G. Hardin, "Tragedy of the Commons," *Science*, **162**, 1243 (1968). See also G. Hardin and J. Baden, eds., *Managing the Commons* (Freeman, 1979).

26. S. van den Bergh, "Summary Paper," in *Light Pollution, Radio Interference, and Space Debris*, D. L. Crawford, ed. (Astronomical Society of the Pacific, San Francisco, 1991) p.329.

7: Effects of Space Debris on Commercial Spacecraft – The RADARSAT Example

H. R. Warren and M. J. Yelle

RADARSAT Program Office, Canadian Space Agency, Ottawa, Ontario, Canada

INTRODUCTION

The purpose of this paper is to describe, for a commercial low earth orbit satellite, the work accomplished to assess the hazard of the space debris environment and to incorporate protective measures. RADARSAT is typical of remote sensing spacecraft – operating in near polar orbit at an altitude well populated with debris. The lessons learned from our investigation may be relevant to other such spacecraft.

RADARSAT is a Canadian remote sensing satellite carrying an advanced Synthetic Aperture Radar (SAR) sensor which operates in a variety of modes to optimize swath widths, incidence angles, and resolutions for a number of commercial applications. NASA and NOAA are participating in RADARSAT through an international memorandum of understanding, with NASA contributing the Delta II launch (see Reference 1 for details of the mission). To help ensure economic viability, the satellite is designed for a mission life of five years, 1995 to 1999. To optimize the ground coverage patterns of the SAR image swaths, the planned altitude is 790 km, in a dawn-dusk sun-synchronous orbit of 98.5° inclination.

The 3000 kg spacecraft, shown in Figure 1, is three axis stabilized, oriented so that the SAR antenna is aligned with the direction of flight. In terms of exposure to orbital debris, the main body presents a frontal surface of 5 sq. m. Such an area, together with a five-year exposure at the end of the century in an orbit that carries a high population of debris, led us to give serious consideration to the risk of damage that could threaten the success of the mission.

The following sections describe the analysis of this threat and the test program, carried out with the cooperation of NASA Johnson Space Center, that led to a number of design modifications. It should be noted that in our case, the spacecraft design was already well advanced when it was determined that protective measures were necessary. Furthermore, the snug dimensions of the launch vehicle shroud placed an additional constraint on the designers.

DETERMINING THE VULNERABILITY OF A SPACECRAFT

In order to quantify the threat of orbital debris, spacecraft designers must undertake an analysis of the orbital debris environment specific to the spacecraft's orbit, and an analysis of

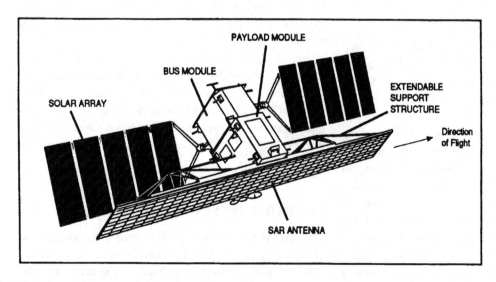

Figure 1

the spacecraft's vulnerability to that environment. As a complement to these analyses, spacecraft components can also be subjected to hypervelocity impacts tests thereby simulating space debris hits.

Defining the environment

The flux of orbital debris on a spacecraft is largely dependent on size of the debris, the orbit altitude and inclination, on the size and attitude of the spacecraft, as well as on the duration of the mission and the solar activity at the time. An orbital debris environment model has been defined by scientists at NASA Johnson Space Center. This model along with its assumptions are described in NASA TM 100 471 [2]. The EnviroNET database, available from NASA Goddard Space Flight Center, can also be used to compute the cumulative flux of space debris of a given size on a spacecraft surface [3].

The number of impacts on a spacecraft surface can be calculated using the flux of the debris, the duration of the mission, and the size of the spacecraft surface. Once these parameters are entered in the flux equation, the number of collisions with the surface becomes a function of the debris size only. Meanwhile, the impact velocity distribution of the flux, also defined in NASA TM 100 471, is a function of the spacecraft velocity and orbit inclination. This impact velocity typically varies between 0 km/s and twice the spacecraft velocity, approximately 15.4 km/s (head-on collision between spacecraft and debris, each travelling at 7.7 km/s) for a low earth orbit. The angle of impact is a cosine relationship between the actual impact velocity and the maximum impact velocity (here, 15.4 km/s). Therefore, the impact angle distribution can be derived from the impact velocity distribution.

Once the orbital debris flux and the impact angle distribution have been determined, the number of hits on specific spacecraft components can be predicted. This involves determining the location of each component relative to one another and to the incoming space debris. This analysis includes considering how components shield one another.

Vulnerability of individual components

In parallel with the above analysis, the investigators also need to determine how resistant each of the components are to impacts. This can be accomplished by using already derived hypervelocity impact equations or by subjecting some components to actual laboratory impacts.

Numerous organisations have been involved in hypervelocity impact research for several years to develop accurate hydrocodes which enable analytical simulation of the impact of particles on simple components. A number of equations have been developed to approximate the damage resulting from impacts on relatively simple structures [4,5,6,7,8]. These equations can therefore be used to determine the ballistic limits, or failure points, of some components. In our case, we would want this ballistic limit expressed as the size of a particle with the predicted in-flight impact velocity which

would perforate the component. Once the particle size to generate failure has been calculated, the probability of an impact with that size of particle can be determined from the flux equation. Thus, the probability of failure of that component due to an orbital debris impact can be determined.

As mentioned earlier, another method of determining the ballistic limit of a component is to subject it to real impacts. Hypervelocity impacts can be achieved in the laboratory using two-stage light-gas guns which are capable of accelerating particles to speeds as high as 12 km/s. This velocity is not quite as high as the expected in-flight impact velocity, therefore the test results need to be extrapolated to the predicted impact velocity using an equivalence of kinetic energies. Thus, these tests can be used to determine the ballistic limit of complex components, or of combinations of components. Since the accuracy of hydrocodes is limited mostly to simple components, laboratory tests must be used to provide information on samples with more complicated structures.

Vulnerability of the spacecraft

The vulnerability of the spacecraft can be determined by combining the probability of failure of its various components. This would include taking into account the redundancy of components, as well as their criticality for the life of the spacecraft. If the vulnerability of the spacecraft is found to be unacceptable, action should be taken to increase the shielding of the spacecraft as a whole or of at least its more vulnerable components.

SHIELDING TECHNIQUES

A number of shielding techniques could be considered to protect the spacecraft from space debris hits. One technique would involve using a bumper, known as a Whipple shield, to break up the debris particle before it hits the spacecraft. A variation of this would be to use a series of bumpers to achieve a more effective debris break-up [5,9]. These two methods however involve increases in size, mass, complexity and cost of the spacecraft.

Another option, although less effective, would be to augment the shielding characteristics of existing components by, for example, thickening the walls of components or by adding a layer of particle breaking material to thermal blankets covering the exterior of the spacecraft. This option would involve an increase in weight, with minimal increases in size, complexity and cost.

THE RADARSAT SPACECRAFT AS AN EXAMPLE

The RADARSAT spacecraft will be used to demonstrate the methodology involved in determining the threat of orbital debris as well as shielding techniques. The actual analyses along with assumptions are described in Reference 10.

Defining the environment for RADARSAT

Using the RADARSAT orbit parameters as well as the spacecraft's configuration the flux of space debris on the spacecraft was determined from EnviroNET [3]. The expected number of hits on the leading (+X) face of the spacecraft with respect to the debris size is shown in Figure 2. The orbital debris velocity distribution, Figure 3, was determined using NASA TM 100 471. The impact angle distribution, with 0° being normal to the leading face, is presented in Figure 4.

The above mentioned figures indicate that the spacecraft leading surface is expected to be impacted by one 1 mm particle during the spacecraft's 5-year mission and by a larger number of smaller pieces. Also, the expected impact velocity is in the 13 to 15.5 km/s range, with the impact angle in the 5° to 30° range. It should be noted that the two peaks of Figure 4 would have been more widely separated if the orbit inclination were at say 28° rather than 98.5, and the altitude were at 500 km.

Defining the vulnerability of RADARSAT

From the above environment information, spacecraft components were examined to determine their vulnerability to impacts. The analysis of the payload module of the spacecraft, with its electronic components protected by honeycomb shear panels, consisted of determining if the shear panels provided sufficient shielding to the sensitive electronics. Hypervelocity impact equations from References 5 and 9 indicated that the components were adequately shielded.

The analysis for the bus module proved to be more complicated due to its configuration. Most of the sensitive equipment of the bus module is mounted on the outside of the bus honeycomb shear panels and is therefore protected only by multilayer insulation (MLI) thermal blankets. These thermal blankets stand off the shear panels by 15 to 25 cm, thereby possibly acting as Whipple shields. The analysis consisted of determining the vulnerability of each component individually, and then combining these numbers to determine an overall bus module vulnerability. This analysis included not only electronic equipment, but also cable harnesses between boxes and propulsion subsystem hardware. A bus survivability against space debris of 0.5 was calculated.

Impact tests on RADARSAT components

To complement the above analyses, some spacecraft components were subjected to impacts at the Hypervelocity Impact Research Laboratory at NASA Johnson Space Center. The tests were conducted under the RADARSAT Memorandum of Understanding between NASA, NOAA and the Canadian Space Agency. The tests were used to verify the assumptions used in the analyses and to determine the effectiveness of various shielding techniques.

The tests articles included different configurations of honeycomb shear panels, different thicknesses of electronic component walls, hydrazine lines, synthetic aperture radar waveguides and wire bundles. Multi-layer insulation blankets were also tested, with and without reinforcement, to verify their shielding effectiveness. The impact angles were varied to simulate impacts on components on both the front and the sides of the spacecraft.

The test results can be summarized as follows:

- An MLI blanket 63.5 mm from a plate provided significant shielding against small projectiles (Pictures 1 and 2). The addition of Nextel* (Picture 3) or aluminum mesh to the blanket improved the shielding even further.
- Wire bundles of 24 gauge suffered considerable damage from a 1 mm aluminum projectile when unprotected (Picture 4), but negligible damage if protected by an MLI and Nextel shield 63.5 mm in front of sample.
- On the basis of equivalent added mass (ie: 1 Nextel layer = 3 aluminum mesh layers, for our samples), Nextel and aluminum mesh performed similarly. However, due to the difficulty of adding three layers of aluminum mesh to the thermal blanket, one layer of Nextel was chosen as baseline reinforcement. These materials were tested as a result of suggestions by NASA Johnson Space Center.
- Honeycomb shear panel substrate caused debris from an impact on the front surface to be funnelled through the core to the rear surface, even at oblique angles. This occurred on various samples with differing thicknesses, materials and cell sizes. This phenomenon created more damage than the Whipple equation had predicted.
- An impact on a honeycomb shear panel with graphite epoxy facesheets produced delamination of the second facesheet.
- Hydrazine line with 0.51 mm thick stainless steel walls was perforated by a 0.40 mm projectile at 8 km/s.
- Synthetic aperture radar waveguides suffered little damage at high oblique angles of impact.

Shielding improvements made to the RADARSAT spacecraft

A number of design changes were made to the spacecraft to increase its chances of survival against the predicted space debris environment. These included adding a layer of Nextel to the MLI blankets of the bus module, thereby increasing the protection of the electronic boxes and the wiring harnesses mounted on the outside of the bus shear panels.

The bus module components which were considered more vulnerable had their walls thickened. A gap which existed between the bus module and the payload module was closed off, thereby protecting a number of hydrazine lines from incoming space debris. Shields were also added to some lines to decrease the probability of direct hits. Finally, the bus forward cornerpost radiators were thickened and widened to give them a double purpose of radiating heat from the spacecraft and shielding some electronic components from incoming debris. All changes are represented on Figure 5.

A total spacecraft mass increase of 17 kg resulted from all of the shielding design improvements. These modifications,

EXPECTED NUMBER OF HITS VS DEBRIS DIAMETER
10% GROWTH, AREA = 4.16 m squ., 5 YEARS LIFE

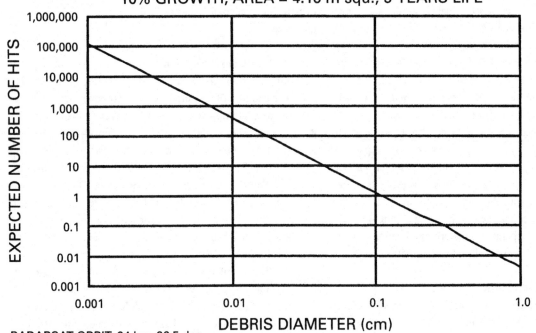

RADARSAT ORBIT, 94 km, 96.5 deg.
Num Hits = 4,11E–3 × d^(–2.493)

Figure 2

ORBITAL DEBRIS VELOCITY DISTRIBUTION
794 km Altitude, 98.5 Deg. Inclination

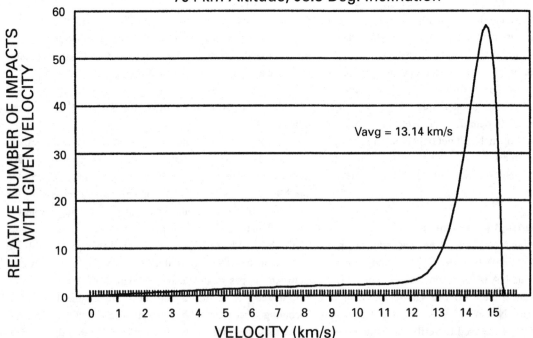

Vavg = 13.14 km/s

Figure 3

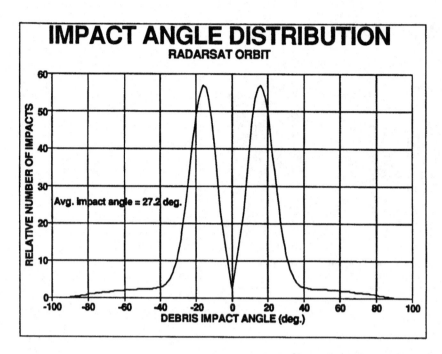

Figure 4

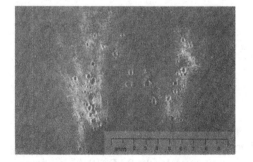

Picture 1: Impact of 1 mm projectile on bare wall.

Picture 2: Impact of 1 mm projectile on box wall shielded by MLI blanket.

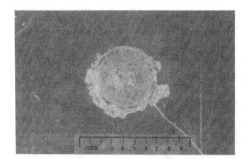

Picture 3: Impact of 1 mm projectile on box wall shielded by MLI reinforced with Nextel.

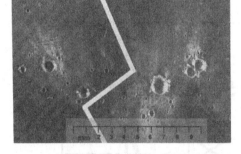

Picture 4: Impact of 1 mm projectile on unprotected wire bundle.

along with some other changes resulting from the evolution of the design, have increased the survivability of the spacecraft against debris from 0.5 to 0.87.

OTHER CONSIDERATIONS

Spacecraft re-entry

One way to limit the constant build-up of man-made debris in orbit would be to de-orbit spacecraft after their useful life. Most three-axis controlled spacecraft use a propulsion system for maintenance of altitude and orbit inclination. Could the residual fuel at the end of the mission be used to lower the perigee altitude enough that the higher atmospheric drag would cause the satellite to re-enter?

Figure 6 shows the result of an analysis of the feasibility of such a plan. As a function of the initial, circular altitude the

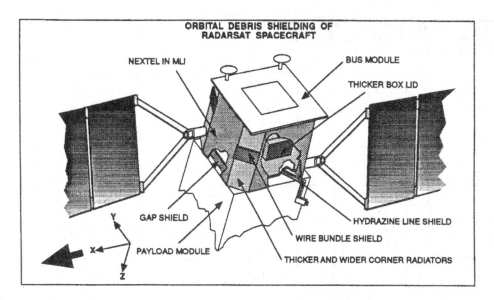

Figure 5

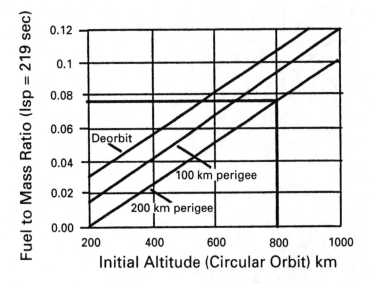

Figure 6

curves indicate the fraction of spacecraft mass that would be needed for fuel (assumed to be hydrazine). The choices are either to de-orbit directly (i.e., re-enter in half an orbit) or to conduct a transfer orbit to bring the perigee just low enough to cause the drag build-up to lower the satellite for eventual re-entry.

As an example, to transfer from an 800 km circular altitude to an orbit with perigee at 200 km and apogee at 800 km, about 7.6% of the spacecraft mass would need to be devoted to hydrazine. In the case of RADARSAT, this would mean carrying over 200 kg of fuel for this manoeuvre. As the normal orbital operations require only 65 kg of fuel, it can be seen that it would require a major design modification to accommodate the extra fuel, its tank and support structure. Furthermore, the large mass increase would be difficult to include in the launch vehicle mass margin.

In terms of time to re-enter, for the RADARSAT configuration and the 200 km perigee case, the re-entry time would be of the order of 530 days. If we had used additional fuel to transfer to a 100 km perigee this time would decrease to 12 days.

Our conclusion from this preliminary study is that for an altitude as high as 700–800 km, which is typical for sun-synchronous satellites, it would not be practical to incorporate the capability to lower the perigee enough to bring about re-entry.

Jettisoned hardware

Many debris objects may be generated during spacecraft separation from the launch vehicle and sequences to deploy appendages. For the RADARSAT program the practice was that there would be no hardware items cast loose into space during either the spacecraft separation or release of the extendible SAR antenna and solar arrays.

It was not found to be difficult to achieve this and it is recommended that all spacecraft designers adopt this practice.

CONCLUSION

This paper has summarized the work carried out by the Canadian Space Agency, with assistance from NASA to evaluate, by analysis and test, the hazard of orbital debris for the RADARSAT spacecraft and to devise protective measures to bring the risk to an acceptable level.

The analysis technique to determine the debris flux, using EnviroNet, is straightforward and readily available. Thus the particle size likely to impact the spacecraft as well as its velocity and direction can be determined. In our experience, the testing of actual spacecraft components conducted by NASA JSC was invaluable in determining how vulnerable they were to damage and how best to protect them.

It was found that shielding modifications that made a substantial improvement in spacecraft survivability could be achieved with relatively small impact to the design, cost, spacecraft mass and schedule of the program, even though the spacecraft configuration was well established when this effort commenced.

Based on the RADARSAT case we would encourage designers of commercial spacecraft to include in early design activities a preliminary analysis of the risk of damage from space debris. By so doing, they will have more scope to implement shielding concepts, will be better able to arrange the layout to place vulnerable units in less exposed locations and they could select materials that would double as effective barriers to incoming debris.

ACKNOWLEDGEMENTS

The authors wish to acknowledge and thank the staff of the NASA Johnson Space Center's Hypervelocity Impact Research Laboratory for their cooperation in performing the tests, compiling the data and contributing shielding ideas. Also to be acknowledged is the initiative and effort of the team at Ball Space Systems Division who implemented the modifications to their bus with a minimum of increase in cost or weight.

ENDNOTE

* Nextel is a 3M product made of woven ceramic fibres.

REFERENCES

1. Ahmed S., Warren H.R., Symonds M.D., Cox R.P., July 1990, *The RADARSAT System*, IEEE Transactions on Geoscience and Remote Sensing, Vol. 28, No. 4.

2. Kessler,D.J., Reynolds,R.C. & Anz-Meador,P.D., April 1989, *Orbital Debris Environment for Spacecraft Designed to Operate in Low Earth Orbit*, NASA Technical Memorandum 100 471, NASA Johnson Space Centre, Houston, Texas.

3. Lauriente,M. & Hoegy,W. ,April 1990, *EnviroNET: A Space Environment Data Resource*, AIAA 90-1370, NASA-Goddard, Greenbelt, MD.

4. Kessler,D.J, Anderson,C.E., Tullos,R. & Cour-Palais,B.G., March 19–22 1990 and May 21–24 1991, *A Short Course on Dealing with the Growing Challenge of Orbital Debris*, Presented by the Southwest Research Institute, 6220 Culebra Rd., P.O. Box 28510, San Antonio, Texas.

5. Cour-Palais,B.G., 1969, *Meteoroid Protection by Multiwall Structures*, AIAA Paper No. 69-372, Presented at the AIAA Hypervelocity Impact Conference, April 30–May 2, 1969, Cincinnati, Ohio.

6. Christiansen,E.L., 1990, *Advanced Meteoroid and Debris Shielding Concepts*, AIAA Paper No. 90-1336, Presented at the AIAA/NASA/DOD Orbital Debris Conference: Technical Issues & Future Direction, April 16–19, 1990, Baltimore, MD.

7. Cour-Palais,B.G., Apr 1979, *Space Vehicle Meteoroid Shielding Design*, Published in " The Comet Halley Micrometeoroid Hazard", ESA SP-153, Proceedings of a workshop held at ESTEC, Nordwijk, Netherlands.

8. Cour-Palais,B.G., 1986, *Hypervelocity Impact in Metals, Glass, and Composites*, Int. J. of Impact Engineering, Vol 5, pp 221–237, 1987.

9. Cour-Palais,B.G & Crews,J.L., 1990, *A Multi-Shock Concept for Spacecraft Shielding*, Int. J. of Impact Engineering, Vol. 10.

10. Terrillon, F., Warren, H.R., Yelle, M.J., 1991, *Orbital Debris Shielding of the RADARSAT Spacecraft*, IAF-91-283, 42nd Congress of the International Astronautical Federation, Montreal, Canada.

8: Potential Effects of the Space Debris Environment on Military Space Systems

Albert E. Reinhardt

US Air Force Phillips Laboratory, Space Survivability Division (PL/WSS), Kirtland AFB, NM 87117-6008

ABSTRACT

The US Department of Defense is in the first phase of a research effort to characterize the space debris environment and establish the potential threat level debris represents to current and future DoD space systems. The US Air Force Phillips Laboratory is acting as the technical lead for this research program. The Phase 1 effort, characterization of the debris environment, is scheduled to complete at the end of Fiscal Year 1993. Phase 1 emphasizes both the measurement and modeling efforts needed to characterize the current debris environment down to 0.1 cm and to project the future environment and its impact on the survivability of current and planned systems. In parallel, policy efforts have led DoD space designers to implement cost effective debris minimization measures for future systems. A joint effort with NASA to develop a Space Debris Minimization and Mitigation Handbook is also geared to fielding "debris clean" space systems. The threat, however, must be well defined and significant before more costly measures are implemented. This paper will discuss the background of the DoD space debris research program, the status and goals of the Air Force Phillips Laboratory measurement and modeling efforts, and the objectives and status of the Space Debris Minimization and Mitigation Handbook.

BACKGROUND

On February 4, 1987, the DoD Space Policy signed by the Secretary of Defense, the honorable Caspar Weinberger, recognized the potential of a space debris problem: " ... DoD will seek to minimize the creation of space debris in its military operations. Design and operations of space tests, experiments and systems will strive to minimize or reduce debris consistent with mission requirements" [1]. As shown in Figure 1, this in turn led to the National Space Policy, signed by President Reagan on January 5, 1988, which reflected the new DoD space policy: "... all space sectors will seek to minimize the creation of space debris. Design and operation of space tests, experiments and systems will strive to minimize or reduce accumulation of space debris consistent with mission requirements and cost effectiveness" [2].

The 1988 space policy also mandated that an Interagency Group (IG) on Space provide recommendations for implementation of the policy.

In response to the National Space Policy, the IG (Space) developed a report on space debris which was endorsed by the National Security Council in February 1989 [3]. Two main recommendations in the report called for joint DoD/NASA studies to develop plans for conducting studies and experiments: (1) to further our knowledge of the current space debris environment; and (2) to address methods for reducing the debris hazard (including minimization techniques) and for making spacecraft more survivable to debris.

As a result of the IG (Space) Report, the Air Force was formally tasked as the lead service and the Air Force Space Technology Center (now the Phillips Laboratory) was made the technical lead for DoD to develop the program plan. NASA selected Johnson Space Center as their technical lead. To carry out the IG (Space) report recommendations, a DoD/NASA study was performed, which provided more detailed program goals for the research program. This orbital debris research program plan was then submitted to the National Space Council and approved in July 1990.

There were two objectives in the study. The first objective was to characterize the debris environment in Low Earth Orbit (LEO), down to debris sizes of 0.1 cm. In addition, the level of confidence associated with the characterization was required. The second objective dealt with the identification of candidate technologies for minimizing debris and enhancing the survivability of orbiting assets. However, the implementation activities under this objective would depend on the environmental characterization to be performed under the first objective [4].

The organization of the Air Force Phillips Laboratory Space Debris Research Program is shown in Figure 2. Measurement and empirical modeling efforts of the debris environment are headed by the Phillips Laboratory Geophysics Directorate Modeling and Remote Sensing Branch. Breakup Modeling efforts and tests are run by the Defense Nuclear Agency's (DNA's) Shock Physics Branch. The Phillips Laboratory's Space Kinetic Impact and Debris Branch supports complex

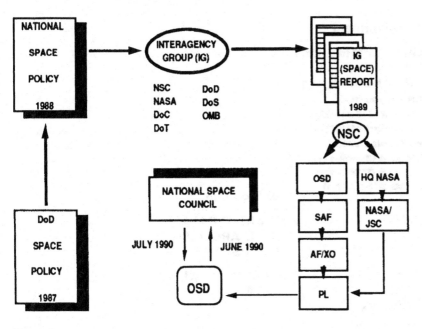

Figure 1

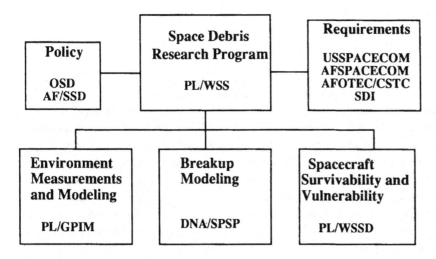

Figure 2

modeling and survivability tool development. Each of these areas are discussed in more detail in this paper.

SPACE DEBRIS MEASUREMENT PROGRAM

Both NASA and the DoD recognized the importance of conducting a measurement program for the characterization effort. As pictured in Figure 3, there is a lack of data and thus great uncertainty in the debris environment for debris sizes between 0.1 and 10 cm [5]. NASA pursued the use of the Haystack radar and development of the Haystack Auxiliary radar for their measurement program, while the Air Force Phillips Laboratory opted to use existing optical systems. AFSPACECOM required that, in addition to detection of small debris, orbital element sets were needed for the debris objects to identify debris orbits (events) of con-

sequence. This is in keeping with AFSPACECOM's Space Surveillance Network's mission of detecting, tracking, and maintaining a catalog of all man-made objects. As discussed later, this small debris tracking goal is driving technology requirements and methods for the ground-based optical systems.

Although the research plan was approved in 1990, actual funding for the Air Force Measurement portion of the program did not become available until April of 1991. The original plan envisioned a three year effort to perform the first objective of the study, that of debris characterization for LEO. NASA meanwhile began measurement of LEO using the Haystack radar in 1990 and then with the Air Force GEODSS telescope at Diego Garcia in 1991. The following sections describe the NASA effort, then provide the initial results and plans for the Air Force optical measurement program.

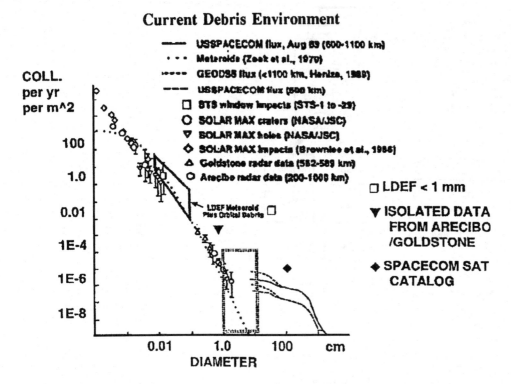

Figure 3

NASA Haystack measurement program

During 1990–1991, NASA collected 1015 hours of data using the Haystack radar (an X-Band radar) in a stare mode. To their credit, NASA also commissioned an independent panel of experts in radar technology to review their methodology and results. The AF Phillips Laboratory participated in this review effort [6]. The NASA findings indicated that there were on the order of 150,000 objects down to slightly below 1cm in size up through an altitude of 1500 km. The flux of 1 cm and larger debris detected at space station altitudes was 30% higher than that nominally predicted by NASA's orbital debris model, but it was within the model's confidence limit of +/- 60% [7].

The Review Committee found that NASA's approach and methodology were sound. The most challenging part in the data analysis was to assess the error bars or uncertainty of the size distribution of the data. Further analyses are focused on the uncertainty issue; however, additional data from the planned Haystack Auxiliary radar (a K-Band radar) will be required to bound the size distribution more precisely.

The NASA Haystack Radar data provided the first comprehensive quantitative assessment of the space debris LEO environment in the size regime of slightly less than 1 cm and greater. Although the methodology used does not readily describe the inherent altitude dependence of the space debris problem, it does take an important first step in providing a macroscopic view of debris in LEO. Additional measurement and modeling techniques are required in order to define the debris population by altitude segments and thus forecast more precisely the threat environment to specific systems.

Air Force measurement program

In April 1991, a site selection review was conducted to select optical sites for the Air Force debris measurement program. Factors such as hardware capabilities, resident expertise, and night sky darkness were considered. Though three sites were preferred by the review team, funding constraints permitted the selection of only two sites: MIT Lincoln Laboratory's Experimental Test System (MIT/LL ETS) in Socorro, New Mexico, and the Air Force Phillips Laboratory's Advanced Measurement Optical Site (PL/AMOS) in Maui, Hawaii.

MIT/LL ETS employs two 80 cm telescopes with selectable 0.5, 1.0, or 2.0 degree fields of view (FOV). These nearly identical telescopes are capable of tracking an object in tandem or independently. PL/AMOS uses the 1.2 m AMOS telescope in tandem with the 40 cm, 6 degree FOV Ground-based Electro-Optical Deep Space Surveillance (GEODSS) Auxiliary telescope for their debris measurement efforts.

On May 1, 1991, the Nimbus 6 Delta rocket body in Sun-synchronous orbit apparently exploded, producing more than 200 catalogued fragments. Both optical sites began twilight searches of this cloud within a few days after the event. Twilight is chosen in order to maximize detection sensitivity, with the objects being illuminated by sunlight while the optical site is in darkness. In order to detect objects fainter than those initially catalogued, the pseudo-track technique was employed. In this method the telescopes are slewed across the sky at the same rate the debris cloud is passing overhead. Thus fixed stars appear as streaking images while debris objects should appear as a steady or blinking (if the object is tumbling) light.

The 1991 measurement results are summarized in Figures 4 through 7, with a representation of the Nimbus 6 cloud shown in Figure 6. In twenty-five hours of observation, MIT/LL ETS was able to detect over 200 probable low earth orbiting objects. More importantly, eight of the uncorrelated targets (UCTs) were tracked over sufficient arcs to determine orbital element sets. Uncorrelated objects are those not in the Space Surveillance Network's Catalog of tracked objects. The faintest object tracked was 14th magnitude which corresponds to 8–10 cm size objects. In forty-four hours of observations, the PL/AMOS site detected approximately 239 objects, down to 15th magnitude (roughly 5 cm) with approximately one-half of the objects being uncorrelated with the catalog. As shown in Figure 7, Air Force personnel at PL/AMOS also began analyzing the data collected for NASA at the GEODSS site in Diego Garcia. Initial analysis of the Diego Garcia data, which was collected in a stare mode, also indicates that roughly one-half of the objects detected did not correspond to the catalog. Current, ongoing analysis of the catalog with this data may reduce the uncorrelated count by 25%.

Discussion of measurement results

The results from 1991 clearly demonstrated that optical sensors could detect, track, and determine orbital element sets for previously uncataloged debris objects. The limitations of the current optical systems were recognized, such as the detection capability being approximately 5 cm at the PL/AMOS site using the GEODSS Auxiliary telescope in a slew mode. In order to push the optical detection sensitivity further, the Phillips Laboratory Geophysics Directorate (PL/GP) began an effort to study the potential application of a Charged Coupled Device (CCD) Camera combined with a telescope for space debris measurements. In theory, magnitudes of 20.5 (mm diameter objects) are achievable when a CCD Camera is used in conjunction with advanced processing techniques.

PL/GP CCD effort

In 1991, PL/GP built an off-the-shelf 512 x 512 pixel array CCD camera system and mounted this device onto a large 100 inch staring telescope at the Air Force Wright Laboratory (AFWL). The AFWL telescope was built in the 1960's for other purposes and operates only in the stare mode. Though this was not the first application of CCD cameras on an optical system, it was the first attempt to use this combination to detect space debris. Data were collected in February 1992 with objects detected down to 13th magnitude. There were viewing problems due to a large temperature gradient between the building housing the telescope and the outside air which limited detection sensitivity. Confirming experiments are scheduled for June and July of 1992.

GEODSS Detection Capability

The experience gained in the PL/GP effort is intended to transition to the PL/AMOS measurement program specifically for either the GEODSS Main or auxiliary telescopes. Figures 8 and 9 show the current limiting diameter detection sensitivity for the GEODSS Main and auxiliary telescopes for the limiting magnitudes shown. A sensitivity of 16.0 magnitude has been demonstrated for the Maui GEODSS Main telescope, while 14.5 limiting magnitude is more the operational norm. Detection sensitivity of the stare and pseudo-track modes are also compared. A CCD properly configured for the FOV of either of these systems should increase the detection sensitivity by a factor of two. Thus, from Figure 9, a CCD camera mounted on a finely tuned GEODSS main should be able to detect objects of 2.5 cm diameter at 1000 km in a stare mode.

Debris population estimates from optical sites

When space debris detection data are collected in a stare mode, such as under the NASA program, the calculation of a population estimate is relatively straight-forward, assuming that the debris population is distributed in the Poisson form [8]. Since the beam area is known, detections are readily converted to flux or number of impacts per year for each square meter of spacecraft cross-sectional area.

This population estimate is not appropriate, however, when the pseudo-track method is employed. Since the telescopes are slewing over large volumes of space while sacrificing coverage to match the relative velocity of known debris clouds, extrapolation of detections to debris population estimates becomes quite complex. The equations and methodology to convert debris detections into population estimates in the slew mode are under investigation by the researchers involved in the Phillips Laboratory program. Based on the complexity of the issue, and on the current and planned capability of the GEODSS Main telescopes, clearly more data is required using the GEODSS Mains in both stare and pseudo-track modes. This additional data will provide the ability to confirm the estimates obtained by the Haystack radar and demonstrate a dual phenomenological measurement capability for the orbital debris population. Unfortunately, additional stare data from GEODSS will not provide insight into the debris distribution with altitude as discussed below.

Debris population distribution with altitude

In order to be able to project accurately the potential threat of space debris to future and planned systems, an accurate measurement of the smaller debris population distribution vs altitude is required. Figure 10 depicts the altitude distribution of the catalog objects for 1987 [9]. Objects catalogued are typically > 10 cm. Clearly, for systems planning to use nearly circular orbits, there are areas of LEO, such as 800 km, which have a higher probability of collision than say 1200 km. As the measurement program develops tools such as the Haystack auxiliary radar and the CCD camera for GEODSS, techniques must be used to describe altitude distribution for the smaller debris sizes detected. If stare techniques are used for debris

Object	i	D	e	ω	M	n	epoch
90006	74.024		0.0016			12.644	
90014	99.758	13.119	0.0209	338.319	337.173	13.6551	91:158.2
90025	99.834	71.987	0.0682	150.691	257.416	12.0172	91:220.2
90026	99.408	70.311	0.0631	130.141	114.672	12.1783	91:220.2
90027	82.586	83.053	0.0073	328.998	184.941	12.7999	91:220.2
90029	101.073	71.002	0.0898	315.114	354.074	15.2385	91:221.2
90030	83.482	98.968	0.0038	234.906	151.404	13.8171	91:178.0
90036	101.210	105.661	0.0579	170.528	343.610	12.5435	91:277.15

Figure 4: Lincoln Laboratory debris anode elsets.

Object	Estimated size	Comments
90006	>1 m	Payload sized object in popular Soviet LEO (74°)
90014	≤15 cm	NIMBUS-6 DB candidate, retrograde from original RB, reacquired 14 rev. later
90025	≤8 cm	NIMBUS-6 DB candidate, posigrade, good Gabbard fit
90026		NIMBUS-6 DB candidate, posigrade, good Gabbard fit
90027	≤30 cm	Similar orbit to Cosmos 1275 DB, but very high Δv
90029	≤100 cm	Very short period, 94 min, sun-synchronous inclination
90030	≤40 cm	Near center of Cosmos 1275 cloud, Gabbard analysis non-definitive
90036		NIMBUS-6 DB candidate, posigrade, good Gabbard fit

Figure 5: Lincoln Laboratory analysis of acquired objects.

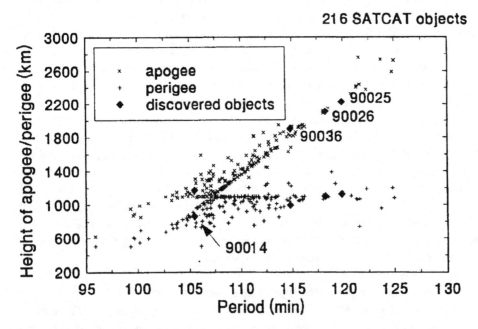

Figure 6: Gabbard diagram of Nimbus 6.

characterization, then additional steps, such as using the signal return time from radar measurements, are required to approximate altitude. Optical stare data, however, does not provide a direct measure of altitude distribution. Optical stare data only provides a rough estimate of the orbit plane. Thus in order to characterize debris vs altitude distribution using optical systems, tracking techniques must be employed. As orbital element sets are developed, then altitude and orbital eccentricity are defined. If desired, a correlation of the event associated with the debris can also be made.

Air Force debris measurement program 1992 status and plans

Observations are continuing in 1992 around new moon periods at both MIT/LL ETS and PL/AMOS. Observations using the GEODSS main at Maui in the stare mode are under considera-

Table I. Debris Observation Results

Day 1991*	Instrument	Hourly Detection Rate	
		Satellites	Meteors
134.0	Maui Aux	11	12
136.0	"	27	7
137.0	"	16	6
138.0	"	22	4
167.0	DG Main	2	8
167.5	"	1	3
168.0	"	3	14
169.0	"	2	6
169.5	"	2	5
170.0	"	3	5
170.5	"	0	1
171.0	"	4	0
171.5	"	0	12
173.0	"	5	14
174.0	"	8	7
184.5	"	4	2
185.5	"	5	3
188.0	"	7	22
196.0	"	1	4
197.0	"	1	13
201.0	"	0	10
202.0	"	3	7
272.0	Maui Aux	5	7
273.0	"	15	3
276.0	"	2	5
"	"	8	9
304.0	"	12	6
306.0	"	5	3
"	"	7	2
307.0	"	5	2
312.0	"	10	6
"	"	6	7
313.0	"	5	7
"	"	5	7
326.0	"	9	9
"	"	8	10
327.0	"	9	7
"	"	5	5
337.0	"	7	8
"	"	10	7
338.0	"	6	10
"	"	2	10
345.0	"	7	2
"	"	5	5

*GMT. NNN.0 = evening twilight; NNN.5 = morning twilight.

PL/AMOS Results

Figure 7

tion by AFSPACECOM for late summer 1992. Two CCD designs have been proposed and are also being considered by AFSPACECOM for installation on the GEODSS Main at Maui. An estimated 18 months from approval is required to build, install and demonstrate either design. If this proposal is approved, then the technique would be to collect data from the Main in the stare mode for characterization purposes and in the track mode for orbital element set deter-mination. An interface between the GEODSS mounts and the PL/AMOS telescopes is under development and planned for demonstration in 1993. This interface would allow the GEODSS to hand-off a detection to the AMOS mount which would then obtain the observations necessary to generate the orbital element set.

Meanwhile, the GEODSS Main would continue searching in a detection mode to continue characterization of the debris environment. MIT/LL ETS will also continue measurements in 1993 and is working to implement an automatic debris acquisition and tracking system for their telescopes to reduce operator errors. Studies are also underway to investigate the range of albedo values of debris objects.

SPACE DEBRIS MODELING PROGRAMS

In order to meet the goals of characterizing the environment to 0.1 cm, models are required. Both NASA and the DoD have several models under development which are attempting to

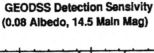

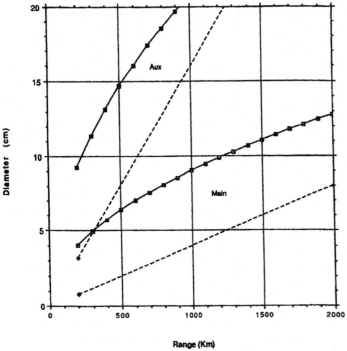

GEODSS Detection Capability at 14.5 Mag

(0.08 Albedo, 11.5 Auxiliary Mag)

⌐·⌐·⌐·⌐ = Stare Mode

······· = Pseudo-Track Mode

Figure 8

simulate various aspects of the debris problem. Such models can be broadly classified as empirical, semi-analytic, or complex codes. The former are typically fast running PC based codes while the latter can be first-principle, physics based codes requiring Cray computers.

At a meeting in January 1992, researchers from NASA and the DoD agreed that the greatest uncertainty in modeling the near-term space debris environment concerned breakup models due to the lack of data for small particles. Establishing the initial conditions of the small particles is the technical challenge. Specifically, accurate measures of mass and velocity distributions are required for both collision and explosive events over the range of events anticipated on-orbit. It will also be necessary to analyze previous events to determine initial conditions that prevailed at the time of the event.

As shown in Figure 11, the smaller particles on orbit still have sufficient kinetic energy to cause serious or catastrophic damage to an orbiting system.

DNA breakup modeling program

DNA recognized the lack of data for debris from breakup events and the need for a modeling capability and established

a test program. Though the military for decades has tested various weapons for lethality effects, a review of the literature found that no one had collected or analyzed data of the smaller particles resulting from an impact or explosive event. It simply was not part of the weapon designer's mission. For the four tests conducted to date, DNA was very careful in providing capture cell material within the test chambers to allow for a statistical estimate of the mass and velocity distributions of the smaller particles. In November 1991, three satellite mockups were impacted with 10 gm aluminum chunks at 10 km/sec. In January 1992 an old OSCAR satellite was impacted with a 150 gm aluminum sphere at 6 km/s. All four tests were complete fragmentation events. Data analysis is ongoing and expected to be complete this fall for these impact tests.

Characterizing the environment

There are many methods to model the space debris environment. A simple approach developed by NASA is the NASA Space Debris Engineering Model which is intended to provide a back-of-the-envelope calculation approach for spacecraft design engineers responsible for mitigating against debris

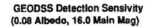

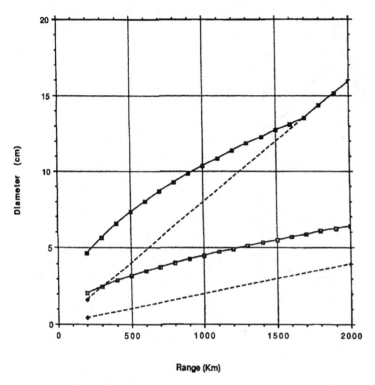

GEODSS Detection Capability at 16.0 Main Mag

(0.08 Albedo, 13.0 Auxiliary Mag)

¬·¬·¬·¬ = Stare Mode

------- = Pseudo-Track Mode

Figure 9

effects in LEO. It provides a quick and conservative estimate of the debris population so that various options concerning spacecraft design can be readily evaluated [10].

NASA EVOLVE model

Another approach is to build a more complex model which consists of various functional subprograms to model launched orbital traffic, debris generating events, and long-term orbital propagation of debris. In this approach, a historical data file is used in conjunction with an empirical breakup model to model the environment. The large number of debris fragments modeled (down to a limiting 0.1 cm diameter) is empirically derived yet is based on the information available on the 132 breakup events which have occurred since 1961. Propagation models then are used to propagate the fragments forward in time to the present. If the breakup models allow it, various debris population sizes can be described and then defined in terms of altitude distribution.

Flux levels are then readily calculated for an altitude of interest. To project the debris flux level forward in time, traffic models and projections of debris growth are used. Until the last few years, a projection worldwide of 120 launches per year was standard. That level has decreased to 80 launches per year with the enormous changes in the Commonwealth of Independent States. Thus care must be taken when projecting the growth of the potential debris population forward in time. In addition, as noted in the January DoD/NASA meeting, more research is required to develop more accurate models of decay effects for long-term predictions.

This complex yet rigorous approach to developing a long-term model of the debris environment is exactly the approach used by NASA in their EVOLVE model [11]. In 1991, the Air Force Phillips Laboratory performed an assessment of the NASA model [12]. We found the code to be extremely well defined and thorough. Most of the recommended changes from our review were already under consideration by NASA.

Air Force Phillips Laboratory debris analysis workstation

Based on our review of the NASA models, the need for a DoD long term model became apparent. The strategy employed for the DoD long term model is similar to NASA's approach with EVOLVE. However, there are two main differences: (1) the Debris Analysis Workstation (DAW) will be hosted within an expert shell system so that non-debris

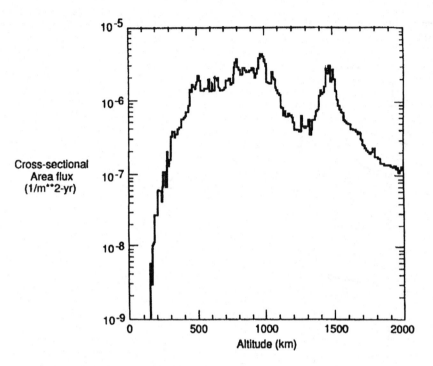

Figure 10: 1987 catalog vs altitude.

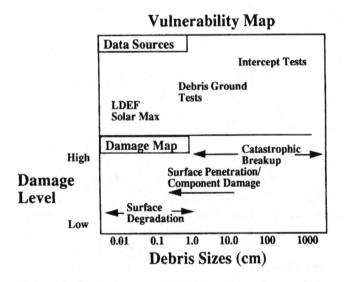

Figure 11: [Source: Tedeschi, DNA]

experts can perform various analyses; and (2) the subprogram models, such as the breakup and propagation models, the historical data file, and traffic models will be different. Of these, perhaps the most critical difference will be in the breakup models. We are currently comparing three empirical breakup models: the Aerospace IMPACT Version 2.0; the fragment model contained within EVOLVE; and Kaman Sciences' FAST model. Results of this study are planned for publication this fall at the 1992 World Space Congress.

DAW status

Figure 12 provides an overview of the subprograms planned

for DAW, and the three broad categories of analyses which will be performed. Currently the breakup and short-term propagation models have been hosted. The first prototype was delivered in January 1992 to the Air Force Phillips Laboratory. Enhanced prototypes are scheduled for delivery to the Air Force Consolidated Space Test Center (AFCSTC) and the Air Force Operational Test and Evaluation Center (AFOTEC) this fall.

The most challenging problem for DAW is that of on-orbit event analysis. In this case, initial conditions of a breakup event will have to be assumed as data is collected from the Space Surveillance Network. In addition, there is a severe time constraint to quickly assess the event and ascertain its threat

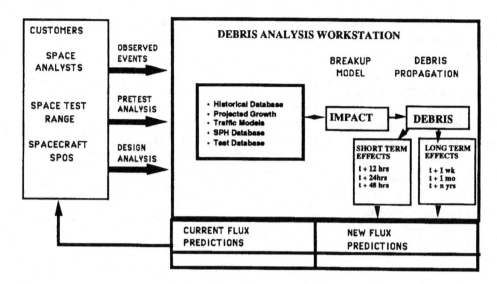

Figure 12

potential to operational systems. Finally, the use of probabilistic models requires a precise and clearly stated assessment of the uncertainties associated with the models in order to bound the real-time threat for operational systems. Thus the on-orbit event analysis drives several requirements for DAW.

Also shown is the planned use of DAW by space test ranges to approve proposed sub-orbital intercept tests. In this way, the potential threat of the few hundred fragments which might achieve orbital velocities will be carefully defined and constrained so as not to impact an operational system. In addition, the predicted ground footprint of fragments will also be studied to determine hazard levels.

Air Force Phillips Laboratory complex modeling efforts
The Phillips Laboratory Weapon Systems Survivability Space Kinetic Impact and Debris Branch is developing a complex computer program known as the Smoothed Particle Hydrocode (SPH) to investigate the physics of breakup events from first principles. An exciting development is that SPH can be used to help the empirical breakup models assess critical issues such as energy partitioning, especially for explosive events which are not yet part of the ground-based test program. Currently, efforts are focused on increasing the efficiency of SPH and on developing algorithms to provide empirical mass and velocity distributions of breakup events. In this way, a database can be established and ported to DAW which will provide a comparison tool against the empirical breakup models for various scenarios.

Air Force Phillips Laboratory debris characterization strategy
Our 1991 and 1992 efforts have focused so far on analysis and development of modeling tools. Current work at the Phillips Laboratory Geophysics Directorate is to develop a model of previous events using the IMPACT Version 2.0 breakup model for debris sizes down to 0.1 cm, then propagate the

debris forward using one of three identified propagation models. A classified traffic model will be developed which will project current and planned DoD space systems launches. Then a constellation-by-constellation flux analysis will be performed, with various debris growth rates overlaid. An assessment of potential losses due to debris will be developed and compared to planned system losses due to reliability failures. In this way, a debris growth rate can be identified to Air Force management at which debris losses exceed planned system failures. Depending upon the level of growth identified, and the probability of that occurring, then the Air Force and DoD management will be in a position to recommend the level of effort to expend via the minimization and mitigation handbook for future systems.

SPACE DEBRIS MINIMIZATION AND MITIGATION HANDBOOK

NASA and the DoD have initiated an effort to provide a handbook to spacecraft designers and launch operators ways to minimize on-orbit debris. Both NASA and the DoD have already implemented cost effective measures. Examples include the venting of all pressurized containers of upper stages, as in the case of the Delta II rocket-body, and moving away from the use of explosive bolts to spring release mechanisms. The handbook is also envisioned to provide shielding data and projected growth of debris. This handbook was established as a goal under the joint DoD / NASA Space Debris Program Plan of July 1990.

The most obviously effective yet most costly measure to implement from a debris viewpoint would be to require end-of-life disposal of all upper-stage rocket-bodies and payloads. The cost in terms of providing residual propellant for such a requirement could be enormous to an individual system. Currently, most DoD space systems are functioning well beyond their anticipated lifetime. This has been achieved through the careful conservation of on-board propellant for station keeping. If this same propellant were required to deboost the payload, then mission lifetimes would suddenly be cut by several years, requiring replacement launches and the increased probability of added debris from launch events.

The DoD will continue to work with NASA to develop "debris clean" space systems and promote cost effective operational debris mitigation measures. An assessment of the debris growth required to generate losses higher than planned from system failures, and the probability of that debris growth occurring, will be required before more costly mitigation measures are implemented. This Phase 1 assessment effort is planned to complete in September 1993.

SUMMARY

The Air Force Phillips Laboratory is acting as the technical lead for the DoD's space debris research program. The Phase 1 effort, characterization of the debris environment, is scheduled to complete at the end of Fiscal Year 1993. Phase 1 emphasizes both the measurement and modeling efforts needed to characterize the current debris environment down to 0.1 cm and to project the future environment and its impact on the survivability of current and planned systems. In parallel, policy efforts have led DoD space designers to implement cost effective debris minimization measures on future systems. A joint effort with NASA to develop a Space Debris Minimization and Mitigation Handbook is also geared to fielding "debris clean" space systems. The threat, however, must be well defined and significant before more costly measures are implemented.

REFERENCES

1. DoD Space Policy, 4 Feb 1987
2. US National Space Policy, 5 January 1988
3. "Report on Orbital Debris," Interagency Group (Space), Feb 1989.
4. DoD / NASA Program Plan for Orbital Debris, July 1990.
5. "Sizing/Projecting the Small Debris Population in LEO," NASA/DoD Technical Coordination Workshop, NASA Johnson Space Center, 22–24 Jan 1992.
6. Orbital Debris Radar Measurements, Peer Review Committee Notes, NASA Johnson Space Center, 22–23 Nov 1991.
7. Stansberry, E., "Characterization of the Debris Environment Using the Haystack Radar," Space Surveillance Workshop Paper, MIT/LL, Apr 8, 1992.
8. Sinew, W.P., ed., "Space Population Study – Final Report, MIT/LL, 11 Mar 92.
9. Kessler, D.J., "Orbital Debris Environment for Spacecraft in Low Earth Orbit," Journal of Spacecraft and Rockets, Vol 28, Number 3, May-June 1991, pgs 347–351.
10. Kessler, D., and Reynolds, R., "Orbital Debris Environment for Spacecraft Designed to Operate in Low Earth Orbit," NASA Technical Memorandum 100741, April 1989.
11. Reynolds, R.C., "Documentation of Program EVOLVE: A Numerical Model to Compute Projections of the Man-Made Orbital Debris Environment," Systems Planning Corp Report, OD91-002-U-CSP, 15 Feb 92.
12. Yates, K.W., "Assessment of the NASA EVOLVE Long-Term Orbital Debris Evolution Model," Orion International Technologies, PL-TR-92-1030, 30 Apr 92.

III. Mitigation of and Adaptation to the Space Environment: Techniques and Practices

9: Precluding Post-Launch Fragmentation of Delta Stages

I. J. Webster and T. Y. Kawamura

McDonnell Douglas Space Systems Company, 5301 Bolsa Avenue, Huntington Beach, CA 92647

I. INTRODUCTION

In early 1981, on-orbit occurrences of satellite fragmentation documented by John Gabbard at NORAD/ADCOM were correlated to breakups of Delta second stages remaining on-orbit after their mission was completed. This correlation was made by Don Kessler of NASA-JSC, and is documented in Reference 1. On 29 May 1981, McDonnell Douglas Space Systems Company (MDSSC), who design, build and launch the Delta launch vehicle, was notified by its customer, NASA-GSFC, of these breakups and was tasked to determine the cause. We immediately mounted an investigation into potential causes of the breakups and corrective action to preclude their occurring on future mission hardware. Early analysis indicated a need to deplete the propellants after completing the primary mission. This was done on Delta 156, just four months later, and has been done on all subsequent Delta missions. There have been no recorded breakups of Delta second stages launched since implementing this corrective action.

II. VEHICLE DESCRIPTION

The Delta vehicle is flown as a two-stage vehicle for low-earth orbit missions and with a third stage for higher energy missions (i.e., GPS, GTO and escape missions). These vehicle configurations are shown in Figures 1 and 2.

The Delta vehicle stands approximately 130 feet high and weighs just over 511,600 pounds (including spacecraft) when ready for launch. The vehicle is made up of four major assemblies: (1) first stage, including the booster, solid rocket motors and interstage; (2) second stage; (3) payload fairing; and as needed, (4) third stage.

The Delta first stage uses a LOX/RP-1 fueled booster with 211,600 pounds of propellant and nine strap-on solid rocket motors (SRMs) with approximately 26,000 pounds of propellant each. The booster main engine and six SRMs are ignited at lift off producing approximately 680,400 pounds of thrust.

The second stage uses a pressure-fed engine that burns nitrogen tetroxide and Aerozine 50. It carries approximately 13,000 pounds of propellants, and the engine peak thrust is approximately 10,000 pounds. The propellant tankage is 410 stainless steel and uses a common bulkhead (Figure 3) between the two propellants. Propellant feed pressure is provided by preflight pressurization of the tank ullages supplemented, in flight, by additional helium from the high pressure spheres carried on the stage. Pitch and yaw attitude control are provided during engine burn by gimballing the engine using hydraulic actuators. Roll control, during all of second stage flight, and pitch and yaw attitude control during coast flight are provided by our redundant attitude control system (RACS) using high pressure nitrogen from the high pressure nitrogen sphere on the stage. Nitrogen is also used by the propellant settling jets to support engine restart. Guidance for both first and second stage is provided by our strap-down inertial guidance system mounted in the second stage guidance section.

The third stage is a spin-stabilized, unguided, solid propellant stage with 4,430 pounds of propellant that burns in 85 seconds. It produces a peak thrust of approximately 17,500 pounds. The third stage carries six pounds of hydrazine for control of coning during motor burn and immediately after burnout. After removal of any post-burn coning, all remaining hydrazine is burned to provide repetitive capture of all available velocity and to depressurize the system totally. All three stages contain ordnance type destruct hardware required by Range Safety.

Delta mission sequence

Typical Delta mission sequences are shown pictorially in Figures 4 and 5. The booster main engine and six SRMs are ignited at liftoff. Ignition of the remaining three SRMs and SRM drop (jettison) are sequenced as shown. The booster burns until depletion of one of its propellants is sensed at which time it is shut down and first-to-second stage separation is accomplished. The spent booster, the interstage, and the burned out SRMs return to earth in the preplanned impact area down range of the launch site.

The second stage engine is ignited shortly after the second stage is separated from the first stage. During the early part of second stage burn the payload fairing is jettisoned and also returns to earth in the predefined impact area. For two-stage missions, the second stage delivers the spacecraft to the

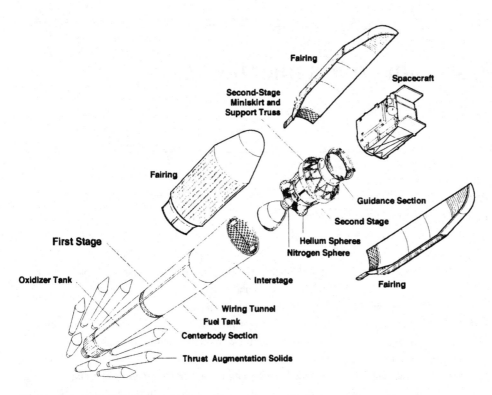

Figure 1: Delta II 7920 Launch Vehicle, 10-foot diameter fairing.

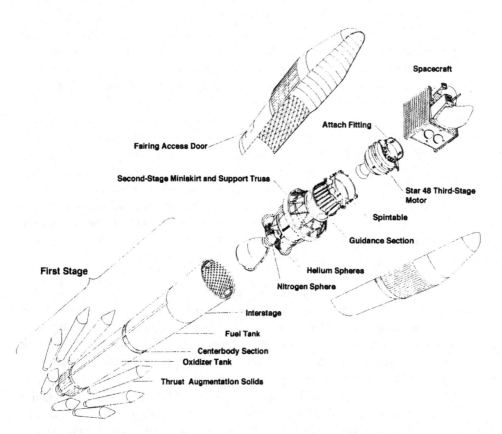

Figure 2: Delta 7925 Launch Vehicle, 9.5-foot diameter fairing.

planned final low-earth orbit (LEO) or medium-earth orbit (MEO). The spacecraft is then separated from the second stage completing the two-stage Delta mission. This leaves the second stage in a long-life LEO or MEO orbit. For most three stage missions, the second stage delivers the spacecraft and third stage to a low-energy elliptical parking orbit with a perigee nominally between 80 and 100 nmi. The on-orbit life of the second stage in this orbit is about one month.

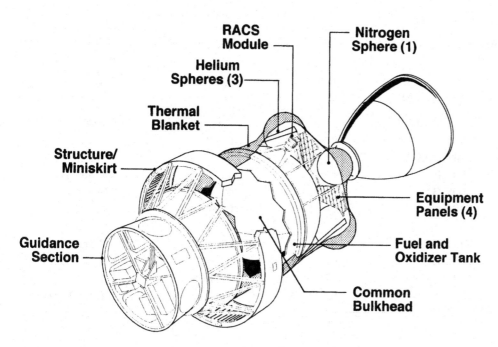

Figure 3: Delta second stage assembly.

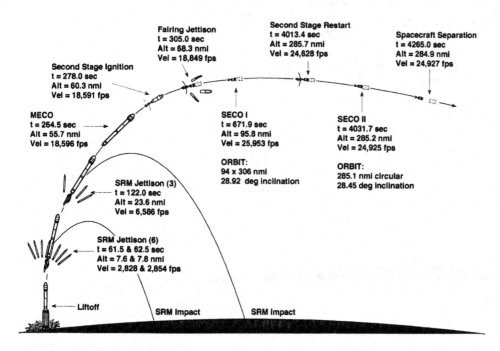

Figure 4: Two-stage mission profile.

The third stage is ignited at or near the parking orbit apogee, imparts approximately 8,000 ft/second velocity to the payload, and establishes the desired transfer orbit (e.g., GPS or GTO). During the post-third-stage burn coast period the NCS propellant and pressurant are completely expended. The spacecraft is then separated from the third stage which completes the three-stage Delta mission. The on-orbit life of the third stage in this orbit is about five years. The depleted third stage is a passive stage and will remain intact while on-orbit. Seventy eight of the last 100 Delta missions have been three-stage missions.

Causes for stage breakup

In 1981, Don Kessler, NASA-JSC Space Environment Office, was able to correlate space debris from satellite breakups recorded by NORAD/ADCOM to Delta second stages left on orbit after completion of their mission. At that time ten Delta second stages were known to have fragmented unexpectedly after completing their missions. Table I lists the missions included in the McDonnell Douglas study (Reference 2).

Five potential causes of orbital breakup were analyzed. They were:

1)   Propellant tank overpressurization

Table I

Delta Sequence Number	Launch Site	Payload Name	Date of Launch	Date of Explosion	Time in Orbit Prior to Explosion
44	ESMC	INTELSAT II F2	01–11–67	Late 1971	5 Years
71	ESMC	INTELSAT II E	07–26–69	07–26–69	1 Day
89	WSMC	LANDSAT A	07–23–72	05–22–75	2 Years, 10 Days
98	WSMC	ITOS-F	11–06–73	12–29–73	1.5 Months
104	WSMC	ITOS-G	11–15–74	08–20–75	9 Months
107	WSMC	LANDSAT	01–22–75	02–09–76	1 Year
				06–19–76	4 Months
109	WSMC	GEOS-C	04–09–75	03–78	3 Years
126	WSMC	ITOS-H	07–29–76	12–24–77	1 Year, 5 Days
132	ESMC	GMS	07–14–77	07–15–77	1 Day
139	WSMC	LANDSAT-C	03–05–78	01–27–81	2 Years, 11 Days

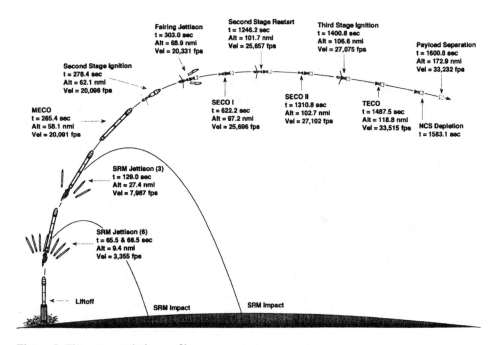

Figure 5: Three-stage mission profile.

2) Propellant decomposition
3) Corrosion of propellant tank common bulkhead
4) Accidental initiation of Range Safety destruct system
5) Overpressurization of nitrogen sphere

Initially, overpressurization of the fuel tank was considered a prime failure mechanism candidate. This was eventually ruled out because, for a given temperature, the vapor pressure of fuel will be lower than for the oxidizer, and because in all instances of exploding stages, there was substantially more residual oxidizer than fuel. Accordingly, oxidizer tank pressures would be higher than fuel tank pressures and would, therefore, be more critical.

Oxidizer tank overpressurization was shown to be the most probable cause of the failures. Thermal analysis of the second stage exposed to orbital environments predicted oxidizer tank temperatures as high as 269°F. At 177°F, the oxidizer tank pressure produced by the heated nitrogen tetroxide and residual helium pressurant reached the predicted burst pressure for the common bulkhead. Fracture of the common bulkhead would allow mixing of the residual propellants that would most probably result in an explosion capable of producing the observed fragmentation. Depleting the residual second stage propellants, before leaving the second stage on orbit would eliminate any possibility of overpressurization of either propellant tank and eliminate the most probable cause of these explosions.

Since the Delta second stage was designed as a restartable stage and since a large percent of our missions had been flown with a second stage restart, accomplishing the depletion burn, from a second stage hardware standpoint, was only a question of ensuring that the dynamic environment caused by a propellant depletion shut down would not destroy the stage. Fortunately, this had already been investigated and successfully tested in several demonstration depletion burns.

The second stage depletion burn was implemented on Delta 156, just four months after the investigation was started. No second stage flown since that time has fragmented on orbit. The fragmentation of Delta 111's second stage on 1 May 1991, after 16 years on orbit, continues to support the study conclusions since it was not flown with a depletion burn.

Post-mission conditioning of upper stages

Today, MDSSC intentionally conditions the second and third stage hardware to preclude on-orbit fragmentation after their mission is complete. In addition to accomplishing a second stage depletion burn to deplete both propellants and the helium pressurant from the tanks, we also actuate the second stage propellant settling jets to deplete the residual nitrogen in the nitrogen sphere, and accomplish a third stage NCS depletion burn to deplete the residual hydrazine and its pressurant. The second stage depletion burn requires careful mission design to preclude contamination of the spacecraft.

For a three-stage mission the second stage primary mission objectives are complete after the spin stabilized third stage is maneuvered to the required burn attitude, is spun up and is separated from the second stage. Second stage depletion burn is delayed until after the third stage burn has been completed and significant separation distance has occurred between the spacecraft and the second stage (Figure 6). For the depletion burn, the second stage is pointed normal to the velocity vector in the local horizontal plane. At this burn attitude most of the energy is used to change the inclination of the orbit, while increases in orbit altitude are limited, thereby preserving the short second stage orbit lifetime (approximately one month). Also at this attitude, the effect of uncertainty in the amount of remaining propellant is small. The depletion burn is normally designed to reduce the inclination, thereby placing the orbit track over less densely populated latitudes. Deorbiting the stage is avoided due to the uncertainty of available propellants.

The Delta two-stage depletion burn design is based on a very detailed relative motion/contamination study completed for one of the early Delta two-stage missions. The contamination limit for that spacecraft was deposition of less than 10 angstroms on the spacecraft (worst case) during the depletion burn. The analysis established the need for a very short evasive burn, prior to the actual depletion burn, for two stage missions and set forth the criteria shown in Table II.

Both the plume angle (the angle between the line from the second stage to the spacecraft and the plume centerline) and the burn time are specified for the evasive burn. By orienting the

Table II *Two-stage depletion burn criteria.*

Evasive burn period	< 6 seconds
Evasive burn minimum plume angle	> 60 degrees
Evasive burn separation distance	> 0.6 nmi
Depletion burn plume angle	> 67 degrees
Depletion burn separation distance	> 19.5 nmi

second stage so that its velocity is reduced by the evasive burn it is placed in a lower orbit than the spacecraft. This positions the second stage below and behind the spacecraft which allows the needed separation distance for the depletion burn to be developed. The depletion burn plume angle is also controlled.

The following second stage sequencing (Figure 7) has been used successfully in completing two-stage depletion burns without contaminating the spacecraft.

- Immediately following spacecraft separation (separation +0.5 seconds) the cold gas retro system is initiated to back the stage away from the spacecraft.

- Following second stage retro, a minimum of fifty seconds of settling jet action are used to provide the needed propellant settling for second stage restart.

- A short evasive burn is performed to the criteria in Table II when the spacecraft-to-second-stage separation distance exceeds 0.6 nmi.

- The depletion burn is performed with the second stage oriented to keep the resulting post-burn orbit lower than the spacecraft orbit.

The actual depletion burn design is unique for each mission. A factor in tailoring the design of each depletion burn for a two-stage mission is the vehicle orientation at spacecraft separation (normally a spacecraft requirement). Since the second stage maneuvers away from the spacecraft, forward facing spacecraft result in retrograde maneuvers by the second stage. In this case, the relative motion of the two bodies places the second stage initially below and behind the spacecraft and the evasive burn assures that the second stage remains below the spacecraft. In the case of an aft facing spacecraft at separation, the second stage maneuvers away in a posigrade direction. The relative motion of the two bodies will place the second stage above the spacecraft. The evasive burn is used to drive the second stage below the spacecraft before the second stage is depleted. Time histories of separation distances and plume angles are established as part of the mission design and any additional contamination analysis needed for that mission is completed. All second stage depletion burns are accomplished in view of one of the TM ground stations.

III. CONCLUSION

Although less than 25% of Delta second stages have long-life orbits, all Delta second stages and third stages are conditioned

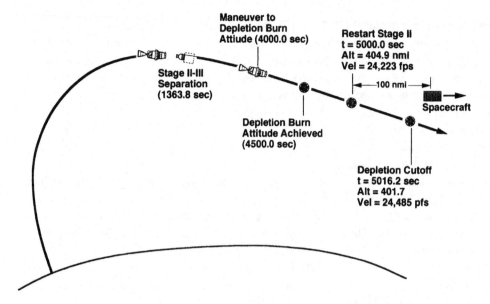

Figure 6: Three-stage post-mission separation profile.

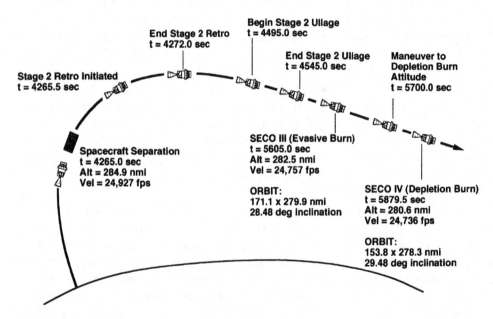

Figure 7: Two-stage post mission profile.

to preclude on-orbit fragmentation prior to being left on orbit. The depletion burn and nitrogen system blow-down implemented on Delta second stages in 1981 have been 100% successful in eliminating fragmentation and have been successfully accomplished without contaminating the spacecraft. The practice of conducting a third stage NCS blown-down as part of our standard three stage mission design was implemented at the time the NCS was introduced.

REFERENCES

1. "Inventory of Orbiting Hypergolic Rocket Stages," technical memorandum SN-3-81-55, Technical Planning Office, NASA-JSC, March 1981.
2. Investigation of Delta Second Stage On-Orbit Explosions," Report MDC-H0047, McDonnell Douglas Astronautics Company, Huntington Beach, CA, April 1982.

10: U.S. International and Interagency Cooperation in Orbital Debris

Daniel V. Jacobs

International Relations Division, NASA Headquarters, Washington, D.C.

EARLY ACTIVITIES

Cooperation between U.S. Government agencies on matters involving orbital debris began with the inception of manned space flight. Since the Mercury program, the North American Defense Command (NORAD) has conducted collision warning analyses for NASA for all manned space missions, using the catalog of objects in orbit, many of which were debris objects. In the late 1970's, at NASA's request, NORAD conducted tests to detect the presence of uncatalogued objects in orbit, and the agencies worked together to analyze the data. In 1976, Don Kessler of Johnson Space Center and John Gavert of NORAD worked both independently and together on understanding the behavior of upper stage explosions in space. Informal exchanges between NASA and DOD scientists on orbital debris have continued since that time.

In 1981, the American Institute of Aeronautics and Astronautics (AIAA) pulled together the first truly interagency group on orbital debris to produce the landmark position paper, Space Debris. This activity involved individuals from NASA, DOD, and NOAA, with private industry representation, and resulted in the first broad consensus that the issue of orbital debris was real and deserved serious attention.[1] The AIAA workshop led to a second workshop in 1982 at Johnson Space Center, involving sixty participants from across the U.S. Government and industry. The workshop ended with the recognition that formal U.S. policies and international agreements would eventually be needed.

Although no formal interagency or international relationships were subsequently formed, several events occurred to keep the issue in the public eye. The first international meeting on orbital debris was held by COSPAR in Graz, Austria, in 1984. Due to a British proposal, the International Consultative Committee for Radio (CCIR) began to discuss debris in 1984. In addition, there was considerable publicity about a large number of satellite break-ups during this time.

U.S. experiences with explosions of the upper stages of the Delta launch vehicle, and the subsequent modification of the upper stage to prevent future explosions, led to a series of informal discussions with other launching nations. In 1985, NASA and Japanese officials informally met to discuss their Delta-derived vehicles, which were modified in 1986. After the explosion of the Ariane 16 upper stage, NASA, the European Space Agency and CNES, the French space agency, formed a study group to examine the event. This resulted in an international conference on upper stage break-ups was held at Johnson Space Center in Houston in 1987.

The French later modified their Ariane 4 vehicle to prevent future explosions. (The Ariane 4 modifications have only been effected for sun-synchronous and LEO launches.)

In 1986, DOD conducted the P-78 anti-satellite test and produced a large number of debris fragments, many of which stayed in orbit for a long time. Upon becoming the head of the Strategic Defense Initiative Organization (SDIO), Lt. General James Abrahamson ordered that all future in-space tests be conducted so as to minimize debris by using low-perigee orbits. The subsequent 1987 Air Force policy requiring all tests to be conducted in a manner that minimized the production of debris made this policy formal. The Delta-180 test followed these guidelines and the resultant debris reentered quickly, with little or no objects remaining in orbit for a significant period of time.

Between 1984 and 1987, additional observations of the debris population were obtained. Observations using ground telescopes detected a population of 1–10 cm objects, which was several times larger than that contained in the NORAD catalog in that size range. In addition, samples of materials exposed to debris impacts were obtained and studied. The Solar Maximum Satellite was captured and refurbished from the Shuttle, and the replaced surfaces were returned for study. A Shuttle window was also struck by a piece of debris, causing its replacement. The results of the analysis of these objects was consistent with the optical observations. All of this new data confirmed earlier models of the debris environment developed at Johnson Space Center, which had predicted that the actual population of debris objects was much larger than indicated by the NORAD catalog.

The debris models, and therefore the predictions of the models, gained considerable credibility from this exercise. New modeling activities by Kessler also began to show that it was only a matter of time before collisions between orbiting objects would begin to occur. It became more difficult to ignore the possibility that the debris population was growing at a rate that might cause operational implications for future spacecraft and missions. This development, together with the Air Force policy and continued international attention, set the stage for the inclusion of a statement on orbital debris in the National Space Policy of 1988.

1988 NATIONAL SPACE POLICY

In January 1988, President Reagan issued a new National Space Policy. The Policy included the following statement:

"All sectors will seek to minimize the creation of space debris. Design and operations of space tests, experiments and systems will strive to minimize or reduce accumulation of space debris, consistent with mission requirements and cost effectiveness."

This was the first formal recognition on the part of the U.S. Government that orbital debris was a real issue deserving the attention of government agencies. The Policy went on to say that "...using NSC staff approved Terms of Reference, an IG (Space) working group will provide recommendations on the implementation of the space debris policy..."

Terms of Reference were developed for the working group which directed it to: 1) present a characterization of the extent of the debris problem now and in the future, 2) review existing agency policies and practices dealing with debris, 3) develop and evaluate technical options for debris management and hazard minimization, 4) describe options for approaching other governments, 5) identify areas for additional research, and 6) recommend language for future policies.[2] The working group was co-chaired by NASA and DOD, and included representation from the Departments of Transportation, Commerce, Treasury, and State, as well as the Office of Management and Budget, the Federal Communications Commission, and the Arms Control and Disarmament Agency. The group was directed to complete a report within 180 days.

With the formation of this group, the orbital debris issue in the U.S. changed character and entered a new phase. Prior to this time, orbital debris was considered by many to be a special interest of certain individuals in specific agencies, but not a significant issue potentially affecting a broad constituency. However, as the group members studied the technical information available at that time and presented the interests and perspectives of their respective agencies, that perception changed.

It became clear that: 1) the effects of orbital debris cut across civil, military, and commercial sectors, and therefore across the interests of the agencies and organizations representing those sectors; 2) although the space assets of all three sectors will be affected by debris, the approaches to dealing with the problem are different; 3) because of the different

interests and approaches, continuing formal interagency coordination would be necessary for actions or policies which went beyond scientific research or internal agency policies; and 4) international implications were significant and international interactions which went beyond scientific research should be coordinated within the U.S. Government.

The working group completed its studies and delivered the *Report on Orbital Debris* to the National Space Council (the successor to the Interagency Group [Space]) in March 1989. The overriding factor driving the recommendations of the *Report* was the lack of knowledge about the debris environment. The first finding listed in the *Report* stated that:

Limitations on debris measurements and the consequent limitations in debris environment modeling create uncertainty as to the urgency for action and the effectiveness of any particular mitigation measure. The need for enhanced measurement capability has been universally recognized.[3]

This uncertainty prevented the working group from making recommendations for specific policies, standards or practices. Four of the recommendations focussed on meeting the need for additional research and studies, and asked NASA, DOD, and DOT to work together to develop research plans and to coordinate research activities.

INTERNATIONAL INITIATIVES

The *Report* also recognized that international interactions and considerations are essential for effectively studying and dealing with orbital debris issues.

One of the recommendations of the *Report* called for the findings and recommendations of the *Report* to be presented to other space-faring nations. NASA took the responsibility for this, and in 1989, met with ESA and with space agencies in Germany, France, Canada, Japan, and the Soviet Union. (Presentations were also given in Washington to the Chairman of the Indian Space Research Organization and to Australian Embassy staff.) At these meetings, information was exchanged by both sides on their current understanding of the orbital debris environment, the causes and sources of debris, projections of the future environment, and implications of the projections on future space activities. In addition, ongoing and planned orbital debris research activities were described and opportunities for potential cooperation were explored.

These initial meetings resulted in a heightened awareness and understanding of the debris issue around the world. A number of ongoing relationships and cooperative activities have evolved from those exchanges.

ESA

The meeting with ESA was especially timely. The ESA Space Debris Working Group had written the report *Space Debris* during the same period the U.S. *Report* was being drafted, and came up with similar findings in most areas.[4] (The major difference was that ESA recommended more immediate action than did the U.S. working group.) Both sides were thus able to report the results of their studies.

Building on a meeting in 1987 in which they shared information on debris, ESA and NASA decided to formalize their relationship through the biannual Orbital Debris Coordination Meetings, held since 1989. At these Coordination Meetings, the results of research activities since the last meeting are shared, along with plans for future activities. Issues covered include debris measurement and detection programs, environment modeling, database management, hypervelocity impact testing, shielding concepts, risk analyses, and international activities. At the last meeting, held in February 1992 at the European Space Technology Center (ESTEC) in Noordwijk, the Netherlands, almost 40 separate presentations were made by over 20 individuals.

These meetings have become the primary means for exchanging data between the two agencies. They have been very successful and have grown in scope and number of participants. The two agencies are now exploring means for reorganizing and coordinating the meetings so as to make them more manageable.

Also under consideration is expansion of the meetings to involve space agencies from other countries where orbital debris research is taking place.

Some specific areas of cooperation have resulted from these exchanges. For example, NASA electronically transmits to ESA the two-line-element database, the USSPACECOM database of orbital elements of space objects which has been provided by the U.S. to the world in paper form for over 30 years. ESA uses this information to develop a larger database called DISCOS, which includes information gained from European sources and special analysis software. This database is used by ESA researchers in their modeling activities. In exchange, NASA has been given access to DISCOS.

ESA and NASA have also developed standards for conducting their hypervelocity tests. These include the identification of benchmark tests to calibrate their facilities, the development of standard reporting formats, and the development of a shared database of test results.

ESA is currently analyzing the data from the cooperative NASA/ESA Infrared Astronomical Satellite (IRAS). In 1983, IRAS scanned a large portion of the sky for astronomical purposes, and in the process, detected a large number of debris objects in GEO which were discarded as noise. ESA is reanalyzing that portion of the database to gain a snapshot of the debris environment in GEO at that time.

Preliminary results from the analysis of this IRAS data include the detection of some 300–400 objects greater than one meter in size in geostationary orbit.

Finally, NASA and ESA are working together to analyze materials and structures from the Long Duration Exposure Facility (LDEF), which was recovered by the Space Shuttle in 1990 after six years in orbit. The data gained from this analysis will provide the space community much needed information on the population of certain sizes of debris which cannot be detected from the ground.

Germany

Both NASA and ESA are working with researchers in Germany on orbital debris research. At the Technical University of Braunschweig (TUBS), the Institute for Spaceflight and Reactor Technology, led by Professor Dietrich Rex, is well-known as a leader in debris environment modeling capabilities. TUBS is conducting several modeling activities for ESA. In addition, NASA and TUBS are conducting parallel modeling activities, using agreed-upon input and parameter standards, to test and validate the results of each other's modeling efforts. This is extremely important, because the NASA and TUBS models are very different, using different approaches to the problem. However, this exercise has demonstrated that they produce strikingly similar results. Because the NASA and TUBS models are the two primary debris environment models used in Europe and the U.S., this type of validation exercise is invaluable.

TUBS is also cooperating with NASA to do conceptual studies of other aspects of the debris issue. For example, one researcher at TUBS is working on concepts for removing debris objects from space with active-maneuver satellites using long tethers for momentum scavenging. Although this not a near-term solution, this is one of the few ideas for debris removal that appears to be potentially feasible and affordable. Another individual is working with NASA researchers to examine collision avoidance systems, such as could be placed on the international Space Station Freedom. This system could potentially warn spacecraft of approaching objects sufficiently in advance of a possible impact so that an avoidance maneuver could be performed. Finally, a TUBS researcher is looking at the Space Station Freedom configuration and examining potential shadowing effects caused by the placement of various components, in order to reduce the amount of shielding required.

Another German institution is also studying orbital debris. FGAN, a German radar research institution, is working with NASA to study the characteristics of objects in space. NASA and FGAN have jointly identified a list of ten objects, which they will separately characterize using their own facilities, both radar and optical. Results will be compared to learn more about the size, shape, behavior, radar cross section, and brightness of the objects.

France

CNES, the French space agency, and NASA have held discussions about orbital debris in the past, as well. A cooperative agreement has been signed between the two agencies, in which NASA electronically transmits the two-line-element database to CNES, as it does with ESA. In exchange, CNES analyzes NASA-provided LDEF samples, using an electronic microscope to determine the type of objects which have pitted the surface of LDEF. NASA and CNES will soon be meeting again to explore other, more extensive opportunities for cooperation.

Japan

The Japanese space agencies NASDA and ISAS have met periodically with NASA for some time, and have recently attended the NASA/ESA Coordination Meetings as observers. After the formal meeting with NASA in 1989, Japan formed an interagency Space Debris Study Group under the Japanese Society for Aeronautical and Space Sciences (JSASS) to study orbital debris. An interim report was delivered to the Japanese ministries in May 1992. The final report of the Group is scheduled for release in January 1993. The study identified opportunities for international cooperation with Japanese institutions on orbital debris. NASA and STA are concluding an agreement to meet on a regular basis to discuss debris research programs and to explore specific cooperative activities. These meetings may occur separately or in conjunction with the NASA/ESA meetings.

Russia

There has been a series of meetings between NASA and Soviet (now Russian) organizations about orbital debris. Since the 1989 meeting, a number of specific cooperative activities have been identified. For example, an agreement has been reached for the two sides to exchange Satellite Situation Reports, or their equivalents. This is a lesser set of the two-line-element database, and will aid each side in its study of the debris population. Other proposals, such as the exchange of objects exposed in space, the flight of debris capture panels, and joint studies of methods for detecting debris, are also under discussion. It is expected that several other activities will also be considered.

China

In June 1992, a meeting was held between NASA and Chinese space officials to discuss orbital debris issues in general and the explosion in orbit of a Chinese Long March upper stage in particular. During these discussions, the Chinese identified a method for preventing such explosions in the future, although they have not yet adopted this procedure. Further discussions are planned at the time of the World Space Congress in Washington in August 1992.

International Organizations

Orbital debris has received considerable attention in international organizations. In April 1990, the AIAA, at the request of NASA and the U.S. Department of Defense, sponsored an international Orbital Debris Conference. This highly successful conference was attended by scientists and space officials from around the world and produced three volumes of proceedings and presentations.

Organizations like COSPAR and the IAF have included Orbital Debris Technical Sessions in their conferences for the past several years. Every year, scientific papers by researchers from various space organizations and universities are presented on orbital debris issues.

Several countries have proposed adding debris to the agendas of the subcommittees of the Committee on Peaceful Uses of Outer Space (COPUOS) in the United Nations for a number of years. The U.S. has consistently maintained that it is premature to do so until there is a more detailed understanding of the debris issue. There are still gaps in the knowledge base about the debris environment and about effective and affordable options for dealing with it. Those issues are being dealt with in national research activities and through bilateral cooperation. Until further progress has been made in those arenas, it would be premature to address the orbital debris issue in the formal and legalistic environment of U.N. organizations.

INTERAGENCY WORKING GROUP ON ORBITAL DEBRIS

During 1991, two separate ideas began to emerge. First, additional research at Johnson Space Center led Kessler to the conclusion that the critical densities necessary to support the cascading effect he had described earlier were being reached far sooner than had been predicted. The level of urgency felt by the scientists at Johnson to deal with the debris problem increased proportionately. In addition, new debris measurements were reducing the level of uncertainty about the size of the debris population from approximately 300% to around 70%. This increased the confidence level about Kessler's conclusions significantly.

In addition, it became obvious to many that several issues with debris implications were maturing and would have to be addressed. For example, it had been assumed that because the major launching nations had voluntarily taken actions to prevent exploding upper stages in orbit, that problem was under control. The explosion of the Long March upper stage in October 1990 demonstrated otherwise. With new entrants into the launch market, this issue needs to be addressed more broadly. Other situations, like increasing international pressure to deal with GEO debris through the use of "graveyard" orbits and commercial plans to develop constellations of LEO satellite systems, also need to be examined.

For these reasons, NASA and DOD agreed that it was time to reconvene the Interagency Working Group on Orbital Debris (IWG-OD). The IWG-OD had met only sporadically since the publication of the *Report*. An annex to the original Terms of Reference was developed, and the group began meeting again last Fall. In February, the IWG-OD completed a review of progress made since the publication of the *Report*. It will soon consider candidates of specific issues to be which could studied in greater detail by the IWG-OD.

CONCLUSION

The various elements of the orbital debris issue such as measurement, modeling, protection and mitigation are being actively pursued by researchers in the major space-faring nations. These national, bilateral and multilateral research activities are expanding as the effects associated with orbital debris become more apparent. It is these efforts that will fill the gaps in our current understanding of the debris environment and our options for dealing with it. This knowledge base will form the basis for the IWG-OD and its international counterparts in their consideration of approaches for dealing with orbital debris issues.

REFERENCES

1. AIAA Technical Committee on Space Systems, *Space Debris: An AIAA Position Paper*, May, 1981.
2. National Security Council, "Terms of Reference for IG (Space) Working Group on Space Debris," June, 1988.
3. Interagency Group (Space), *Report on Orbital Debris*, February, 1989.
4. European Space Agency, *Space Debris*, ESA SP-1109, November, 1988.

11: ESA Concepts for Space Debris Mitigation and Risk Reduction

H. Klinkrad

Mission Analysis Section, European Space Operations Centre, ESA/ESOC, Darmstadt, Germany

ABSTRACT

In the present earth particulate environment, for hazardous objects larger than 1mm, man-made space debris are already prevailing over the natural meteoroid and cosmic dust background. Most of these debris were caused by about 120 on-orbit fragmentations of upper stages of rockets, or by spacecraft break-ups. Following these events, some altitude regions have already attained critical concentrations which could ultimately trigger collision chain reactions and make portions of the near-earth space unsave for manned operations. This paper will describe concepts of the European Space Agency for the mitigation of space debris which could assist in the preservation or improvement of the space environment. The outline of ESA debris mitigation measures will cover the formulation of related design and operation policies for space systems under ESA control, and give examples of the implementation of such measures for the reorbiting of GEO spacecraft and for the passivation of Ariane upper stages after their mission completion.

INTRODUCTION

Currently, some 7,000 trackable space objects are orbiting the earth. This population is the remainder of a total of 22,000 objects which have been tracked operationally by NORAD and USSpaceCom (its successor for space surveillance) since Sputnik-1 was launched in 1957. The other 15,000 objects have mostly reentered and burnt up in the earth atmosphere, or have been retrieved by controlled reentry or by the Space Shuttle. Almost 50% of the observable space population can be correlated with more than 120 historic on-orbit fragmentation events, each of which typically generated 120 to 200 trackable debris objects.

Most of these fragmentation events are related to the explosion of upper stages of launch vehicles (mostly detonations of cryogenic fuel remnants), sometimes months or years after the launch. To a lesser extent, spacecraft explosions were contributing to this class of debris (accidental explosions due to propulsion or battery failures, or intentional explosions of military spacecraft). Another 45% of the USSpaceCom Catalogue

consists of defunct spacecraft, upper stages, and mission related objects. Less than 5% of the observable space object population is due to operational spacecraft.

The current operational Space Surveillance Network (SSN) of USSpaceCom has detection thresholds of about 10cm in LEO and 1m in GEO altitudes. The corresponding 7,000 correlated objects are believed to represent only 1/18 to 1/5 of the debris larger than 1cm[1], which is the maximum shieldable object diameter with today's on-orbit shielding technology (e.g. for Space Station Freedom, SSF). Hence, there is a range of particle sizes between 1 cm and 10 cm which can cause catastrophic spacecraft fragmentations, which do, however, pass current ground-based surveillance capabilities undetected.

Apart from technologically and cost demanding improvements in surveillance and/or shielding techniques, the most efficient means to reduce the risk of catastrophic on-orbit collisions is the reduction of the space debris creation per se (because it is improving the debris environment rather than improving mission survivability). The urgency of such measures is emphasised by debris environment predictions which indicate that within the next decade the spatial object densities in some altitude and inclination bands could reach levels which may be sufficient to trigger a self-sustained chain reaction of debris-debris collisions which could render parts of the LEO environment unsave particularly for manned space activities. Space debris prevention is the most efficient means of on-orbit risk reduction, since it is directly affecting the problem source, rather than curing the symptoms. The prevention of space debris, or space debris mitigation in general, can be addressed source-wise (e.g. upper stages or spacecraft), or according to the particular orbit environment (e.g. Space Station, sun-synchronous orbits, or geostationary orbits). In the following, the policy issues, related design criteria, and additional research activities of the European Space Agency in the different areas of space debris mitigation will be summarised.

ESA ACTIVITIES IN AN INTERNATIONAL FRAME

The near-earth space is a common resource for mankind. Its

pollution by man-made particulates is hence a problem which must be addressed at an international level. There are regular discussions on this issue at meetings and symposia organised by the International Astronautical Federation (IAF), by the International Academy of Astronautics (IAA), and by the Committee on Space Research (COSPAR), a sub-organisation of the International Council of Scientific Unions (ICSU). Regulatory and legal aspects are dealt with by the International Institute of Space Law (IISL), and the Unitied Nations Committee on the Peaceful Use of Outer Space (UN COP-UOS) within two sub-committees is investigating scientific, technical, and legal aspects of space debris. Special problem areas related to astronomical observability conditions (light pollution), and related to the reduction of operational risks of communication spacecraft are addressed by the International Astronomical Union (IAU), and by the International Telecommunications Union, respectively (ITU providing inputs to WARC, the World Association for Radio Communication, via the CCIR study group).

The European Space Agency has recognised that space debris constitutes a serious long-term problem. Since a number of years ESA has been actively involved in space debris related discussions on an international level, and has funded and coordinated space debris research activities within its Member States. In 1986, following a request by ESA's Director General, the Agency established a Space Debris Working Group composed of ESA staff and of experts from the Member States. The SDWG, chaired by Prof. D.Rex (TU Braunschweig, FRG), issued a debris status report and a set of recommendations under the title 'Space Debris' in 1988. The findings and a refined set of recommendations of the group were later submitted to the ESA Council and approved in the form of a 'Resolution on Space Debris' in 1989. This Council resolution defines the following ESA objectives:

- minimisation of the creation of space debris
- reduction of debris induced risks for manned and unmanned spacecraft
- improvement of debris environment models with consolidated data
- investigation of legal aspects of space debris

The European Space Operations Centre (ESOC) has been entrusted with the coordination of all ESA activities which are on-going and planned in order to accomplish these goals. An advisory body of leading European experts, the Space Debris Advisory Group (SDAG), has been formed to advise the Agency in all debris related matters. Chairman of SDAG is Prof.Rex (TU Braunschweig, FRG).

ESA is conducting regular meetings for the coordination and harmonisation of debris research activities throughout its establishments (ESA HQ, ESTEC, ESOC, and ESA Toulouse). On an international level regular meetings with NASA have been held since 1987 (under the coordination of NASA JSC). More recently, also formal cooperation agree-

ments with NASA have been established. Since 1991, also Japan (NAL, NASDA, CRL) has joint this regular NASA/ESA forum for information exchange in space debris matters. On a non-regular basis ESA has also held meetings with USSR/CIS scientists. These meetings were mostly concerned with re-entries of risk objects (e.g. Salyut-7).

SPACE DEBRIS RISK ASSESSMENT

A good knowledge of the space debris environment is a prerequisite for an efficient implementation of risk reduction and mitigation measures. It is known from ground-based observations (USSpaceCom, Millstone Hill, Goldstone, Arecibo) and from returned spacecraft surfaces (LDEF, Solar Max, Shuttle windows) that man-made particulates are prevailing over the natural meteoroids and cosmic dust in all size regimes larger than 1mm. The most up-to-date information for this size range can be obtained from the ELSET Catalogue of USSpaceCom. The associated SSN tracking network consists of radars (mainly for LEO observations), and electro-optical telescopes (GEODSS, mainly for HEO and GEO surveillance). The operational SSN detection threshold is on the order of 10cm in LEO and 1m in GEO. Experimental observation campaigns (radar as well as optical, and mainly of statistical nature) allowed an extrapolation of the debris population, and thus a bridging between operational ground-based observations and on-orbit surface impact data, across the cm size range. An analytical formulation of the space debris and meteoroid environment based on these data has been produced by D.Kessler and co-workers of NASA/JSC. These models, which provide particulate flux and velocity distributions versus flight azimuth of a target spacecraft, are very compact and convenient to use in engineering applications (e.g. in conjunction with damage equations for SSF impact and puncture risk assessments), and the modelled flux rates are known to fit quite well with in-flight data. The compact formulation of the Kessler model could, however, only be achieved by introducing a number of simplifications such as circular target and debris orbits, and by low resolution analytical fits of altitude and inclination distributions. ESA works on an alternative debris model with improved spatial resolution within a study contract with Battelle/Europe and TU Braunschweig. This will lead to an ESA Space Debris and Meteoroid Reference Model by 1993. As was the case for the Kessler model, also the ESA model will initially focus on the LEO environment. The modelling of the small size debris population (> 0.1mm) will be supported by on-ground explosion tests on simplified, scaled models of upper stages.

The current LEO debris environment is dominated by particulates of upper stage and spacecraft fragmentations. Most of these events were due to the explosions of residual fuel, and most of them occurred on near-polar, sun-synchronous orbits between 900km and 1,500km altitude. These explosion fragments contribute to about 50% of the trackable LEO population (USSpaceCom Catalogue), and almost exclusively determine the particulate environment below sizes of 10cm. For Space Station Freedom (SSF), assuming a shieldable debris diameter of 1cm, the associated risk of puncture of a manned module for a 30-year mission duration is on the order of 2%, with a most likely collision velocity of 10.5 km/s. For a remote sensing satellite on a near-polar orbit around altitudes of 800km, the most likely collision direction is almost head-on with relative velocities of 15 km/s. Due to this higher collision speed and due to a higher population density at this altitude, the debris flux per unit surface and year is about 6 times larger than for Space Station. Assuming an unshielded spacecraft (e.g. ERS-1 with a 3 year mission duration), there is a 10% risk of a collision with an object larger than 1mm. Such particles may cause major damage on sensitive equipment.[2]

Outside the LEO region, only the GEO ring of near-equatorial orbits around 35,785 km altitudes is exposed to a considerable collision risk. In contrast with the LEO environment with its self cleansing mechanism due to airdrag, the GEO and near-GEO objects stay within a confined altitude and inclination band for an infinite period of time. Assuming a GEO population of 200 operational (controlled) spacecraft, and 1000 debris objects, there is a 1% collision risk over a 10 year time span. Due to the limited number of longitude slots in GEO, some spacecraft operators share a standard size window of $\Delta\lambda = 0.1°$. In case of 4 uncoordinated spacecraft sharing the same slot (e.g. near 19° W), there is a 60% risk that proximities closer than 50m can occur within less than one year. Non-operational spacecraft and mission related objects (e.g. apogee kick-motors, upper stages, lens covers) tend to build up orbital inclination up to 15° within a 53 year cycle. At the same time, solar radiation pressure is distorting the orbit shape (increase in eccentricity). Both effects lead to a build up of the transient velocities during GEO transitions up to 800 m/s, and hence lead to a larger debris flux on the operational spacecraft.

Single, intact spacecraft or upper stages on highly eccentric orbits, including the GTO transfer orbits to GEO, are of minor importance for statistical risk analysis. This is due to their low resident probabilities during LEO passage, and due to their low spatial densities in the GEO region. It is, however, likely that explosions of upper stages on GTOs have occurred unnoticed (due to poor SSN observation geometry), and that the GTO population is considerably larger than currently assumed. In this case, particularly after a break-up event on GTO, the effect of a related debris cloud deserves special attention in risk assessment studies. In LEO, such debris approach Space Station from the anti-flight direction at velocities around 3 to 4 km/s. These impact speeds are known to constitute a worst case scenario for shield penetration, and the expected hits will be from directions which are presently considered of low risk potential. In GEO, a GTO debris cloud causes an object flux from due East with relative speeds of about 1.5 km/s.

There are feasibility studies on-going to equip Space Station with on-orbit collision warning sensors which allow to forecast potential collision events based on previous sightings of a debris object. For GTO debris, due to their peculiar approach geometry and due to no previous sightings by the sensors, evasive manoeuvres or temporary shield deployments will be impossible in view of the short reaction times.

The particulate environment situation for μm-size objects (both man-made debris and meteoroids) which has been described by the models of Kessler for debris, and Gruen, Zook, Giese, and Fechtig for meteoroids, was largely confirmed for LEO altitudes by the post-flight analysis of LDEF surface impact data. Some time-correlated impact data of the first year on orbit also show indications of debris clouds on highly eccentric orbits (of Molniya type). A number of European institutes is actively involved in the analysis of LDEF samples (e.g. University of Kent, University of Munich, EMI Freiburg, MPI Heidelberg, ONERA Toulouse).

In the domain of Catalogue-size objects, the German Research Institute for Applied Science (FGAN, Wachtberg-Werthhoven) is using a high performance L-band/Ku-band radar with a 34m dish antenna for tracking and imaging, respectively. In a joint study with NASA, the tracking data of selected targets have been processed by FGAN to determine the orbits, radar cross-sections, and masses of the objects.

ESA SPACE DEBRIS MITIGATION AND RISK REDUCTION POLICY

In order to reduce the space debris related risk imposed on manned and unmanned space missions, ESA intends to implement a number of related operational and design criteria. The following non-exclusive list summarises the most immanent measures with only moderate impact on cost, manufacturing, and operations:

– reduced release of mission related objects
– safing (venting or depletion burn) of upper stages and satellites at EOL
– selection of GTOs with reduced orbital lifetime
– re-orbiting of GEO spacecraft to safe disposal orbits after EOL

Such procedures and further reaching recommendations will be contained in revised ESA safety regulations and associated design and operation criteria. They will be summarised in ESA-PSS guidelines (procedures, standards, and specifications) which constitute high level applicable documents for the design and operation of ESA space systems and associated equipment. PSS-01-40 (Safety Assurance of ESA Spacecraft and Associated Equipment) is a level II document of the ESA Standardisation Policy and Architecture (PSS-00-0) which also addresses space debris related issues. The existing PSS-01-40 document (clauses II-1.x.x) includes the following requirements:

– "Means shall be provided to prevent the hazardous descent of debris as the result of a launch vehicle mission abort, or of the controlled de-orbiting or orbital decay of spacecraft or space system elements that are likely to survive reentry."

- "The creation of space debris in orbits that are repeatedly intersecting orbital paths used by space systems shall be avoided."

- "Residual propellants contained in spent or aborted stages shall be safely dispersed."

IMPLEMENTATION OF ESA SPACE DEBRIS MITIGATION AND RISK REDUCTION

Some of the foregoing recommendations have already entered into force in a number of spacecraft and launcher design and operation aspects under ESA control. Precursor activities in the field of space debris mitigation and risk reduction were re-orbiting operations of ESA GEO spacecraft after their end-of-life (EOL), and Ariane third stage passivation after delivery of LEO spacecraft into orbit.

ESA recommends to re-orbit GEO spacecraft to altitudes at least 300km to 400km above the geosynchronous ring. The associated amount of propellant is on the order of a few percents of the station keeping fuel budget, and less than 1% of the overall propellant at GTO injection ($\Delta V = 3.65$ m/s for each step of +100km of uniform altitude raising). While the amount of required fuel for re-orbiting at EOL is small, the accuracy with which the remaining fuel supply can be determined is also poor. In case of very accurate book-keeping over all manoeuvres (including thruster calibration), and with additional indirect gauging via temperature and pressure measurements of pressurants (PVT method), uncertainties in the estimate of the remaining propellant may be ± 50% of the median value (as was the case for OTS-2). For Olympus, which underwent complex recovery operations during an accidentally induced drift around the earth, the estimates of the remaining fuel are likely to have even larger error bounds.

Recognising the uncertainty in the available ΔV for the re-orbit manoeuvre, ESA procedures call for a 3-burn orbit raising operation with 12 hour time separations. The ΔV's of burns 1 through 3 are commanded such that 25%, 50%, and again 25% of the estimated propellant[3] is used up to depletion. Application of this concept for GEOS-2 (Jan.1984), OTS-2 (Apr.1991), and Meteosat-F2[4] (Dec.1991) resulted in super-synchronous altitudes at Δh = +270km, +320km, and +542km/+334km. In the latter case, the circularisation burn no.3 was incomplete (the available fuel was at the low end of the 1σ uncertainty), leaving the spacecraft in a slightly eccentric orbit.

Similar operational procedures for removing end-of-life spacecraft from the valuable GEO region have been adopted by the US, by the USSR/CIS, by Japan, and by several international operators (e.g. INTELSAT). While the validity of the re-orbiting concept has been widely accepted, different GEO spacecraft operators aim at different super-synchronous alti-

tudes for their EOL re-orbiting. An internationally accepted code of conduct or agreement in this matter would be highly desireable.

Apart from end-of-life reorbiting, also collision risk minimisation during operational periods can be important for GEO spacecraft. Due to constraints from GEO coverage geometry, some longitude regions are of higher commercial and/or scientific interest than others (e.g. at European or US longitudes). Such regions are consequently more densely populated. They may, however, be efficiently used by a concerted operation of a number of spacecraft within a common longitude slot (typical slot width: 0.1°). Such a concept has been implemented for 4 European satellites near 19°W: TDF-1, TDF-2, Olympus-1, and TVSAT-2, which are in position since Oct.88, Aug.90, Jul.89, and Aug.89, and which are controlled by CNES, ESA/ESOC, and DLR/GSOC. They use assigned longitude positions in between 18.8°W and 19.2°W. During phases of their operation, some of the assigned longitude slots were overlapping, and the spacecraft operators devised a coordinated control scheme with maximum spatial separations by optimal selection and control of the individual eccentricity and inclination vectors. Without coordinated control, the common longitude assignment for TDF-1 and TDF-2 (at 18.8°W) would have led to a statistical estimate of a proximity closer than 50m (100m) within one year (4 months, respectively). The corresponding risk of proximity passes or even collisions is virtually zero in case of coordinated orbit maintenance.

Most of the observable debris and almost all of the small size debris population has its origin in the fragmentation of upper stages of launchers, and in the break-up of satellites. Most of these events were due to explosions of residual propellants, and due to propulsion system failures, respectively. ESA has so far had one entry into the history of fragmentation events due to the explosion of an Ariane-1 third stage 9 months after the delivery of SPOT-1 into a sun-synchrounous orbit of 825km altitude. The fragmentation of the third stage of flight V16 occurred on an ascending orbit pass over Africa on Nov.13, 1986. The maximum count of V16 fragments in the USSpaceCom Catalogue reached 488 within less than 2 years (current count: 40). Jointly with the fragmentation of a Titan 3C-4 on 15.Oct 1965, this was the worst break-up event in space history. In the aftermath of this event, ESA and Arianespace devised strategies for the passivation of Ariane upper stages.

The explosion of the Ariane-1 V16 upper stage was most likely due to the detonation of a near-stoichiometric mixture of the gaseous H_2/O_2 propellants remnants. The mixing of the fuel components may have been the result of a rupture and/or reversal of the common bulkhead which seperates the two tanks, and which is sensitive to overpressure from the oxygen side (could be caused by failure of the pressure relief valve in case of differential heating from the LOX side[5], or by pressure drop in the LH2 tank due to particle impact or tank rupture). Turbulent mixing of the two components, subsequent ignition by O_2 reaction with exposed metallic aluminium, and finally the onset of a deflagration with possible transition to a high energy detonation may describe the chain of events that led to the V16 break-up.

Arianespace has studied possible failure scenarios and concluded that depressurisation of both the LH2 and the LOX tank will be the safest measure to counteract bulkhead reversal or rupture as the most likely break-up cause.

The depressurisation of the tanks necessitated a change in the pyrotechnics command chain. The implementation of Ariane upper stage passivation on LEO missions has been practiced since flight V35 (SPOT-2, 22 Jan.90) and also for ERS-1 (V44, 17 July 91).[6]

For the passivation of GTO upper stages, further design changes were necessary, and the first complete depressurisation of a third stage on GTO will be performed on the Ariane launch V60. Thereafter, all GTO and LEO missions of Ariane will terminate with an inert upper stage. Concerning the non-passivated Ariane stages on GTO orbits through flight V49, 8 have decayed, 41 are still tracked (by USSpaceCom), and 1 stage (L8, launch vehicle of Intelsat 5-F8) has been lost. It cannot be excluded that this object has fragmented.

Upper stage passivation procedures similar to those of Arianespace (depressurisation or depletion burn after end-of-mission) have been implemented or are in preparation for Delta II launch vehicles, and for the Japanese H-1 and H-2 upper stages (which will perform depletion burns). While upper stage passivation provides a short-term improvement of the debris environment, long-term predictions indicate that subsequent on-orbit collisions will nullify such actions. The only ultimate solution would be a de-orbiting of the stages, which entails considerable design and cost implications, and which requires an agreed code of conduct among launch vehicle manufacturers and operators.

As a means of risk reduction to populated areas on earth, ESA has since 1983 offered a reentry prediction service to its Member States in case of uncontrolled orbital decays of risk objects. This service was triggered in view of the reentry of Kosmos-954 over Canada on 24 Jan.1978 (with a nuclear reactor on board), and in view of the reentry of the massive Skylab (75t) over Australia on 11 July 1979. The ESA long-term prediction of decay date, and short-term forecast of geographic reentry area (with associated uncertainy) was first performed for the decay of Kosmos-1402 parts A and C (Jan./Feb. 1983). More refined methods were then applied recently during the reentry prediction campaign for Salyut-7/Kosmos-1686 (a 40t USSR space station) which decayed over South America in Feb.1991. The good prediction accuracy which was achieved during these campaigns was largely due to the support of ESA by NASA/USSpaceCom, and by the German FGAN tracking station. For Salyut-7, ESA was also supplied with pre-processed orbital element sets from the USSR (via IKI/Moscow).

In order to prevent on-orbit collisions of ESA controlled spacecraft with deterministically known debris objects from the USSpaceCom Catalogue, one could devise collision or proximity prediction procedures. In case of passes within a given threshold distance (e.g. 10km), these predictions in conjunction with related orbit prediction uncertainties could be used for the planning of evasive manoeuvres of LEO spacecraft (e.g. ERS-1 or Polar Platform), and for the selection of collision free ascent or descent trajectories of Ariane or Hermes. The operational feasibility of such techniques has already been demonstrated. No decision has, however, been taken yet as to its implementation for on-going or future ESA missions.

SUMMARY AND CONCLUSIONS

Man-made space debris is already prevailing the space particulate environment in the near-earth space. Its current growth rate entails a degradation of the space environment which will make manned and unmanned activities unsave within the next decade, unless preventive measures are taken.

The most effective means of space debris risk minimisation is the avoidance of debris creation, which according to the historic debris sources can be achieved with moderate cost and design impact by the passivation of upper rocket stages and of spacecraft after their useful operational life. ESA in cooperation with Arianespace has initiated design modifications of the upper stages of future launchers of the Agency. These hardware changes are aimed at a depressurisation of the cryogenic fuel tanks (by controlled propellant release), and hence a passivation of the stage after its end-of-mission. This passivation of Ariane upper stages was first performed on launch V35 (Jan.1990, SPOT-2), and will be adopted for all launches (including GTOs) as of V60. Long-term forecasts indicate, however, that only the de-orbiting (or retrieval) of upper stages and spacecraft will guarantee a stable space debris environment.

A further means of debris mitigation is avoidance of on-orbit collisions in densely populated orbit altitude regions, since collisions are known to release large amounts of debris which could on-set collision chain reactions if a critical particle concentration is exceeded. ESA has applied this mitigation philosophy to most of their GEO spacecraft at end-of-life (GEOS-2, OTS-2, Meteosat-F2) by re-orbiting them to super-synchrounous altitudes well above the geostationary ring. ESA is also considering the implementation of collision-free trajectories for Ariane ascent, Hermes ascent/reentry, and LEO satellite missions by proper selection of launch/reentry time windows, and by the planning of evasive manoeuvres, respectively.

All debris related activities of ESA are coordinated by its operations centre ESOC. Backed by resolution C/LXXXVII/ Res.3 of the ESA Council, and following recommendations which were formulated in 'Space Debris' (The Report of the ESA Space Debris Working Group, 1988), ESA has prepared a long-term Space Debris Research Programme (ESA/C(89)24, Rev.1) which was approved by the Council at its 87th meeting.

The main objectives of this Debris Research Programme are:

–   to gain a better understanding of the space debris problem (environment definition, hypervelocity impact studies)
–   to assess the level of debris induced risks for ESA's space programmes
–   to identify methods for active and passive debris risk reduction (debris mitigation, collision avoidance, shielding)
–   to support the establishment of an overall ESA policy on space debris

In order to attain these goals, ESA will establish a set of operational and design guidelines in the form of ESA-PSS documents as part of the ESA Standardisation Policy. These high level documents take direct influence on future system designs, and they are formal, applicable design rules.

ESA's debris mitigation policy, debris protection developments, and debris environment definition studies are established jointly with the Space Debris Advisory Group (SDAG). Regular information exchange with NASA, Japan, and the USSR/CIS in space debris matters also assist in directing the available resources in an efficient manner.

ENDNOTES

1. these numbers account for the Henize correction factor, but do not fully consider recent Haystack measurements
2. note that given collision risk estimates do not fully consider recent Haystack measurements (suggesting an under-representation of the small-size objects of the USSpaceCom Catalogue well beyond the Henize correction factor)
3. normally, the best estimate less a 1σ uncertainty margin is used as baseline for the remaining propellant mass

4. Meteosat-F1 was operated until fuel depletion as a matter of operational expedience; it is now drifting at GEO altitude
5. this was identified as cause of the early Delta break-ups
6. upper stage passivation also performed for TOPEX/Poseidon (launch V52)

REFERENCES

1. ESA Space Debris Working Group, **Space Debris**, ESA SP-1109, Nov.1988
2. US Interagency Group (Space), **Report on Orbital Debris**, For National Security Council, Washington/D.C., Feb.1989
3. D.G. King-Hele, D.M.C. Walker, A.N. Winterbottom, J.A., Pilkington, H. Hiller, G.E. Perry, **The RAE Table of Earth Satellites 1957–1989**, RAE, Farnborough/Hants, England, 1990
4. N.L. Johnson, D.J. Nauer, History of On-Orbit Satellite Fragmentations (4th Edition), TBE Technical Report **CS90-TR-JSC-002**, Jan.1990
5. N.L. Johnson, D.S. McKnight, **Artificial Space Debris**, Orbital Book Comp., Malabar/Fla., USA, 1987
6. D.J. Kessler, Orbital Debris Environment for Spacecraft in Low Earth Orbit, **Journal of Spacecraft and Rockets**, Vol.28, No.3, May/June **1991**
7. P. Eichler, D. Rex, Chain Reaction of Debris Generation by Collisions in Space – A Final Threat to Spaceflight?, 40th IAF Congress, paper **IAA-89-628**, Malaga, Spain, Oct.7-13, 1989
8. R. Jehn, Dispersion of Debris Clouds from In-Orbit Fragmentation Events, **ESA Journal**, Vol.15, No.1, p.63–77, 1991
9. H. Klinkrad, R. Jehn, The Space Debris Environment of the Earth, **ESA Journal**, Vol.16, No.1, p.1–12, 1992
10. N. Longdon (editor), **Re-Entry of Space Debris**, ESA SP-246, 1986
11. B. Battrick (editor), **The Re-Entry of Salyut-7/Kosmos-1686**, ESA SP-345, Aug.1991
12. W. Flury, Activities on Space Debris in Europe, 42th IAF Congress, paper **IAA-91-589**, Montreal, Canada, Oct.5–11, 1991

12: Space Debris: How France Handles Mitigation and Adaptation

J.-L. Marcé

CNES, Toulouse, France

1. INTRODUCTION: SOME WORDS ABOUT THE *"CENTRE NATIONAL D'ETUDES SPATIALES"* (CNES)

The *"Centre National d'Etudes Spatiales"* (CNES), a Government body, commercial and industrial in character, is the national agency responsible for the development of French space activities. Created in December 1961, CNES exists in effect since 1st March, 1962.

1.1 Mission

CNES mission is:

- to analyze the long term issues and future course of space activities and to submit proposals to the French government regarding the actions and facilities required to enable France and Europe to participate in their development;
- to conduct, in pursuance of the French government's space policy decisions, the major development programmes undertaken as a national effort as well as some European Space Agency activities delegated to France, such as ARIANE development.

1.2 Role

In discharging its mission,CNES plays a multi-faceted role:

- in association with the scientific community, it implements a fundamental research programme in the space area, involving the laboratories of the *"Centre National de la Recherche Scientifique"* (CNRS) and of the Universities.
- it develops supplier-customer type relationships with French users of Space: France Telecom, the Broadcasting Authority (*Télédiffusion de France*), the Ministry of Defence, the Meteorological Office and Earth Observation users;
- it fosters the competence and credibility of French exporting firms by awarding them, whenever possible, programme prime contractorship and execution responsibility and having them conducting a Research and Technology programme;

- in the operations area, CNES plays an important role by providing services for satellite in-orbit control and commissioning and for the running of the Kourou launch range;
- it strives to capitalize on the technical know-how acquired through space programmes by initiating the setting-up of commercialization companies (e.g ARIANE-SPACE and SPOT IMAGE) whenever the space applications market warrants it and when no structures geared to take on such activities exist;
- in association with the Ministry of Foreign Affairs, CNES represents France on the various bodies of the European Space Agency (ESA).

1.3 Organization

CNES staff stands at about 2400 persons, working at one of the following locations: CNES Headquarters in Paris, Launcher Directorate in Evry (near Paris), the Toulouse Space Center – programme preparation and development, spacecraft's operations, test facilities, Hermès Directorate (joint team ESA/CNES), the balloon launch range of Aire sur Adour and the Kourou launch range located in French Guiana.

The chairman of CNES Board is Mr. Jacques-Louis Lions and the Director General is Mr. Jean-Daniel Lévi.

1.4 Activities

CNES activities encompass French participation in the European Space Agency's programmes, as well as projects undertaken in a national or cooperative framework.

National programmes include Research and Technology activities, the SPOT programme (with participation of Belgium and Sweden) with two operational spacecraft SPOT 1 and SPOT 2 and two spacecraft under development , SPOT 3 under storage and SPOT 4 to be ready for launch in 1995, the participation to the development, in cooperation with France Telecom of TELECOM 2 (two spacecraft launched in 1991 and 1992), in cooperation with TELEDIFFUSION DE FRANCE of TDF (two spacecraft launched in 1988 and 1990) and in cooperation with the French *"Délégation Générale de l'Armement"* the future military observation satellite, HELIOS (the first spacecraft to be launched in early 1994).

Bi-lateral cooperation covers:

- scientific projects with the United States of America, (in particular the TOPEX-POSEIDON oceanographic project to be launched next summer, as well as MARS Observer) and with the ex-Soviet Union (inter alia manned flights such as ANTARES, the SIGMA programme launched in December 1989 and the Mars 94/96 project).
- location and data collection with the ARGOS programme carried out in cooperation with the United States of America
- search and rescue programme SARSAT-COSPAS carried out in an international cooperation framework

ESA programme with major role of CNES covers:

- ARIANE (delegation of development to CNES)
- HERMES (joint team ESA-CNES)

1.5 Budget

In 1992, the CNES budget stands at about FF.10,730 Millions apportionated as follows:

- 45.4%   for participation in ESA programmes,
- 5.4%   for bilateral cooperation,
- 22.3%   for the national programme,
- 15.3%   for programme support
- 11.6%   for general operating costs.

2. CNES ACTIVITIES AND PRACTICES RELATED TO SPACE DEBRIS

2.1 Generalities

Space debris refer to man-made earth-orbiting objects which do not serve a useful purpose. As part of the more general problem of pollution of space that includes radio-frequency interference and interference to scientific observations, this field was considered as a major concern for the design and the operations of space system. The critical areas which required our attention were:

- designing systems which minimize space debris
- designing systems with adequate protections against space debris
- implementing current practice which minimize the production of space debris in the operation of space systems.

Since its creation, as French Space Agency, CNES has payed attention to the problem of space debris in its various aspects and led various activities relevant to this multidisciplinary domain.

In this paper, I will present a summary of these activities, define the current CNES practices in this field and outline the relevant organizational aspect.

2.2 Design practices for space systems in order to avoid production of debris

2.2.1 Separation and deployment systems

In order to reduce the production of space debris, current design practices of separation and deployment systems have always used self-trapping items such as bolt catchers; no free belt were used in particular.

2.2.2 Despin systems

Yo-yo devices for spacecraft are no more used since long.

2.2.3 Pressurized spacecraft systems

Adequate design margins including fatigue considerations due to thermal cycling are current practice for pressurized vessels (propellant tanks, battery...) on board spacecraft such that the risk of catastrophic failure leading to production of space debris by explosions is negligeable.

2.2.4 Launcher's upper stages

As design authority of the Ariane Launch Vehicle, by delegation of the European Space Agency, CNES has studied and implemented practices as soon as the problem was evidenced by the V16 third stage explosion (SPOT 1 Launch on a Sun-Synchronous Orbit) in November 1986, i.e ten months after the launch.

The most probable cause of this explosion has been identified as tank rupture due to structural fatigue resulting from repeated thermal cycling. Therefore, as corrective action, we have defined a passivation of this third stage consisting in tank depletion and depressurisation after separation from the payload.

This practice is presently currently used for ARIANE 4 launch on Sun-Synchronous Orbit. It was used for the first time for the V35 launch (SPOT 2) in 1990 and repeated for the V44 launch (ERS 1) in 1991; no break up of the concerned third stages was observed up to now which demonstrates the efficiency of this practice. It will be also used for the next V52 launch on Low Earth Orbit of TOPEX in July 1992.

The implementation of this practice for ARIANE 4 launch on Geostationary Transfer Orbit was delayed due to technical difficulties and was anyhow considered as less critical. Studies, hardware and software modifications, will be completed this year; therefore, the passivation will be implemented for all Geostationary Transfer Orbit launch from V60 (presently scheduled for end 1993) onwards.

For ARIANE 5 the problem is only relevant to the L9 upper stage because the H150 cryotechnic stage will never be left in orbit and will naturally fall after cut off, whatever the mission is. As for ARIANE 4, the L9 upper stage will be left inert by using an adequate passivation method; in case of use for an orbit close to the space station, it will be removed from the orbit of the space station before passivation.

2.3 Design space system with adequate protections against the effect of space debris

This item will not be developed lengthly in this paper, being not fully relevant to the topic of the Symposium. For spacecraft design, the design environment can no more be limited to the natural meteroid debris left from the formation of the solar system, but shall also consider man-made orbiting object; the consideration of manned flight such as HERMES has also put emphasizes on this topic in CNES.

For many years, CNES has conducted study of natural and artificial environment, with the support of French laboratories such as ONERA/DERTS.

In particular, various experiments were performed such as GIOTTO/HALLEY dust sensor, LDEF/FRECOPA and MIR/ARAGATZ; additionaly, ground tests on hypervelocity impacts are being studied in available facilities such as the one of ETCA/GRAMAT. New experiments are planned for EURECA 1 this year, and for some shuttle flights mid-1992, and are under investigations for EURECA 2 planned for 1994/1995.

The objective is to achieve modeling of Low Earth Orbit particle population and to perform hazard assessment for space missions (manned and unmanned).

Various theoretical studies have also been performed:
– dealing with collision probability assessment of geostationary satellites
– dealing with probability assessment of low earth orbit satellite collision with space debris
– dealing with definition of possible orbital manoeuvers in view of collision avoidance of a manned vehicle with space debris
– dealing with the impact on the launch slot of the taking into account of the risk of collision with space debris

Computer modelling of hypervelocity impacts are also part of the investigations.

2.4 Current practices for the operations of space systems

2.4.1 Operations of space systems

Since the first time it was involved in the operation of Geostationary spacecraft, CNES has proposed to reorbit geostationary spacecraft at its end of life.

This was currently implemented for the French-German spacecraft SYMPHONIE: SYMPHONIE 1 was reorbited (increase of the semi-major axis by 80 km) in August 1983 and similarily SYMPHONIE 2 in December 1984.

The reorbiting procedure for TELECOM 1A, which is no more in service since the successful launches of TELECOM 2A and B (Spacecraft belonging to FRANCE TELECOM which subcontracts the platform operations to CNES) is scheduled for August 1992 (increase of the semi-major axis by 160 km).

In order to make this reorbitation in an economical way, i.e typically to know one year beforehand with a precision of one month, the date of propellant shortage, in order to prepare the launch of the replacement satellite, CNES is continuously developing various methods in the frame of its Research and Technology programme.

The major running activities are:
– development of a thermal flow meter for the measurement of the mass flow rate of propellants (contract with *Société Européenne de Propulsion*)
– development of an improved P/T method (Pressure and Temperature measurements in the propellant tanks) by adding a KALMAN numerical filter (contract with AEROSPATIALE).
– study of different measurement techniques and of alternative solutions for reorbitation such as additional tank (in house study).

In this context it should also be mentioned that CNES has developed an original method for station-keeping of colocated spacecraft which are currently implemented for the operations of satellites located at 19°W namely TDF 1 and 2 (Spacecraft belonging to *Télédiffusion de France*, which subcontracts its platform operations to CNES) OLYMPUS operated by ESA/ESOC and TVSAT-2 operated by DLR/GSOC ; this coordination between three different spacecraft operating centers for four satellites is unique in the world according to our knowledge.

2.4.2 Surveillance of space

In the frame of its activities as spacecraft operator, CNES as an operational service dedicated to space surveillance which is used to monitor re-entry such as the one of SKYLAB, COSMOS 954, 1402, 1900 and SALIOUT 7 in particular. Due to funding limitations, CNES has presently no dedicated monitoring means such as radar, and has therefore signed various agreements in order to use data from NORAD or from the French Defence facilities.

3. INVOLVEMENT IN VARIOUS COMMITTEES AND GROUPS DEALING WITH SPACE DEBRIS

CNES has appointed representatives in various international groups dealing with space debris:
– NASA group dealing with LDEF experiment
– ESA Space Debris Advisory Group (SDAG)
– Space Debris ad'hoc group from the International Astronautics Academy
– United Nation Committee for Peaceful Use of Outer Space (UNCOPUOS)
– CCIR/ITU (indirectly dealing with space debris)
– ISO

CNES is also having discussions in this field with other National Agencies.

4. CONCLUSION

I hope that this paper has demonstrated that CNES, as French Space Agency, has devoted , since the early phase of its activities, major attention and has implemented numerous activities relevant to the problem of space debris.

CNES has also demonstrated that it is ready to implement adequate practices and measures in order to reduce debris production provided that they are based upon reasonable assessments.

The best demonstration of the high importance brought by CNES to the problem of space debris, is highlighted by the recent settlement of an internal CNES Space Debris Synthesis Group involving representatives of the various Directorates concerned by this matter covering all involved disciplines from space mechanics to material behavior including legal aspects and project oriented engineers; this group has a permanent secretary and is reporting to an Orientation Committee chaired by the CNES General Director.

13: Facing Seriously the Issue of Protection of the Outer Space Environment

Qi Yong Liang

Head of the Team for Study of Space Debris Issue, People's Republic of China

The rapid development of the space technology of the last 20 years has been of great benefit to mankind. In many areas such as communications, navigation, remote sensing, meteorology, and scientific studies, the production activities and the way of human life has entered a new stage. As more and more vehicles have been delivered into outer space, the population of space debris created by spent rocket stages or deactivated satellites that continue to orbit is a situation that is of increasing concern to many international communities and governments. The protection of the space environment has become an important problem which needs to be seriously treated in the development of space technology.

1. PROTECTION OF THE OUTER SPACE ENVIRONMENT – A PROBLEM OF EVER GREATER SIGNIFICANCE

The Outer Space Treaty approved by the United Nations in 1966 points out: "The progress in the exploration and use of outer space for peaceful purposes are related to the common benefits of whole mankind," therefore, protecting the outer space environment will be advantageous to the space activities for mankind, while a seriously polluted outer space environment will have a harmful effect on the exploration and use of outer space. From a long-term point of view, it would be useful for future exploration of outer space and could produce twice the results with half the effort if people would seriously consider methods or measures for protecting outer space as early as possible. We shouldn't wait to protect the space environment until it becomes damaged. The industrialized society on Earth has experienced a process of pollution followed by control; the same disastrous road shouldn't be followed in outer space.

Both the development of space technology and the protection of the outer space environment are not fully opposite and can complement each other. Stressing the development of space technology to the neglect of the protection of the outer space environment would jeopardize increasingly the space activities. However, technical problems related to the protection of the outer space environment could only be solved through the development of space technology.

China has developed space technology to a certain extent, but, as a developing country, its space activities are still limited in size and capabilities. In order to forge ahead with the economy, science, and technology, China will develop space technology further and perform more space activities. Therefore, protecting the outer space environment is also of great importance to China. Although China's few space launches have created only a very small amount of space debris compared to that of space powers, China has paid special attention to the protection of the outer space environment while developing space technology.

2. CHINA'S EFFORTS TO PROTECT THE OUTER SPACE ENVIRONMENT

China's space industrial departments have made a due contribution in that their space activities – ranging from fundamental research and technical design to the management of scientific-technical works – have fully reflected the principles of protecting the outer space environment.

As early as the mid 1970's, the Academia Sinica had made use of the optical and radio observation station in its man-made satellite system for tracking China's satellite launches and performing scientific research of outer space. In the meantime, some space debris created by domestic or foreign satellites were observed and some data were collected for studies of the outer space environment. In 1986, China's space industrial departments had collected a series of materials relating to space law, space contamination, maintenance of launch safety in outer space, effects of space debris on mankind, etc. In 1989, the Team for Study of the Space Debris Issue was formed, in which specialists from the Chinese Academy of Space Technology, China Academy of Launch-vehicle Technology, Shanghai Space Agency and Zijinshan Observatory were organized to study the space debris issue.

In order to expand information exchange, and under the support of the Ministry of Foreign Affairs, China's space industrial departments have invited specialists from the International Association of Space Law, the United States, and other countries to participate in discussions or give lectures. A Chinese delegation sent to the United States in 1989 exchanged wide-ranging views relating to international space law, especially to the space debris issue, with responsible persons in the UN's Outer Space Division, the International Association of Space Law, and NASA. In 1991, China's specialists in space debris issues visited Japan to attend the "91 Symposium on Space Debris" there. In the same year we received a delegation from the US to exchange opinions about the space debris issue.

China's space industrial departments have not only accomplished much in studies of space technology and international information exchanges, but have also made technical efforts in protecting the outer space environment.

China's space industrial departments have insisted on the principles of independence, autonomy, and self-reliance in the development of space technology. Therefore, only limited space objectives have been selected according to the actual needs and the economic status of our country to decrease the number of space launches and increase the rate of success. From 1970 to the end of 1991, only 29 space launches were carried out by China to deliver 33 satellites (two of them were foreign satellites) into outer space.

For the sake of maintaining the outer space environment, China's space industrial departments have adopted some measures in the design of satellites and launch vehicles. For instance, we have redesigned the launching trajectory of some geostationary satellites so that the perigee of the geosynchronous transfer orbit could be lowered from 400km to 200km. In this way, the orbital life-time of upper stages would be substantially shortened. An earlier reentry of the upper stages would lower the possibility of creating space debris. Another example is that the covers of the IR sensors onboard the Chinese meteorological satellites remained attached to the satellites after their jettison, thus avoiding the generation of more debris. These technical measures are favourable to the protection of the outer space environment.

Other technical options being studied by China's space industrial departments for avoiding the creation of space debris are as follows: In order to prevent the upper stages from explosion due to propellant remnants, it is intended to remove the residual hypergolic propellants and high-pressure gases remaining in the upper stages when they are separated from the payloads but still fly in orbit. Moreover, the "deorbit" technique will be used after the orbital insertion of the satellite. In this way, the upper stage would be oriented and restarted into the lower altitude orbit for an earlier reentry into the Earth's atmosphere. Besides the above mentioned options, other technical measures will also be explored.

3. SOME CONSIDERATION CONCERNING THE PROTECTION OF THE OUTER SPACE ENVIRONMENT

In view of the fact that the development of space science and space technology is still in the ascendant, and more and more countries possess the capability to perform space launches and satellite development, it is impossible to hold back the development of space technology merely to maintain the outer space environment. Furthermore, there are many technical and economic problems which are hard to solve at present when people wish to remove the existing space debris as quickly as possible. The problem, hence, is how to protect the outer space environment with feasible measures and techniques on the premise that the rapid development of space technology will be assured.

3.1 Publicizing knowledge of the outer space environment

At the moment, people are aware of space activities and show interest in them, but most of them are not aware of the state of the outer space environment, its effect on, and the hazards to, space activities, as well as how to protect the space environment. So it is necessary to publicize knowledge relating to the space environment through various means such as newspapers, magazines or lectures, in order to attract the attention of all governments and launch organizations.

China is also preparing to publicize further the knowledge concerning the international space law and space debris. Relatively detailed knowledge will be introduced to the organizations and persons directly related to space activities. Our purpose is to impel the related departments in the space industry to participate actively in studies of the outer space environment in conjunction with their own job, and to pay attention to the treatment of space debris, thus generating their own efforts to protect the outer space environment.

3.2 Establishing the file system of space objects

Seeing that the protection of the outer space environment is the common responsibility of mankind, we wish to make a proposal to the UN's Committee on the Peaceful Uses of Outer Space for establishing a filing system of space objects. The system would release information on space objects regularly or irregularly, or provide accurate, timely information for the countries and organizations according to their needs for performing launch activities. Such information would be necessary and beneficial to the development of the space technology in many countries. In this way, man could have an overall understanding of the space environment and make use of the information provided to avoid damage to the space launches. All launch nations and organizations, especially the space powers, would have an unshirkable duty to cooperate closely with the UN's Committee on the Peaceful Uses of Outer Space for establishing the filing systems of space objects and for doing a good job of releasing information.

3.3 Solving the technical problems of space debris control

The primary problem in the space environment is space debris. To solve the problem, the first step is to resolve a series of technical problems including: understanding the mechanism of the creation of space debris, the technical approaches necessary to minimize the creation of space debris, the technical improvements in the design and operation of launch vehicles, the technical options for removing space debris in low Earth orbit, etc. The settlement of these problems needs long-term studies and the accumulation of experience. At the moment, emphasis should be put on preventing the upper stages from exploding and decreasing the orbital life of space debris.

3.4 Promoting international exchange and cooperation

Since the protection of the outer space environment is a problem of global scale, it won't be solved by any one country or organization alone. The protection of the space environment will be effective only through international exchange and cooperation. International exchange and cooperation could be conducted through various channels and in different forms, including: exchange of the results of studies, discussion on special topics, and cooperative studies. There could be bilateral or multi-lateral discussions or international conferences and symposia.

International exchange and cooperation should adhere to the principles of friendship, coordination, promotion and effectiveness. Whether large or small, all countries and organizations should make a joint effort to protect the outer space environment.

In the international exchange and cooperation arena, countries and organizations which possess some techniques for protecting the space environment or knowledge obtained as the result of studies, should actively spread those results and technical achievements, for example: the mechanism of the creation of space debris and techniques for removing residual propellants from the upper stage. Popularization of these techniques is a contribution to the development of the space technology and will be favorable to the international society as well as to the knowledge owners themselves.

Techniques for protecting the space environment should first be seriously studied through international cooperation. On this basis and in line with the goal of paying equal attention to both the promotion of space technology and the protection of the outer space environment, an international agreement or convention should be approached and formulated as the norm that must be followed by all countries in their future explorations and peaceful use of outer space.

14: Space Debris – Mitigation and Adaptation

U.R. Rao

Chairman, Space Commission, Bangalore, India

1. INTRODUCTION

The growing man-made debris deposition in space, because of its cumulative effect, has become a major concern of potentially harmful space environmental pollution. In particular rapidly increasing probability of space debris colliding and damaging functional spacecrafts, the distinct possibility of re-entry of large space debris into habitated areas and the debris causing interference with radio observations have become a real threat to future space activities. It is clear that all space faring nations, in their own interest, have to focus their attention on evolving adequate preventive measures to minimise the space debris hazard and also agree upon appropriate national/international codes for ensuring the safety of future space activities. We have carried out extensive modelling studies, to model debris generation and distribution caused by break up of space bodies which are briefly summarised in this paper. This paper also addresses the possible ways of preventing further deposition of space debris due to fragmentation and discusses the policy which India proposes to follow, in the coming future, towards achievement of this objective.

2. MODELLING OF SPACE DEBRIS ENVIRONMENT

The population distribution of large space debris objects, of larger than 4 cm at 400 km altitude, 10 cm at 1000 km and one meter at geostationary altitude of about 7000 trackable objects using radar and optical means, has been reported. However, the distribution of untrackable objects, below the present tracking limit size, estimated to be about 95% of the total debris, requires to be clearly established for properly assessing the total space debris damage potential. A number of studies[1,2] based on fragmentation model of rocket objects and spacecrafts are available in literature for estimating the untrackable debris distribution. We have carried out an independent study by using a refined mathematical model for fragmentation and established empirical relationships. In order to derive orbital elements of the debris, we have used the incremental velocity distribution along three orthogonal directions assuming an isotropic distribution of frag-

ments in the frame of centre of mass system and have compared our results of mathematical modelling with the observed distribution of 33 major on-orbit fragmentations out of the 90 such fragmentations, as a consistency check of our model. Figure 1 shows the close comparison between the observed altitude profile of trackable debris with that derived using our mathematical model, which clearly brings out the density peaks at 800, 1000 and 1500 km altitudes with corresponding densities of 1.83 x 10^{-8}, 1.64 x 10^{-8}, and 1.2 x 10^{-8} objects/km^3 respectively. Extension of this mathematical model for predicting the statistical density distribution of untrackable debris has been carried out. Figure 2 shows the statistical distribution of trackable and untrackable debris derived from our model, which clearly shows that the untrackable debris density to be 3 to 5 times higher than the trackable ones at different altitudes.

While we also assign a 'most probable' value for the incremental velocity imparted to the debris at the instant of explosion, the significant feature of our model is in the assumption of a double Gaussian distribution for the velocity of fragmented debris particles in the centre mass frame of reference. The details of our mathematical model are given in a separate paper being published elsewhere[3].

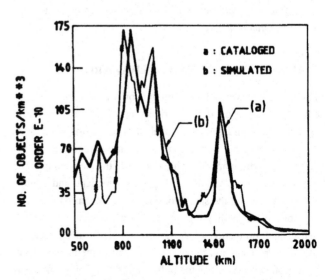

Figure 1: Spatial density of tracked debris: comparison of simulated against cataloged.

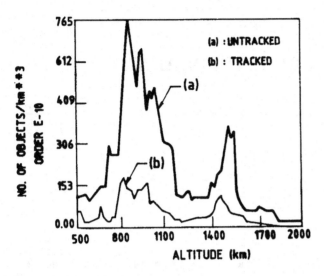

Figure 2: Spatial density.

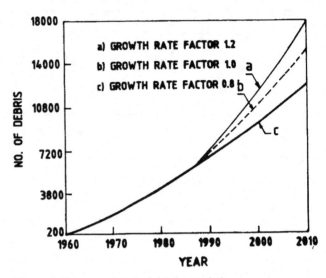

Figure 3: Forecast of tracked debris population.

In order to assess the future damage potential of these debris, the fragmentation profile as well as the debris growth profile for different scenarios of space activities, ranging from 0.8 to 1.2 times the present level of global space activity, have been computed upto the year 2010. Figure 3 shows the growth profile of trackable debris. It is clear that if the debris deposition in space continues unabated, the population of trackable objects can rise to over 18000 with the space traffic reaching 1.2 times the present scale, while the total population of trackable and untrackable objects would exceed 3.4×10^5. Applying appropriate statistical methods, we have computed probability of collision of debris with active space objects at different altitudes and as a function of time using the method of closest approach and orbit intersection. It can be observed from figure 4 that the current collision probability per m^2 per year at 500 km, 800 km, 1000 km and 1500 km are 1.2×10^{-6}, 4.00×10^{-6} and 2.5×10^{-6} respectively for tracked debris. On an average the probability of collision increases almost 3 times by 2010. For untracked debris the collision increases almost 3 times by 2010. For untracked debris the collision probability is almost four times that of tracked debris.

3. MINIMISING THE DEBRIS HAZARD

Three possible ways of controlling debris hazard are undoubtedly prevention of debris, protection of active spacecraft against debris and active removal of space debris. Extensive discussion on each of these measures are available in literature[4,5], the summary of which is that the scavenging for removal of space debris is largely impractical and highly expensive leaving the space communities to essentially evolve a global understanding on the method of prevention of space debris and protection of space hardware. In this paper we restrict our discussion to possible prevention methods and the basic policy issues which the Indian Space Program proposes to adopt, on a voluntary basis, in the conduct of its own activity, notwithstanding the fact that the contribution of India and other

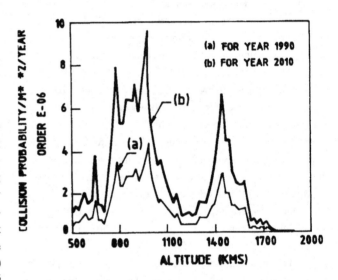

Figure 4: Future collision probability.

developing nations to the debris pollution in space is totally negligible. The basic principle which we have followed is primarily to vent the residual propellant of spent stages of rockets as a precautionary measure against explosion/fragmentation or, as a secondary option, deorbit them, if possible, to make them rapidly enter the atmosphere. In the case of satellites, in addition to taking precautions against explosion/fragmentation, our strategy resorts to deorbiting or orbit raising of the spacecraft at the end of their useful life, in order to protect the two regions of interest to all countries involved in the exploitation of space for peaceful applications, namely, the altitude region of interest to remote sensing satellites and the geostationary arc.

Recognising that the venting of residual propellant in the last stage of a rocket while minimising the hazard of explosion, cannot remove the spent stage from the crowded zone, we have also studied, the possibility of deorbiting the last stage after the separation of the spacecraft. Such a strategy calls for reorientation of the spent liquid upper stage and then deorbiting the same using the residual propellant. Figure 5 depicts the

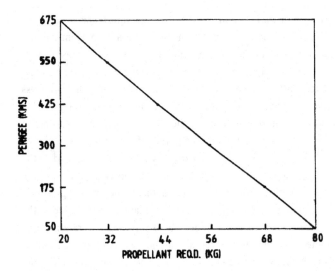

Figure 5: Propellant mass for deorbiting spent stages of polar launch vehicles to different perigee altitudes.

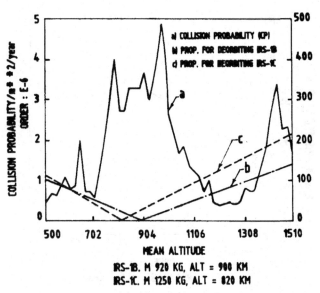

IRS-1B. M 920 KG, ALT = 900 KM
IRS-1C. M 1250 KG, ALT = 820 KM

Figure 6: Collision probability and propellant mass needed for deorbiting IRS class of satellites.

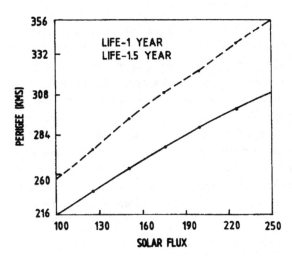

Figure 7: Max perigee height as a function of solar flux to effect complete decay in 1–1.5 year (apogee height: 900 km).

perigee altitude to which the spent stage of a Polar Launch Vehicle like PSLV can be brought down for different values of residual propellant. This method, particularly in the case of low earth missions, is advantageous for reducing rocket debris pollution as it enables the spent stage to have a faster decay and drastically reduces the chances of explosion. In the case of geostationary launches, the residual propellant of the GTO stage can be used to lower the perigee thus ensuring faster re-entry.

With regard to the satellites, our study has concentrated on possible deorbiting strategies of used-up satellites to areas of less debris density. A typical variation of propellant mass required to bring such satellites to different altitude bands is depicted in figure 6. The collision probability at these altitudes are also shown in the same figure. To bring down the mean altitude of a typical polar sun synchronous remote sensing satellite at 900 km to about 700 km (perigee height about 500 km) about 50 kg of propellant is required. An alternative will be to push the spacecraft into 1100 km circular orbit by spending almost the same amount of fuel. This has the disadvantage that continuous debris deposition at this altitude can make this altitude band unusable in future. Additional propellant is needed to enable the spacecraft, after completing its mission life, to re-enter the Earth's atmosphere within a stipulated time of about a year. The extra residual fuel is dictated by the solar activity and the perigee height to which the spacecraft needs to be de-orbited. Typical perigee heights, which can ensure re-entry into atmosphere in a stipulated period of one or one and half year of an IRS class of satellites is depicted in figure 7, as a function of solar activity.

In the case of geostationary satellites, the best strategy seems to be to use the residual propellant of the spacecraft for orbit raising, at the end of its useful life, to ensure that the spacecraft is placed in an orbit which will not interfere with the geostationary arc. The altitude to which the spacecraft is to be raised is dictated by the area to mass ratio of the spacecraft

and the orbital perturbation that acts on it. Figure 8 shows the minimum altitude raise required as a function of area to mass ratio (A/M). For a typical spacecraft with a dry mass of 900 kg. and a cross sectional area of about 23 square meters, this altitude is about 80 km and requires about 1.2 kg of propellant which in terms of operational life is a reduction of useful life by 1 month. However, if the altitude raise needed is 500 km, the fuel requirement will go up to 6 kg which is equivalent of 5 months of operational life.

4. CONCLUSION

While more detailed and focussed research is needed to fully understand the fragmentation mechanism in space and better sensors are needed to monitor the dynamics and distribution of debris population we believe that even with the existing knowledge and technology, certain actions can be taken to

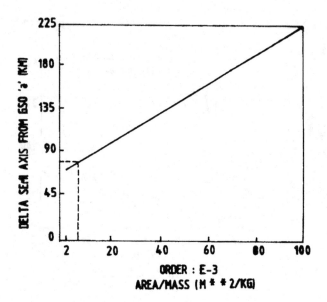

Figure 8: Minimum altitude raise as a function of area/mass ratio for reorbiting geo satellites.

minimise debris hazard in space. On India's part, the voluntary decision to adopt reorbiting strategies for geostationary spacecraft and venting of spent stages of vehicle upper stages, whenever possible, is a step in the right direction. In the long run, a concerted and joint effort by all space faring nations to evolve appropriate and affordable strategies to minimise the impact of space debris on future space missions is called for, which would finally form a basis for international understanding.

If the goal is to minimise and reduce accumulation of space debris, the required strategies must form a part of the very design of rockets and spacecraft in all future space ventures. To avoid getting entangled in the discussion of identification and responsibilities of the polluter nations, we suggest formation on an equal basis, of an inter-governmental committee, consisting of experts from all space faring nations to discuss the technical details and arrive at appropriate strategies. To ensure implementation of these strategies, all information available and technology required, if necessary, have to be freely shared between the nations which alone can lead to the total acceptance of the system by all the present and emerging space nations. It must be noted that none of the presently accepted international space law regulations really address the space debris issue. Acceptance of any specific rational and objective international regime on this issue will emerge only if the voluntary agreements based on technical evaluation and free flow of information for solving this particular issue are agreed upon.

REFERENCES

1. Kessler, D.J.et.al: 'Orbital Debris Environment for Spacecraft Designed to operate in Low Earth Orbit, NASA TM-10047, April 1989.
2. Culp, R.D and Madler, R.A: 'Modelling Untrackable Orbital Debris Associated with a Tracked Space Debris Cloud', AAS 87-442, 1987
3. U.R. Rao: 'Space Debris – Environment and its Control', to be published.
4. P. Eichler and A.Bade: 'Removal of Debris from Orbit', AIAA 90-1366, AIAA/NASA/DOD Orbital Debris Conference: Technical issues and Future Direction, April 16–19, 1990, Baltimore
5. Eric.L.Christiansen: 'Advanced Meteoroid and Debris Shielding Concepts', AIAA 90-1336, AIAA/NASA/DOD Orbital Debris Conference: Technical Issues and Future Direction, April 16–19, 1990, Baltimore.

15: Near Earth Space Contamination and Counteractions

V. F. Utkin and S. V. Chekalin
Central Research Institute of Machine Building, Russia

The level of near earth space contamination with orbital fragments has been evaluated. Basic research lines to solve the "space debris" problem have been analyzed. Definite proposals on international cooperation have been presented.

The problem of near earth orbit pollution with rocket and space technology fragments is of global character and concerns many countries. For the 35 year period of space exploration more than 20,000 orbital man-made objects measuring more than 10–20 cm have been registered using available on-ground tracking systems. Depending on their orbits and masses, some of these objects and fragments entered the upper atmosphere and burnt, while the rest continued their space flight. Specialized optical-electronic and radio systems functionally integrated into a single space monitoring Service observe orbital objects constantly. This Service controls the given systems and performs centralized automated data processing and cataloging of information supplied by the systems.

At the present moment the near earth orbits accommodate about 6700 observed artificial objects: 36% of them represent used spacecraft and rocket stages, 58% separated elements and fragments, produced by spacecraft and rocket stage explosions, and only 6% are active spacecraft.

The bulk of "space debris" is the result of Soviet and American spacecraft launches. The share of both countries is approximately equal. While more used spacecraft and launch vehicle stages fall to the share of the ex-USSR, the US is responsible for two-thirds of trackable fragments, remaining in orbit after space object explosions. Fragments generated by Soviet spacecraft explosions depart orbit sooner, since these objects are destroyed at lower altitudes.

The "space debris" population contains tiny untrackable fragments too. In accordance with expert estimates, the population of small fragments several cm in size produced by objects' destruction already numbers some tens of thousands, while centimeter-size and smaller fragments number hundreds of thousands.

The yearly increase in large fragments in near earth space runs approximately 4–5%, while the rate of increase of small-sized untrackable fragments and man-made particles is supposed to be even higher. So, judging by the number of break-throughs in the aluminum foil samples, 10 and 20mkm in thickness attached to the "Saljut-6" and "Saljut-7" stations' external surfaces, it is possible to state that the yearly increase of high-velocity particles, 0.001–0.003 mm in diameter runs into 6–7% in 350 km orbits.

The highest concentration of "space debris" is observed between 300–1600 km, just within the altitude range where the majority of manned and unmanned space vehicles fly. Even now, collision probability for such large space objects as advanced as the orbital station of the "MIR-2" and "Freedom" type with a dangerous fragment of 1 cm or more may run into several percent for a year.

The presence of nuclear power spacecraft within this zone in "illuminating" orbits, the collision of which with artificial fragments may result in unwanted aftereffects, causes anxiety.

Collision risk with "space debris" fragments in GEO is relatively low as compared with that in low earth orbits. But this orbit, due to its specific factors, is being extensively utilized by many countries. Today there are up to 450 traceable objects in GEO. The yearly increase of objects in orbit amounts to 35–40 pieces, but the orbit capacity is limited. Today, separate segments of GEO are overpopulated already. The contamination problem may become one of many factors reducing the efficiency utility of this specific orbit in the near future.

Evaluation of the current near earth orbit contamination situation and the forecast of its continuing development point to the necessity for undertaking technological and legal steps to solve the "space debris" problem.

First of all, it would be necessary to go on with evaluation research of the space contamination level, particularly with the tiny fragments undetectable by ground observation systems. Information on small particles and estimates of the population of small man-made particles are obtained from the effects of their collisions with target surfaces, attached to a long service-life orbital station. The information can be useful for modelling the generation and population of these particles. Under investigation is the possibility of registering microparticles in space using a small satellite launched as an extra payload, together with series spacecraft. Particle collisions are registered at the moment they break through inflatable satellite shells which serve as condenser-type film transducers, while particle mass and speed are measured by ion sensors.

Technical and organizational steps, aimed at preventing near earth space pollution in future are other important fields of activity. The following steps are under consideration:

- reduction of space vehicle launches at the expense of active service life increase, integration of missions and revision of the launch program as a whole (the annual launch number has been diminished by 20–30% on the average);
- exclusion of in-orbit object destruction, beginning with deliberate spacecraft explosions (since 1982, satellite interception experiments have not been conducted; developments to prevent spacecraft self-explosions have been carried on);
- exclusion or maximum reduction of the number of spacecraft elements and SLV stages separated in orbit (today each spacecraft launch is accompanied, on average, by the separation of three standard components);
- refusal to utilize orbital propulsion plants using propellants which leave solid particles while burning (for example, one-third of the solid-propellant engine combustion product falls on aluminum oxide particles of 0.0001–0.01 mm).
- use of materials and coatings for orbital objects production least exposed to erosion emission caused by near-earth space factors.
- controlled spacecraft removal upon termination of its active service life from the operating orbit into the upper atmosphere or in a "burial" orbit (as for GEO spacecraft, spacecraft transfer to an outer orbit with an increase of the altitude up to 250–300 km).
- elaboration of launch programs without an SLV upper stage orbit insertion, while the spacecraft is inserted in orbit by an apogee kick motor (such a launch pattern has been realized for the SLV "Energiya" and is provided for a number of new spacecraft to be launched by "Proton" and "Zenith" SLV).
- in future, changeover to utilization of completely reusable space transportation systems (STS) would contribute to reduction of the level of space contamination.

As for removal of "space debris" from the near earth space, it would be quite substantial to begin, first of all, with removal of large passive objects being potential generation sources of small fragments. Reusable vehicles of the "Buran" and "Shuttle" type and orbital transfer vehicles (OTV) equipped with robots-manipulators could be used for the purpose. This is a very expensive job, but gathering and removal of small fragments from space seem to be a much more complicated and expensive mission. And once again we become convinced that getting rid of environmental waste contamination is much more difficult than preventing it. And it is necessary to move forward in order to find the solution.

Solution of this world-class problem of "space waste" needs an international legal basis and close partnership. Working out of legal norms and provisions must be aimed not so much at limiting rocket and space technology use as at its perfection in order to prevent further contamination of space, at first. Therefore, all aspects are to be analyzed by technical experts first. Such working group meetings have been conducted by Russian and American specialists.

The following may be considered to be definite proposals on cooperation:

1. Agreement in principle of maintaining regular tracking and data exchange for small space objects, including data about orbital objects, objects departed from orbit or exploded objects.
2. Constitution of a common database on objects presenting a high level of risk (large satellites approaching the upper atmosphere, uncontrolled objects, etc.).
3. Exchange of samples subjected to the actions of microparticles in space, as well as the analysis of the results of these collisions. Performance of a joint experiment on registering man-made particles in space by "MIR" onboard systems.
4. Development of near earth space pollution models, assessment and prediction of collision rate for different objects. Simulation result comparison.
5. Elaboration of standard requirements for rocket and space technology in order to lessen or prevent further pollution of space.
6. Working out of legal bases for regulating space activities in relation to the space waste problem.

It seems expedient to set up an international working group on orbital fragments, to comprise technical specialists and lawyers, which would focus on solutions to the problem of lessening the near earth pollution rate and providing orbital flight safety. Cooperation in solving technical and legal problems would permit a more objective formation of policy in this field with increased understanding of the complexity of the problem and the limitations of technological capabilities.

In May, 1992, a book, *Space: Tomorrow's Troubles*, by S.V. Chekalin (V.F. Utkin, a reviewer) will be published in Moscow. The book could be presented at the Symposium with the author's brief speech. The book's Table of Contents follows.

1. Modern cosmonautical achievements.
2. New purposes – new problems.
3. Space technology. How does weightlessness operate? New materials from space. Space biotechnology. Steps of space production.
4. Orbital missions. Large structures. Assembly operations in space. Spacecraft maintenance and repairs in space. Means of orbital maintenance. Role of man and robot.
5. Space and commercialization. "Space market" capabilities. Foreign launch vehicles. GLAVKOSMOS offers: Experiments in orbit. Use of satellites. Commercial L.V.

6. "Energy Buran." A system is known to be in comparison. Problems of the development. Priorities of application. Contribution to the economy.

7. Space transport of the future. Reusability and problems of technology in the search for advanced STS Aerospace plane. Orbital transportation vehicles. STS with an extravehicular power supply.

8. Power supply in space and space energy transfer to the Earth. Electrochemical energy sources. Perfecting of solar batteries. Space power station. Orbital mirrors. Radioisotope and nuclear power facilities.

9. Lunar forepost. "Return to the Moon." Priorities of the exploration. Problems of deploying a base. Works program and organization.

10. Martian mission. Dreams and realities. Mysterious planet. Preconditions and problems of the mission. Possible versions of the flight. Phases of the program.

11. Ecological aspects of cosmonautics. If space becomes overcrowded, what is to be done? Impact of LV launches on the atmosphere. The problem of reducing the number of fall zones and their areas. Ecological experiments and control. Possibility of nuclear waste removal.

12. On the way to far-out space. Interplanetary flight energy consumption. Stellar routes and antimatter. Projects of galaxy space vehicles.

16: The Current and Future Space Debris Environment as Assessed in Japan

Susumu Toda

National Aerospace Laboratory, Japan

1. INTRODUCTION

Space debris is considered to be a problem that all space faring nations must work together in order to maintain a safe environment for the future space development.

One of the world's earliest warnings on this issue was made in Japan in 1971 by M.Nagatomo and his colleagues of the Institute of Space and Astronautical Science (ISAS)[1]. Since that time, independent research on this topic has been carried out by various organizations in Japan. However, the quality and quantity of the research is still kept low, and very limited significant international contributions have been made. It should also be pointed out that no national guidelines on this problem have yet been formulated .

In light of this situation, the Space Debris Study Group was founded by the Japan Society for Aeronautical and Space Sciences (JSASS) in September 1990. Its objectives are to promote overall space debris related research, to stimulate public awareness of this issue and to provide guidelines to cope with it. The Interim Report was published in January 1992[2]. In this paper the space debris environment and the related technologies assessed by the group are described.

2. SPACE DEBRIS ENVIRONMENT

Trackable objects in earth orbit are steadily increasing with small fluctuations due to periodical atmospheric drag increment at the maximum solar activity and occasional launch activity stall. The history of object number provided by the United States Space Command (USSPACECOM) is shown in Fig.1. This figure also shows major breakup events in orbit. These breakups are the main source of about half of objects being tracked. However, the breakups are believed to have also created numerous untrackable smaller objects, many of which are large enough to be hazardous to space systems. The number of these hazardous untrackable objects is estimated to be 3,500,000, which is larger than the number of the trackable objects by a factor of 500. The total mass of objects is estimated to be 3,000 tons[3].

Detection capability of USSPACECOM is 10 cm in Low Earth Orbit (LEO), and decreasing according to the altitude, down to 1m at Geostationary Orbit (GEO). Smaller debris distribution in the 500 km altitude vicinity was estimated through returned materials, for instance, Solar Max exterior materials and LDEF. Because of limited area and exposure time of returned materials, the maximum size of detected debris was limited to less than 1mm. The orbital velocity of objects in LEO is about 7 km/s. The relative impact velocity of debris depends on the angle of orbit crossing and the average is estimated to be 10 km/s. This hypervelocity impact results in severe damage or breakup when collision occurs. For instance, an aluminum sphere of 1 cm in diameter has the equivalent kinetic energy of a compact car running at 50 km/h. Recent studies[4],[5] have pointed out the existence of a critical density, above which collision chain reactions increase the amount of debris even without further launch. It is shown that the present spatial density already exceeds the critical density at altitudes of 1000 km and 1500 km[4].

In GEO, it was once considered that the immediate danger is less severe because the amount of debris is not large and the relative velocity is low. Functioning geostationary satellites are controlled within a narrow ring and the density at the center part of the ring is very high. On the other hand, spent satellites without orbit control pass through the narrow ring twice a day in the north-south direction. The average relative speed at the time of GEO passing is 300 km/s, which is equivalent to the speed of passenger jet planes. Assuming 20 annual launches and no removal of spent satellites, the probability of a collision occurring within the next 20 years is estimated to be 5% as shown in Fig.2[6]. It is important to note that at the GEO no data are available on objects smaller than 1m. Small fragments created in this region would be difficult to be detected and they will pose an eternal hazard to other geostationary satellites since there is no atmospheric drag to lower the orbit.

3. THE STATE OF TECHNOLOGIES

3.1 Observation of space debris

It is generally acknowledged that improvement is needed in ground observation quality. The accuracy of debris size esti-

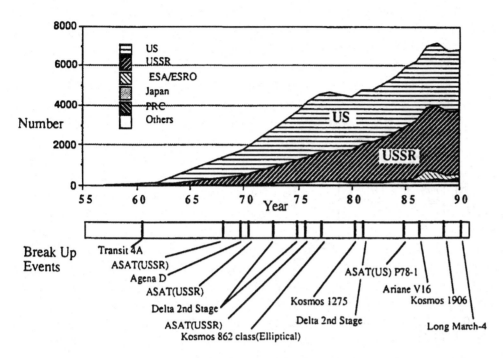

Figure 1: Catalogued object number and major breakup event.

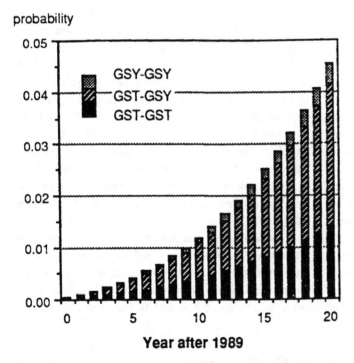

Figure 2: Probability of encounter among GEO objects within 10 m.

mations from radar cross section (RCS) is very uncertain, and present methods provide no data on shape or rotational motion of objects. Density also is uncertain by a factor of 2 to 10.

There are two Japanese efforts to contribute to the knowledge of debris environment. Radar observations have been made by Kyoto University utilizing the MU (Middle and Upper atmosphere) radar, which is a monostatic pulse Doppler radar operating at 46.5MHz. The height distribution of space debris observed by the MU radar is fairly well compared in

Fig.3 with the result determined from the USSPACECOM catalog[7]. It is expected to provide valuable information for RCS comparisons using radars with different frequencies, which is one of the most promising means of assessing the shape and mass of space debris.

Optical observations of geostationary objects have been made by Communication Research Laboratory (CRL), utilizing a 1.5 aperture telescope. Up to present, only known geostationary satellites were detected and no smaller, unknown

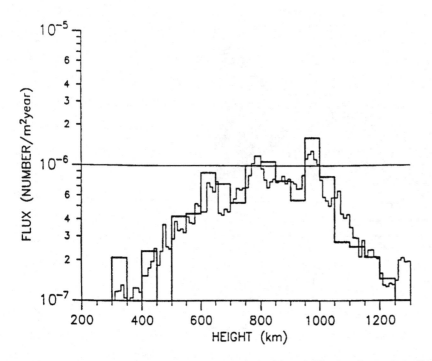

Figure 3: Height distribution of the debris flux derived from MU radar observations (thick line) and from the USSPACECOM catalog.

objects were found. The use of other telescopes in Japan is also found promising. For more detailed and long lasting observation, however, a devoted debris observation system will be required.

3.2 Causes of space debris

Among the different types of orbital objects, the largest number is that of fragment debris created by explosions. These are the result of rocket upper stages explosions, intentional destructions and other unknown factors. The nature of the latter is not quite clear, but is believed to include explosions caused by hypervelocity impacts. Up to present, none of Japanese space projects are known to have created a large amount of debris. However, considering past mission failures concerned with the upper stage motor collision and abnormal motor burning, some debris have been released from Japanese activities. In these cases, it is considered that the debris creation took place at high altitude so that the debris could not be sensed by ground observation, or that the debris did not stay in orbit long enough to be detected. In order to avoid significant accidental debris creation in the future and to maintain Japan's clean record, intensive studies should be performed to determine the debris creation possibilities inherent in the present and future space activities, and to ascertain any appropriate means to eliminate them.

3.3 Modeling technology

Modeling technologies are important in design specification, design cost reduction, future hazard forecasting, and mitigation plan establishment. The environment model describes the collisional environment as a function of altitude, inclination, time and solar activity, even for those areas of space not described by observation data. The LEO environment model which has been extensively used by spacecraft community is obtained by Kessler et al[8]. In Japan, GEO environment has been analyzed in detail in NTT laboratory based on known object data by US sources. The peak density in GEO was found larger than previous analysis by one order of magnitude, when the analysis resolution was made finer[6]. Various fragmentation models have been developed in the USA based on orbital collision/explosion observations and hypervelocity impact tests. The characteristics of debris released as a result of breakup can be estimated by these models and the results may be applicable to estimations of eventual environment change.

The main research area to be pursued at present are reliable estimation of debris ejection initial parameters under various impact parameter, reliable debris distribution data in smaller size region, and the long term effect of orbit evolution by various natural forces such as gravity, solar pressure and atmospheric drag.

3.4 Protection technology

The space debris protection systems have been studied for the design of the International Space Station Freedom. In Japan, the debris related activities of the National Space Development Agency (NASDA) are concentrated in this field. Because data accumulation by the design group itself is vital for design purpose, this activity should be continued. However, its achievements are still very limited compared with those of the USA and Europe.

Hypervelocity impact experiments are important not only for the space station protection system design but also for the development of basic hypervelocity impact science and of understanding debris creation and dispersion phenomena. Tests by use of both a two stage light gas gun and a rail gun are being planned at the National Aerospace Laboratory (NAL) in collaboration with various organizations.

Software for impact analyses is in progress. Such software packages are important design tools and will play even more important role in the future. It will be vital to develop more effective codes for wider applications, and it will be possible to do so by combining work with experimental efforts.

4. CONCLUSIONS

Orbit environment conservation with respect to space debris is indispensable for insuring long lasting and expanding space activities. Various technical and legislative proposals have been made so far and they would necessitate, more or less, additional cost and reduction of space systems capabilities. These potential penalties make space organizations reluctant in considering efficient policies against space debris. However, history will prove that earlier effort results in less expense. What is important now is to evaluate various measures, to define associated penalties expected and to select acceptable cost effect measures. These tasks could be carried out by the respective countries or groups, but would best be dealt with by representatives of all. Execution of the selected measures will only be carried out under international understandings.

REFERENCES

1. M.Nagatomo, H.Matsuo and K.Uesugi, "Some Considerations on Utilization Control of the Near Earth Space in Future," Proc.9th ISTS, Tokyo, 1971, pp.257–263.
2. Space Debris Study Group Interim Report, JSASS, Jan.1992.
3. Report on Orbital Debris by Interagency Group, Jan.1989, Washington, D.C.
4. D.Kessler, "Collisional Cascading : The Limits of Population Growth in Low Earth Orbit," MB 2.2.2, 28 COSPAR, The Hague, 1990.
5. P.Eichler and D.Rex, "Chain Reaction of Debris Generation by Collisions in Space-A Final Threat to Spaceflight?" IAF-89-623, 40th IAF, Malaga, Oct.1989.
6. T.Yasaka and S.Oda, "Classification of Debris Orbits with regard to Collision Hazard in Geostationary Region," IAA-90-571, 41st IAF, Dresden, Oct.1990.
7. T.Sato, K.Ikeda, T.Wakayama and I.Kimura, "RCS Variations of Space Debris Observed by the MU Radar, " Proc. of Space Debris Workshop '91, Sagamihara, JSASS and ISAS, Nov.1991, pp.9–16.
8. D.Kessler, R.Reynolds and P.Anz-Meador, "Orbital Debris Environment for Spacecraft Designed to Operate in Low Earth Orbit," NASA TM100471, April 1989.

17: Orbital Debris Minimization and Mitigation Techniques

Joseph P. Loftus, Jr., Philip Anz-Meador and Robert Reynolds

Lyndon B. Johnson Space Center, Houston, TX 77058

ABSTRACT

Man's activity in space has generated significant amounts of debris that remain in orbit for periods of sufficient duration to become a hazard to future space activities. Upper stages and spacecraft that have ended their functional life are the largest objects. In the past, additional debris has been generated by inadvertent explosions of upper stages and spacecraft, by intentional explosions for military reasons, and possibly by a few breakups resulting from collisions. In the future, debris can be generated by collisions among spacecraft as the number of orbital objects continues to grow at rates greater than natural forces remove them from orbit.

There are design and operations practices that can minimize the inadvertent generation of debris. There are other design and operations options for removing objects from space at the end of their useful service so they are not available as a source for the generation of future debris. Those studies are the primary concern of this paper.

The issues are different in the low Earth orbits and in the geo-synchronous orbits. In low Earth orbit, the hazards generated by potential collisions among spacecraft are severe because the events would take place at such high velocities. In geosynchronous orbit, the collision consequence is not so severe, because the relative velocities are low – less than 1 km/s. But, because of the value of the limited arc and the extremely long orbital lifetime of the satellites, it is necessary to remove any debris generated in the orbit to a different orbit at the end of life if it is not to be a hazard to future operational spacecraft. The issue at present seems to be how high the reboost maneuver must be and what the system design and maneuver strategy should be to ensure its effectiveness.

The most economic removal of objects is achieved when those objects have the capability to execute the necessary maneuvers with their own systems and resources. The most costly option is to have some other system remove the spacecraft after it has become a derelict. Numerous options are being studied to develop systems and techniques that can remove spacecraft from useful orbits at the end of their useful life and do so for the least mass penalty and economic cost.

INTRODUCTION

At the end of 1990, approximately 2,000,000 kg of man-made materials were in orbit, and about two-thirds of which is in low Earth orbit below 5000 km altitude, or an orbital period of 201.31 minutes. Table 1 illustrates the characteristics of all the activities that have taken place in space up to October 1991. The categories are Fragment debris which originate from breakup events, Operational debris which are generated in normal operations and include such items as lens covers and separation devices etc., Rocket bodies which are the upper stages of the launch vehicle used to place the spacecraft in orbit, and Payloads which are the spacecraft. The column Other includes objects in orbits other than those in previous columns, the row Other is miscellaneous objects such as planetary probes and some of the Project West Ford needles. Table 2 illustrates the characteristics of the current Earth orbit population. To date we believe that the breakup events have been due mostly to propulsion or electrical battery explosions or military tests; but there are many cases in which the cause of the event that generated the fragment debris cannot be ascertained.

Figure 1 illustrates the present earth orbit population of satellites as a flux versus altitude. The peaks in the flux indicate the operationally favored altitudes. There are sharp peaks at the geostationary altitude favored by communication satellites and the semisynchronous altitude favored by navigation satellites. The highest density is in the altitude regions characteristic of the sunsynchronous orbits. One of the reasons this region is so densely populated is that it is the altitude band in which the largest number of breakup events has occured.

Table 3 indicates the most recent of the breakup events. The interval from August of 1989 until October of 1990 in one of the longest intervals without a breakup. The most recent event was the breakup of a Delta second stage that had been used to launch the Nimbus 6 spacecraft on July 12, 1975. The suspected cause of the breakup was corrosion and thermal cycling induced failure of the common bulkhead leading to mixing of the residual hypergolic propellants. This phenomenon was first recognized in 1978, and since that time residual propellants have been burned to depletion. There have been no breakups

Summary Catalog of all objects launched into space (8 October 1991)

	Low Earth Orbit (below 5000KM)	Medium Earth Orbit (18K - 25K KM)	Geosynchronous Orbit (30K - 41K KM)	Medium Transfer	Geo Transfer	Molniya	Supersynchronous	Other
Fragments	7158/32.93	0/0	0/0	9/.04	16/.07	32/.14	1/0	19/.08
Operational Debris	5776/26.57	2/0	2/0	112/.51	217/.99	31/.14	1/0	61/.28
Rocket Bodies	2682/12.33	16/.07	111/.51	59/.27	174/.80	187/.86	8/.03	11/.05
Payloads	3597/16.54	69/.31	322/1.48	23/.10	46/.21	184/.84	45/.20	17/.07
NCE/NEM	0/0	0/0	0/0	0/0	0/0	0/0	0/0	529/2.43
Other	0/0	0/0	0/0	0/0	0/0	0/0	0/0	219/1.00

Total number of all cataloged objects 21736
Other consists primarily of interplanetary probes
NCE/NEM = No Current Elements/No Elements Maintained
Cell entries are the number of objects/percentage of the population

Table 1

Summary of On-Orbit Cataloged Objects

	Low Earth Orbit (below 5000KM)	Medium Earth Orbit (18K - 25K KM)	Geosynchronous Orbit (30K - 41K KM)	Medium Transfer	Geo Transfer	Molniya	Supersynchronous	Other
Fragments	2607/37.29	0/0	0/0	4/.05	13/.18	32/.45	0/0	2/.02
Operational Debris	837/11.97	2/.02	1/.01	64/.91	68/.97	21/.30	1/.01	37/.52
Rocket Bodies	612/8.75	16/.22	111/1.58	44/.62	119/1.70	128/1.83	6/.08	7/.10
Payloads	1177/16.83	69/.98	322/4.60	14/.20	40/.57	126/1.80	25/.35	13/.18
NCE/NEM	0/0	0/0	0/0	0/0	0/0	0/0	0/0	338/4.83
Other	0/0	0/0	0/0	0/0	0/0	0/0	0/0	135/1.93

Number of objects currently on orbit 6991
Other consists primarily of interplanetary probes
NCE/NEM = No Current Elements/No Elements Maintained
Cell entries are the number of objects/percentage of the population

Table 2

of stages that have depleted propellants and pressurants. The last previous Delta second-stage explosion was in 1981. Only one unvented Delta stage remains on orbit, the Nimbus 7 rocket booster.

It remains a significant source of concern that, despite the adaptation of a number of mitigation procedures by all the spacefaring nations, there has been no appreciable reduction of the rate of breakup events.

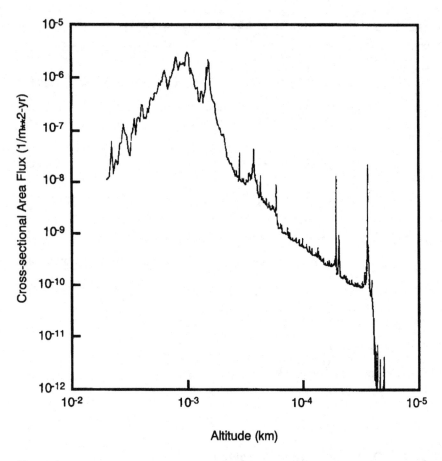

Figure 1

At typical collision velocities, a 3 mm fragment with a mass of 0.04 gm could cause an explosion if it hits a pressurized container such as a propellant tank. A 5 cm fragment with a mass of 0.5 kg or any larger fragment could cause fragmentation of a entire spacecraft. It is this consequence of the high intersection velocities in LEO, characteristically 10 km/s, that makes mitigation procedures urgent. At such velocities, millions of fragments result from a single event and hundreds of these fragments are large enough to generate another such event.

Table 4 indicates the estimates of the size and mass characteristics of the debris in LEO. There is significant uncertainty in these values because they are derived from a limited number of tests. Johnson and McKnight assess that this estimate is high by 50% (4). Efforts are currently in progress to improve our estimates, both by making special radar observations of the objects on orbit and by doing g more extensive explosion and hypervelocity impact testing and modeling. The number of objects, however, is quite large even if the lower values are used for each of the size ranges.

Kessler (5) has analyzed the consequences of the accumulation of satellites in preferential regions and has defined the concept of critical density. Critical density is achieved when a population is of sufficient size in an orbit of sufficiently long life that the population will produce fragments from random collisions at a rate that is increasing and is greater than the removal

rate due to natural processes. Figure 2 indicates his estimate of the critical density value in relation to the spatial density of the cataloged objects as of December 1989. By this estimate the regions at 900 to 1000 km and 1450 to 1650 km are at critical density. An alternative estimate would indicate that the population might increase by a factor of two or three before critical density is attained. Eichler and Rex have reached similar conclusions (2). If the critical condition is not upon us already, it will be in a few more years at present launch rates.

As part of the modeling studies used to define the environment, the Johnson Space Center has developed a model called EVOLVE which is used to predict the future environment. In figure 3 a family of monte carlo runs were made to assess when and how frequently collisions among satellites might occur. The darker line is the mean expectation, the lighter lines particular time histories. Since such collision events are random there can be significant uncertainty as to when the first and each subsequent event will take place.

Figure 4 illustrates the mean value of the expected surface flux for three case of such EVOLVE studies. Case 1 is business as usual with historical explosion rates and launch rates. Case 2 illustrates the effect of having no further explosions after the year 2000 on the basis that mitigation efforts are effective. This delays the onset of population growth due to collisions but not for very long. Case 3 introduces the effect of deorbiting stages after the year 2000 and all spacecraft after the year 2030 and as

Recent Fragmentations
Revised 18 September 1991

Launch Date	Event Date	Parent	Breakup Height (km)	Apogee Height (km)	Perigee Height (km)	Inclination (deg)	Large Fragments Fragments Cataloged	Large Fragments Fragments In Orbit
6-12-75	5-1-91	Nimbus 6 Delta 2nd stage	1088	1102	1093	99.9	226	>226
2-12-91	3-5-91	SL-8 2nd stage	1560	1750	1470	74	60	60
12-29-83	2-4-91	SL-12 propellant tank?	18500	18800	330	52	0	>2
12-1-90	12-1-90	USA 68	850	850	610	99	27	7
10-1-90	11-30-90	Cosmos 2101	210	290	200	65	3	0
9-3-90	10-4-90	Long March 4A 3rd stage	900	900	890	99	75	72
7-18-89	8-31-89	Cosmos 2031	270	370	250	51	8	0
7-12-89	7-28-89	Cosmos 2030	150	220	150	67	0	0

Table 3

Estimated Population of Debris

Particle Size	Number	Percentage	Mass (kg)	Percentage
>10 cm	7,000-15,000	0.2	2,998,900	99.5
1-10 cm	35,000-150,000	2.0	1,000	0.05
0.1-1 cm	3,000,000-40,000,000	97.3	100	0.005
Total		100.0	3,000,000 (~2,000,000 LEO)	100.0

Table 4

can be noted these practices not only protect the environment but over time improve it. In figure 5 case 2 is replicated to serve as a reference. Case 4 is business as usual but with an increase in the launch rate of five launches each year as new entrants develop launch capability. Case 5 illustrates the effect of the increased level of activity but using environmental management measures to protect the operational environment.

Thus far, the objective of the discussion has been to present the data that indicate the need to begin active measures to prevent explosions or breakups in space and to remove from space the large objects, most particularly previously useful spacecraft and upper stages, which, if not removed, will be subject to hypervelocity collision and the source of large quantities of future fragments.

DEBRIS ABATEMENT PRACTICES

There are many actions to reduce debris that can be taken at minimal cost to space operations. Items traditionally discarded can be retained by being attached to the satellite so they do not become independent hazards, e.g., lens covers, etc. Pin pullers can be used instead of cable cutters; bolt catchers and lanyards can retain objects. Retention of the apogee kick motor will have some opportunity cost in the attitude control system, since it will increase the inertial mass of the spacecraft.

Circuit protection can be provided so that batteries are protected against both internal and external short circuits, which could cause the battery to rupture.

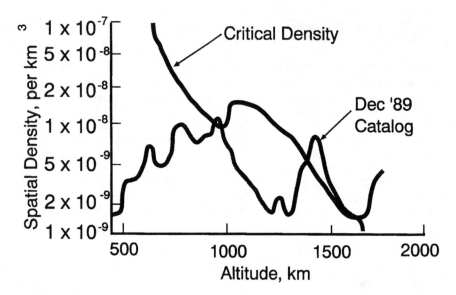

Figure 2

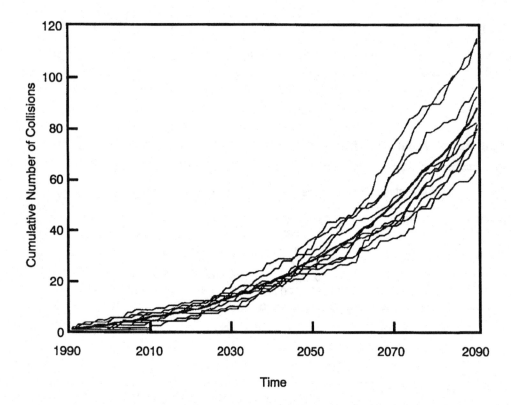

Figure 3

Upper stages can be vented as is now done with the Delta and Ariane stages placed in sunsynchronous orbits. The Delta second stage is burned to depletion after placement of the payload and execution of the collision avoidance maneuver. Ariane uses pyrotechnically actuated vent valves to deplete all propellants and pressurants. The Japanese H-I and H-II upper stages provide for an idle mode firing of the engine to deplete propellants after payload separation. The PRC are reviewing design and operations for the Long March following the October 1990 breakup of their upper stage

Testing of weapons can be done in orbits with sufficiently low perigee so the orbital lifetime of any resulting debris will be brief. Similarly, spacecraft that are to be destroyed for national security reasons can be maneuvered to an orbit with a low perigee prior to detonation so the resulting debris will enter rapidly.

While such practices will prevent adverse near-term environmental effects, they do not reduce the long-term threat. To control the environment for the longer-term objects must be removed from space at the end of their useful life so they do

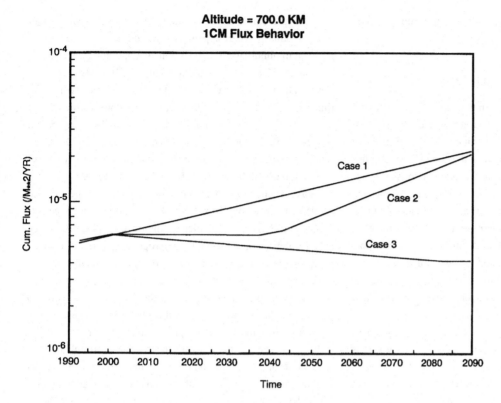

Figure 4

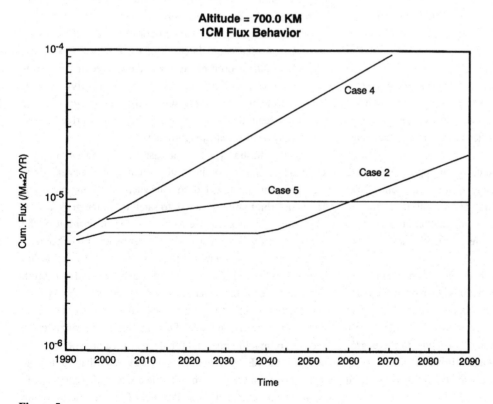

Figure 5

not become the source of debris proliferation through subsequent collisions.

DEBRIS REMOVAL TECHNIQUES

Table 5 lists the methods of exploiting natural forces or spacecraft systems design to remove spacecraft and stages from orbit at the end of their useful life. Since the system used to remove a spacecraft or a spent stage at the end of its useful life is parasitic, that is, not contributing to immediate mission success, it is desirable to minimize both the absolute cost and the opportunity cost of such systems. In assessing each of these options, cost is a primary consideration; but the reliability, or expectation that it will in fact effect its function, is also important. Not all of the techniques listed are of comparable maturity, but some of the more exotic are listed because potentially their costs could be least if the technology were in hand and execution routine.

In all disposal options the criterion measure applied will have a great influence on the methods used. If removal means that the object must reenter on the next passage through perigee, the penalty can be significant. If the criterion is that the object reenter within ten percent of what would have been its orbit lifetime, the reentry initiation will be less of a penalty and this may be an adequate level of precaution but we have not yet concluded the analyses as to what the effect of such variance in the criterion measure signifies it the long term.

SELF DISPOSAL OPTIONS

Any of the techniques listed can be used by a spent stage or a spacecraft for initiating its own early entry. There are clearly different costs for the implementation of any of the options as a function of the design of the particular system to which augmentation is applied.

Propulsive methods are most suitable for upper stages since they have large, efficient propulsion systems and some flight performance reserve residuals. Drag or tether systems may be more suitable to spacecraft that have minimal propulsion capability. The two most useful "figure of merit" measures for assessing a particular system are the marginal mass cost of implementing the system and the direct system acquisition cost. The marginal mass cost is either mission payload or duration displaced to accommodate the removal system. The direct system acquisition cost is that required to develop and apply one of the techniques to a particular spacecraft; to the degree that a "standard kit," could be bought, it becomes less costly and more attractive if the weight is a small fraction of the total system weight.

Propulsive maneuvers. Deorbit with a conventional propulsion system is effective for all orbital altitudes. The direct cost and the opportunity cost are very sensitive to the basic configuration of the system.

For upper stages that have attitude control, large efficient engines, multiple start capability, and are, at the time of disposal, at minimum weight, the propulsion option is generally the best choice. The only capabilities that must be added are additional batteries to assure that the stage has the operational lifetime to initiate its entry into an ocean disposal area, which might not be immediately available after the payload separation sequence. Some additional propellant above the flight performance reserve may be required to assure that the velocity increment to control entry can be achieved before engine cutoff. In a maneuver such as this, one could probably calculate the required propellants on a "depletion" cutoff basis rather than a performance cutoff criterion in order to minimize propellants. Range safety criteria would dictate the appropriate standard. For most systems, the cost of this strategy would be nominal, since the basic systems already exist and are merely augmented. The mass margin cost is less than 1% of the stage mass delivered to orbit, since the penalties are additional batteries, propellant, and mission management time, as opposed to an entire new system.

For a spacecraft that does not have a large propulsion system, the cost can be substantially greater. Figure 6 illustrates the cost to deorbit a spacecraft in LEO with a propulsion system calculated as a fraction of the inert weight to be deorbited. These curves assume that the cost of the disposal system is completely additive to the cost of the basic system; e.g., add a solid rocket motor or a liquid propulsion system solely for the purpose of deorbiting the spacecraft.

Figure 7 shows comparable data for the case of the geosynchronous transfer stage. Because of the effectiveness of the execution of the deorbit maneuver at the apogee of the elliptical orbit, the propellant mass penalty is greatly reduced. However, the mass penalty to sustain the system operation for the hours required to attain that point in the orbit (10–14 hours) may be moderately significant.

Figure 8 illustrates a special case for the objects in geosynchronous transfer orbit. In geosynchronous transfer orbits, the stage can be targeted to subsynchronous or supersynchronous orbit to minimize risk to satellites in the geosynchronous arc. It is also feasible to control the disposition of the upper stages for geosynchronous transfer by control of the time of launch (6). As figure 8 indicates, there is a relationship between the initial Sun angle and the lifetime of the orbit. The influence of the lunar-solar gravity on the orbit is such that it lowers or raises the perigee and, thus, influences the lifetime of the orbit. The perigee of the transfer orbit strongly influences its orbital lifetime, and for stages with a high perigee, the hazard to the LEO environment may be greater than for a stage in a circular orbit at the same altitude because of the longer life of the deeply elliptical orbit. It is not often feasible to accept this constraint in addition to the other constraints, such as range safety, which control the planned time of launch.

It is interesting to note that at altitudes above 25,000 km, the propulsive energy cost to accelerate to escape is less than that for entry; at lower altitudes, entry is a lower cost choice.

Methods To Remove Objects From Orbit

1. Propulsive
   - Upper Stages
   - Payloads

2. Aerodynamic Drag
   - Basic Ballistic Characteristics
   - Enhanced Drag Systems

3. Solar Pressure
   - Solar Sails
     - Escape
     - Entry

4. Tether Techniques
   - Momentum Exchange at Deployment
   - Momentum Exchange at Retrieval
   - Electromagnetic Drag

5. Solar-Lunar Perturbations
   - Time of Launch Constraint
   - Geosynchronous or other deep elliptical orbits

There are no other forces available with which to influence the orbit of satellites.

Table 5

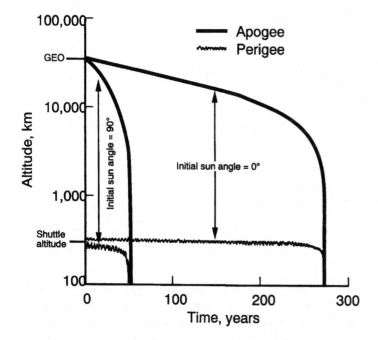

Figure 6

Drag augmentation devices. The most fundamental force that causes spacecraft and space debris to enter the Earth's atmosphere and burn up is the encounter with the rarified atmosphere at orbital altitudes. The forces are small because the mean free path of molecules is large, but the force is always present and its effect integrates constantly over time and grows larger as the altitude declines. The magnitude of the force at any given altitude is a function of the 11-year solar cycle and its differential heating of the atmosphere.

The purpose of drag augmentation devices is to create a much larger drag area for the mass to be decelerated so the effects will occur more rapidly (as shown in figure 9). The incentive is that the less time the "hard body" occupies critical regions, the less risk there is of a bad consequence. By implication, this requires that the drag device be "benign" in any encounter with a functional spacecraft; i.e., the functional spacecraft may be perturbed, but the device will be so fragile that it will experience the bulk of the damage.

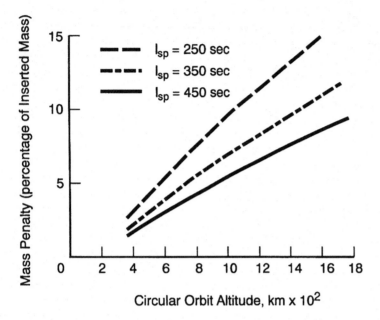

Figure 7

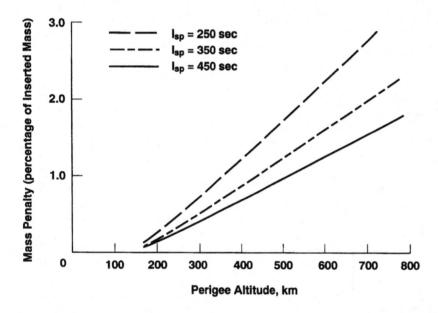

Figure 8

Design for minimal weight and strength helps to achieve this effect.

A number of devices can be considered, for example, a balloon such as the ECHO satellite used as a radio frequency reflector in the early 1960's; or a more efficient drag configuration, which, at a potential increase in fabrication cost, might have a lower mass penalty. A consideration in the configuration and the sizing of such a device is to recognize that the satellite and its augmentation device will "sweep out" a volume of space comparable to that which the spacecraft would have transited over a longer time. While only five percent of the objects the system could encounter are active satellites, it would be well to design the system so that such encounters would have a low probability of damage to a functioning unit.

If there is no propulsion system, then the mass and cost penalty of a drag augmentation device may be near minimum for most applications. A 100 to 1 increase in area might well be accomplished for 2% to 4% of the mass.

Such analyses indicate that balloons would be impracticably large for initial altitudes above 750 km. In all cases, the effectiveness of such drag-augmentation devices varies over the course of the 11-year solar cycle and is greatest during the period of maximum solar activity and least during solar activity minimum.

Tether systems. The use of electromotive tethers may well be competitive with drag devices as passive and low cost systems to deorbit spacecraft. Tethers in their electromotive force application can be used to convert satellite motion into the genera-

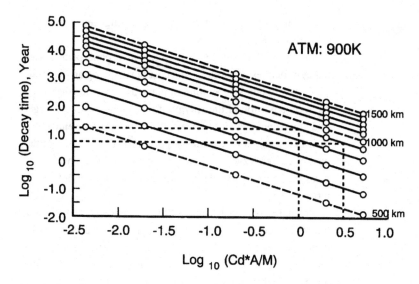

Figure 9

tion of electromagnetic current. A ten km length of 12-gauge wire with an ion gun, hollow cathode, or large area conductor at each end has a mass of about 200 kg. Passive contactors are less efficient than hollow cathodes but are not life limited by the supply of working fluid. One of the attractions is that this system is not dependent upon atmospheric density, so they are effective above 700 km. The efficiency of such electromotive systems is dependent upon the inclination of the orbit relative to the earth's magnetic lines of force. It is useful at low inclinations but not efficient in high inclination orbits. (9)

Further, just as for drag devices there is an orbital debris hazard. A 10 km tether could be severed by a 0.5 mm meteoroid or debris object in less than the time required to effect the deorbit maneuver. In addition, it would represent a collision cross-sectional area of more than 100 square meters to other operational spacecraft.

Comparison of propulsion package and drag devices. Drag devices, both aerodynamic and electromotive, and the propulsion package, have different advantages. Each system would reduce the time in orbit for inoperative satellites and spent stages and thus would decrease the chance of an internal explosion or a random collision. One drawback of the drag device is that the decrease in collision probability resulting from shorter orbital lifetime is offset to some degree by the increase in cross-sectional area. The satellite alone and the satellite with the drag device attached would each sweep out the same volume of space over the course of its time in orbit; however, debris impact on the balloon would not generate hazardous high-density debris particles.

Although there is yet no complete analysis to support such a conclusion, it appears to us that the same considerations would apply to tether systems; that is, with the increased collision area due to the tether, there is no obvious way to make an inadvertent encounter between the tether and a functional spacecraft "benign." The advantage of the drag device is that it is simple,

passive, and requires no attitude control system. For altitudes below about 700 km, drag devices appear to be a lower mass alternative to propulsion packages. The disadvantage of the drag device is that it is a mass penalty to the mission objective performance of the spacecraft. For spacecraft which have orbit maneuver capability, the marginal increase in propellant mass to lower the perigee and limit orbital lifetime may be substantially less than the mass of a drag device. For spacecraft in high altitude orbits, propulsive lowering of the perigee and a small drag device may offer practical advantages.

These analyses are based upon the assumption that the drag device is a spherical balloon. At the cost risk of a more complicated geometry, they could be conic sections similar to aircraft landing deceleration parachutes. The increase in drag efficiency might warrant such cost.

ACTIVE RETRIEVAL AND DISPOSITION

One approach for the removal of large debris objects is to collect them with a maneuverable space vehicle. In the evaluation of this approach, it was assumed that rendezvous would be accomplished with an autonomous or remotely controlled vehicle such as the Orbital Maneuvering Vehicle (OMV), recently under development and study by NASA. Assuming the OMV can grapple the target spacecraft, there are several options for disposition. The OMV can perform a deorbit maneuver, separate from the object, and reinsert itself in orbit while the discarded object enters the atmosphere. Another option is to station the objects at some location where they can be maintained together in a safe orbit and possibly salvaged for spare parts or raw materials.

Figure 10 shows the altitude range for OMV retrieval as a function of apogee and perigee height and the mass of the object being retrieved. This analysis is based on the assumption that the OMV starts in a circular orbit at an altitude of 500 km and returns to that orbit with the retrieved debris. It is also

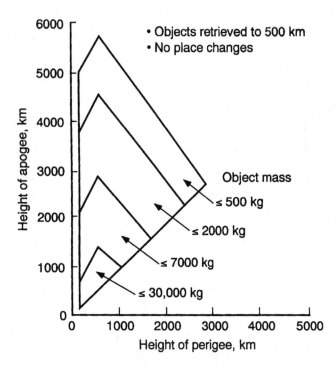

Figure 10

assumed that the OMV makes no propulsive plane changes. The range of the OMV becomes progressively smaller as the mass of the target object increases.

The relative performance cost for deorbit versus collection depends on the mass of the object and its orbital altitude. For objects in LEO having masses less than 2000 kg, collection in orbit is less costly than deorbit in terms of OMV performance. A third alternative is to rendezvous with an object using the OMV and then attach a separate deorbit device to the object, rather than using the OMV for propulsion. The attached device might be a deorbit propulsion package or a drag-augmentation device. Attaching devices rather than maneuvering the objects with the OMV would expand the envelope of accessible objects.

Several concerns about using the OMV for debris recovery should be noted. It may be difficult to grapple uncooperative satellites. The satellites may be tumbling; they may have no convenient points to grapple; and some may contain hazardous material. The mission time required for orbit phasing and rendezvous could overtax the power supply of the OMV. Objects at the same inclination as the OMV may not be in the same orbital plane; therefore, the OMV may have to wait while natural orbital precession brings the respective orbital planes into alignment. Propulsive plane changes of more than a few degrees would be impractical, since the energy required is large – exceeding the amount of energy to raise or lower the orbit by many hundreds of kilometers.

Reducing the population of large debris would require the use of several OMVs dedicated to retrieval missions as well as a large number of launches from Earth to deliver and service the OMVs in specific orbit planes. The magnitude of this oper-

ation illustrates the desirability of providing new spacecraft with devices for self-disposal.

Eichler (3) has developed an alternative concept that uses tether principles to minimize the cost of retrieving and deorbiting derelict bodies. Operating in plane as would the OMV, he recovers a portion of the energy of rendezvous with the target body by using a tether to lower its perigee to assure entry and, at tether separation, acquires a portion of the energy for the next target rendezvous. In such a system, there is a cost for the mass of the tether and the capture system it uses to lower the target object.

An alternative employment of the concept might be to use a tether system for the initial deployment of a spacecraft so the deploying upper stage is displaced to a low perigee to ensure its entry. Such practice would eliminate almost half of the debris mass that now enters orbit.

Geosynchronous orbit

Figure 11 illustrates the population of objects in the geosynchronous arc. The objects on the 0 degree latitude line are the actively controlled satellites. The other satellites are communications satellites no longer active, upper stages and apogee kick motors and other uncontrolled objects. These objects begin to oscillate across the equatorial line due to lunar and solar perturbations. The maximum excursion of 15 degrees has a period of 53 years. Eccentricity of the earth will cause altitude variances of +/- 50 kilometers as a function of longitude and the spacecraft will oscillate east and west under the influence of the geopotential variations

Some operators have begun to reboost spacecraft to a higher orbit at the end of the operational life of the spacecraft. Such a maneuver can be motivated by either operator concern to make a given operational longitude "slot" available or as a measure to protect the long term environment. (1, 7) If such maneuvers are to have a long term beneficial effect there is a minimum increase in the perigee that must be assured. If the boost maneuver does not put the spacecraft beyond the operational torus it may in the long term add energy to an eventual collision.

A reboost maneuver to a perigee 300 kilometers plus 2000 kilometers x the area in meters squared/ mass in kilograms will assure that the spacecraft will not reenter the operational torus. Even this separation will not protect the operational torus if the spacecraft explodes in transit to the higher orbit or at any later time.

A more significant issue is the potential for explosions or other breakup events in the geosynchronous region. Recently two explosion events have been reported. An Ekran satellite was observed to explode during a recharging cycle on a nickel-hydrogen battery in 1978. A Transtage was observed to breakup for an unknown cause in February of 1992. Because of the distance of the geosynchronous arc from the sensors objects less than a meter in diameter are not discernable. Because of the limited field of view of the sensors a breakup event would not be detected unless there were some reason to

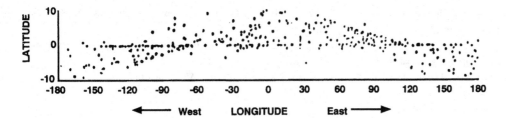

Figure 11

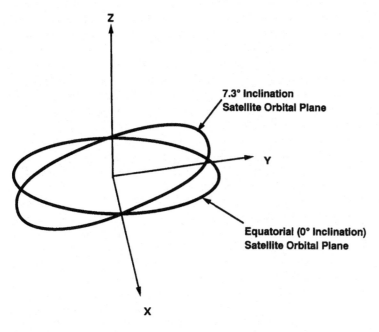

Figure 12

observe the object or one accidentally was examining the location where the event occurred. If breakup events occur in geosynchronous orbit at rates comparable to LEO we would expect that there have been others that have not been detected.

The experience suggests that there need to be the same mitigation measures applied in GEO that are already in use in LEO.

There is an alternative geosynchronous orbit that could be used, figure 12. At 7.3 degrees inclination and a right ascension of the ascending node of zero degrees there is a "stable orbit" in which spacecraft North-South station keeping is not required. Obviously use of this orbit implies that the ground station either must track the satellite or have a sufficiently large aperture that tracking is not required. In this orbit there is essentially no collision risk among spacecraft since the relative velocity is always less that 5 meters/sec as opposed to the 800 meters /second that can develop in the equatorial orbit. Both satellites in the stable orbit and the equatorial orbit would be less at risk if some assets were placed in each.

SUMMARY

Environmental management is as important in space opera-

tions as it is in any other human endeavor. In space, as on Earth, prevention of pollution is much more cost-effective than remediation. The level of space activity is now such that active intervention techniques are necessary. It is important that the participants in space design and operations take actions in design and operations to manage the environment so future operations can continue to be cost-effective. For management of the environment to be effective, all those who operate in space must exercise common and cooperative efforts.

REFERENCES

1. Chobotov, V. and Wolfe M. G., End of Life Disposal of Spacecraft in Geosynchronous Orbit. IAA-89-631, 40th IAF Congress, Malaga SP.
2. Eichler, P. and Rex, D., Chain Reaction of Debris Generation by Collisions in Space – A Final Threat to Spaceflight, IAA-89-628, 40th Congress of the International Astronautical Federation, 1989 Malaga SP.
3. Eichler, P., Orbital Debris Removal Using Tethers: The Advanced TERESA Strategy, NASA/TUBS Joint Orbital Debris Working Group Meeting, April 11–12, 1991.
4. Johnson, N., and McKnight, D. S. (1989), An Evaluation of the Mass and Number of Satellites in Low Earth Orbit. Paper presented at CNES Symposium in Nov. 1989.

5. Kessler, D. J., Collisional Cascading: The Limits of Population Growth in Low Earth Orbit, Advances in Space Research, Vol. 11, No. 12 (in press).

6. McCormick B. in Orbital Debris from Upper Stage Breakup, AIAA, Progress in Aeronautics and Astronautics, Vol. l.121, Washington, DC, Loftus, J. P., Jr., Ed. 1989.

7. Perek L., Safety in Geostationary Orbit after 1988, IAA-89-632, 40th IAF Congress Malaga SP.

8. Report on Orbital Debris by Interagency Group (Space), For National Security Council, Washington, DC, Feb. 1989.

9. Tethers in Space Handbook, NASA, August 1986

IV. Economic Issues

18: In Pursuit of a Sustainable Space Environment: Economic Issues in Regulating Space Debris

Molly K. Macauley

Senior Fellow, Resources for the Future, 1616 P Street, N.W., Washington, D.C. 20036

This paper was prepared for "The Preservation of Near-Earth Space for Future Generations," a symposium for the centennial celebration of the University of Chicago and the International Space Year and under the sponsorship of the John D. and Catherine T. MacArthur Foundation and the Midwest Center for the American Academy of Arts and Sciences. Comments from Ted Glickman, Tom Rogers, Paul Uhlir, and Symposium participants are greatly appreciated. Responsibility for errors and opinions rests with the author.

As the other papers at this conference note, most experts appear to agree that current levels of space debris are presently manageable. The experts caution, however, that the rate of debris growth, estimated to be doubling roughly every decade, could render many orbital locations unusable within the next decades.[1] Because the impact of space debris seems to loom largest in the future, policy towards the environment of space – as the conference theme puts it, "preserving near-earth space for future generations" – must take a long-run perspective. Such a perspective invites comparison of the issue of space debris with the issues of pollution and one of its broader contexts, sustainable development on earth. By reasoning by analogy, perhaps some light might be shed to illustrate how decisionmakers can conceptualize and address debris issues. In particular, innovative approaches to pollution control and sustainable development, including marketable permits for air pollution, debt-for-nature exchanges, and transactions to commercialize tropical biodiversity, suggest a newly emerging social willingness to experiment with economically oriented strategies.[2] This paper echoes this theme in highlighting economic-based strategies for mitigating space debris.

SUSTAINABLE DEVELOPMENT AS A PARADIGM FOR ADDRESSING SPACE DEBRIS

As a concept for preserving earth's resources, sustainable development eludes precise definition. It is generally taken to mean something along the lines of no net loss over time in the global stock of human and natural capital associated with environmental quality, atmospheric integrity, natural resource adequacy, biodiversity, and other desiderata. Such development would "meet the needs of the present, but not compromise the ability of future generations to meet their own needs."[3]

An analogous concept pertaining to the preservation of space for future generations might be a "sustainable space environment" in which present-day space activity is carried out to benefit the current generation, but levels of debris and other environmental impacts are moderated just to the point where future generations are not unduly compromised. Of course, ascertaining that point is quite challenging; perhaps most difficult is that it requires us to presume to know the preferences of future generations and to make judgments involving moral, legal, and economic values about these preferences. Just as sustainable development doesn't require the cessation of polluting activities, however, a sustainable space environment doesn't necessarily require zero debris. That is to say, some amount of debris may be endurable. Moreover, this endurable amount may be large or small depending on whether there are techniques to offset harm. An example is if the present generation were to develop new debris shielding techniques, debris collectors, or other alternatives which could mitigate the effects of debris for future generations, such that on net the present generation's additions to debris do not compromise the future.

Before proceeding, it is key to note that a fundamentally different interpretation of a sustainable space environment (and this interpretation has a counterpart in the context of sustainable development on earth) might argue against any debris generation unless it is fully cleaned up to return the space environment to a pristine form. In other words, debris generation would be permitted only if its effects are fully reversible. This school would contend that it is in some sense immoral, unfair, or both to pollute, to presume to know the preferences of future generations, or to engage in choosing a discount rate with which to link present and future activity.

This paper assumes the former interpretation of a sustainable space environment – that there is a liveable amount of debris, defined as the socially optimal amount given all of the benefits and costs that accrue from debris-generating activities. This approach agrees in spirit with much of the current public debate that argues for debris mitigation, but it extends debate to emphasize that questions such as how much to mitigate and at what cost should also figure prominently in debate. Costs of mitigation include not only the direct costs of the mitigating actions that are taken, but also the costs of administering, enforcing, and monitoring compliance with any mandatory requirements and related regulations. In turn, these costs need to be balanced against the benefits of debris mitigation. Weighing these costs and benefits would also indicate the desirability of alternatives to reducing debris, such as shielding spacecraft, pursuing new techniques to clean up debris or in other ways adapting to debris, or some combination of debris-reduction and -adaptation actions.

OUTLINE OF THE PAPER

After discussing the challenge of specifying a liveable amount of space debris, the paper considers the present status of debris, a "debris or not debris" situation insofar as the probability-weighted expected value of spacecraft loss may not presently be large enough to focus attention on debris. For example, estimates of the probability of a geostationary communications satellite being ruined by debris average about 1 in 1000 during the expected life of the spacecraft. Multiplying this probability by the cost to replace a typical satellite gives $500,000 as one measure of the expected loss. This amount is so small that most satellite owners do not worry about the impact of debris. This section of the paper also argues, however, that while the low expected loss values may accurately represent private losses, they may underestimate social losses because of externalities attributable to the technology of debris proliferation (collisions can beget so-called "cascading" amounts of debris) and other factors. For instance, by the year 2000, the estimated probability in the preceding example may rise to as high as .4, a four-hundred fold increase, due to the proliferation of debris.

Next sections of the paper then address whether an ounce of debris prevention is worth a pound of cure in that some actions other than reducing debris may be effective alternatives, and discuss regulatory options for debris regulation, ranging from voluntary actions to economic-based incentive strategies. Economic-based strategies include debris-related launch fees, "deposit-refund" mechanisms for controlling debris, and performance bonds for debris control. The concluding section discusses a "three musketeers' phenomenon," that is, the likelihood of requiring virtually global consensus on approaches to debris, as activities by one country alone are likely to be insufficient – and perceived as unfair – as a debris control strategy.

A "LIVEABLE" AMOUNT OF DEBRIS[4]

If individuals were asked what the most desirable amount of space debris is, the inclination of almost everyone is to say "zero."[5] But debris is a byproduct of activities that provide benefits, including satellite communications that enhance the quality of life; remote sensing that provides weather information, monitors the quality of the environment, and contributes to national defense; and interplanetary exploration and scientific investigation that augment our stock of fundamental knowledge and understanding. Indeed, unless we are willing to live with some debris, space activities would virtually have to end, insofar as debris can be generated merely by the accidental collision of a single spacecraft with naturally existing debris such as micrometeoroids.[6] In addition, controlling the amount of debris is not free, in that valuable resources are spent on debris-control devices and actions. By permitting some debris, the money not spent on debris control can instead be spent on other space-related research or exploration or other activities.

However, the problem still remains of ascertaining and controlling the "endurable" amount. Generally speaking, in the case of pollution, when it is unregulated, polluters will pollute excessively, enjoying the benefits of polluting activities but not realizing the costs (since they are borne for the most part by third parties – parties other than the polluters). Interestingly, the situation of space debris is somewhat different from this more general pollution problem. The difference is that debris generation is more likely to create mutual harm – that is, not only others but the generator as well may be impacted by the generator's debris.[7] Another aspect of this mutual harm is that shielding or other actions taken to protect a given spacecraft from debris can also serve to reduce the additional debris generation from this spacecraft that could occur if the spacecraft were not shielded. The reduction in mutual harm – or, the resulting "mutual benefit" – is not guaranteed, however, because the vastness of space and the way in which debris propagates and migrates through various orbital planes complicate the prediction of who or what will be affected (in fact, this difficulty in impact prediction is partly at the heart of the problem of controlling space debris). Thus, spacecraft owners are probably likely to shield their spacecraft in an amount determined largely if not solely by self-benefit rather than mutual benefit. For this reason, owners may well have incentives to tend to pollute excessively, contributing to the total amount of debris proportionately more than they may expect to benefit from their own efforts at debris reduction.

Because of the costs associated with debris prevention, is there any situation in which the optimal level of manmade debris might be close to zero? Yes, if the benefits of debris never exceed its cost. Such cases do exist in the case of environmental pollution on earth – for example, as Helfand, 1992, notes, DDT in the United States has been completely banned because the costs were thought to exceed its benefit even at a use level of zero. In space, orbits without the natural cleansing effect of atmospheric drag or orbits of particular usefulness may be examples of settings where the desirable amount of debris is close to zero and expenditures to eliminate manmade debris are deemed "worth it."

At the other extreme, is it ever possible that the socially optimal level of pollution is the unconstrained level? Such a case would exist if the benefits of polluting increase at a faster rate than the costs. As Helfand, 1992, comments, this case may describe the common perception of pollution problems in previous periods in history when pollution was considered a necessary consequence of a growing economy. In the case of space debris, this situation characterized the early days of spacefaring.

In most cases, though, public policy acknowledges at least implicitly that completely eliminating pollution would be too costly, particularly with only small benefits achievable for elimination of the very last units of pollution. Similarly in the case of space debris, decisionmakers have generally recognized the desirability of "minimizing" or "reducing" debris rather than eliminating it.[8]

This paper assumes that some amount of space debris is optimal, but that public policy to regulate the extent of and rate at which further debris is generated is desirable. The paper also assumes that the liveable or optimal extent of debris and its rate of growth can be specified, although the social benefit-cost calculus to do so is left as a topic for future research.

ATTRIBUTES OF A SUSTAINABLE SPACE ENVIRONMENT: ON THE BENEFIT SIDE

In addition to controlling the amount of space debris that is generated, are there other social desiderata that might be associated with preserving the environment of space for the future? An additional objective might be to improve adaptation opportunities – that is, the possibilities for accommodating debris growth. These opportunities may include leaving for future generations a legacy of technological innovations to adapt to debris (spacecraft shielding, debris vacuum cleaners). They may also include ensuring that debris growth occurs gradually rather than abruptly, such that future generations themselves can develop techniques for adaptation.[9]

Improving the ability to specify the parameters of debris – such as its rate of growth, location, and other characteristics – might be another objective of a sustainable space environment. Numerous uncertainties are associated with the current state of the art in modeling, measuring, and monitoring the characteristics of debris and its growth rate. For example, it is reported that only relatively large pieces of debris (exceeding 10 centimeters in diameter) are able to be detected and tracked using present-day technology, but millions of smaller pieces are thought to be in orbit. These smaller pieces can pose risks to spacecraft particularly in lower orbits (the usable volume of orbital space declines geometrically and relative inertial velocities increase in lower orbits compared with higher orbits, increasing the consequences of collision).[10] Perhaps more importantly, the impact of these small pieces can generate additional debris. The probability, size, and economic consequences of impacts with these small pieces are extremely chal-

lenging to model and quantify.[11] Are there ways to improve our understanding of these small pieces? More generally, are there ways to improve monitoring of the random component of the debris growth rate and the monitoring and prediction of the endogenous growth rate? Advances in understanding these parameters could significantly increase the ability of present and future generations to adapt to space debris.

Risk-related priority setting is another possible goal of a sustainable space environment. Ranking priorities on the basis of risk might include the identification of types of activities that are most versus least worst in terms of the hazards they pose, in turn dependent on characteristics such as specific orbits, types of debris (e.g., mass), and other particulars. Presumably remediation of the most egregious hazards should be given higher priority, although an exception to this observation may be a situation in which amelioration of several less damaging hazards could contribute as much or more to overall remediation at lower cost than amelioration of the worst hazards. Priorities may range from the removal of the upper stages of rockets, given their large mass and their potential for contributing significant debris, to the venting of excess propellant from these stages to reduce the potential for chemical explosions and the .severity of possible collisions when they occur (because venting removes additional energy in the object).[12]

Finally, some notion of fairness in terms of who wins and who loses, on net, both now and in the future, in debris impact and mitigation is also a likely desirable attribute of space sustainability. Issues of fairness may arise to divide spacefaring and nonspacefaring nations, or developed and developing countries. They may also divide commercial and government entities if, for example, private industry views debris-regulation requirements as adding unfairly to project costs. Also of importance are the relative burdens of the cost of debris impact and the cost of debris control on the commercial space industry. If commercial launch vehicles, payloads, or both, are harmed by debris, then the company can lose revenue and customer confidence, and face higher insurance rates. At the same time, however, debris-control costs increase the cost of commercial activities. What is needed, then, are policies that adroitly balance the benefits and costs of debris (or debris control). Some strategies – generally arguing for flexible approaches – are discussed below.

ON THE COST SIDE

Taken together, the objectives noted above – improved adaptation opportunities, greater parametric debris control, risk-related priority setting, and equitable access – represent a possible set of benefits of debris mitigation that might be undertaken in pursuit of a sustainable space environment. Presumably, however, balance is desired between achieving these objectives and the costs of so doing. The costs of mitigation include the following direct costs: direct costs of the mitigation activity; the costs of monitoring the activity;[13] and, if it is taken in response to regulation, the costs of enforcing the activity. In addition to these more or less obvious costs, related costs are the likelihood and impact of debris in contributing to the self-propagating nature of debris collisions (cascading effects); and the effect of the costs of the mitigating actions or other strategy on the pace and direction of longer-run technological innovation (e.g., if the actions retard or bias the development of new space technologies). These costs represent a mix of costs borne by one company or government engaged in the activity (the costs of the mitigation action) and the costs borne by society more generally (the costs of cascade effects and technological change). Individual governments or companies are likely to take the first set of costs, the privately borne costs, into account but not the second set, the socially borne costs. If the second set is large enough, then governments, industry consortia, or other centralized entities may want to intervene to regulate debris generation. The costs of intervention – administration, policing, etc. – must be smaller than the social costs of debris, however, for the intervention to make economic sense.[14]

DEBRIS OR NOT DEBRIS?

Before focusing on regulatory options for debris mitigation, the potential economic impact of debris warrants some consideration. How large might this impact be? Table 1 offers rough estimates of the expected monetary loss associated with various activities if they failed catastrophically in orbit due to the impact of debris. Two expected loss values are given: "private expected loss" and "social expected loss." One measure, used here, of the private expected loss is the product of the lifetime average collision probability for spacecraft in the orbit in which the activity takes place and the replacement cost of the activity. This calculation assumes that the collision is fatal to the activity and that the replacement cost is approximated by the original cost of the activity (adjusted for inflation).

One measure of the social expected loss is the sum of the private loss and additional costs which would be imposed on society as a whole by the collision. These might include factors such as the contribution of the impact to debris in various locations, such as different orbits or at various longitudinal locations along the geostationary orbit (some geostationary locations are more valuable and populated than others); the contribution of the debris resulting from the collision to the cascading of additional debris; and in the case of, for instance, loss of the shuttle, any programmatic delay due to special investigations or public concerns.

These estimates of probability and replacement costs reflect a host of caveats, but the size of relative (rather than absolute) magnitudes for private and social losses can be illustrative.[15] The relative sizes suggest that private losses, or those values specific to the agents carrying out the activity (e.g., the corporation, the particular science community, NASA) may be significantly smaller than the losses for society at large. As a con-

sequence, private agents confronting only private expected losses may not find it worthwhile to take mitigating actions. For example, boosting a commercial communications satellite out of geostationary orbit near the end of the spacecraft's operating life is currently estimated to consume from a few months to a year's worth of fuel. An upper bound on the equivalent total foregone operating revenue is around $20 million. Would another spacecraft operator likely be willing to induce an operator to boost the expiring spacecraft out of the orbit and to compensate the operator of the expiring spacecraft for foregone revenue? Probably not, because the operator of the expiring satellite faces a private expected debris-induced loss of only about $500,000.

Similarly, the probability of debris-related loss of a shuttle is probably quite small given the shuttle's brief on-orbit duration (although its large physical cross-section increases the area exposed to debris and this increases the collision probability). While the cost of a shuttle flight, including imputed value-of-life estimates for the crew, is on the order of a billion dollars, the private expected loss can be much smaller. In this light it is understandable why debris risks have generally not been considered to be a significant concern in the shuttle program, and why decisionmakers may not have judged extensive protection or avoidance measures to be cost effective.[16]

The social expected loss values for the activities, however, may be quite larger. These social values are some function of parameters such as location, contribution to cascade effects, and program delay. By way of example, the entries for a geostationary satellite during the 1990s compared with those in 2000 suggest what the value of "g" might be for debris impacts during the 1990s. As the collision probability increases from .001 to .4, private expected losses increase from .5 to 200 million 1992 dollars. The difference in these losses, about 200 million dollars, reflects the costs imposed on the future (year 2000) satellites by the continuation of present (1990s) satellite operators (under current launch rates, operating parameters, and debris potential). Thus some fraction of the $200 million could be ascribed to each 1990s satellite to represent its social loss. The fraction should probably be higher for those geostationary satellites in prime orbital locations (e.g., positions within view of the contiguous U.S. or linking the U.S. with Asia or the U.S. with Japan).

To the extent differences between private and social expected losses are indeed large, if the specific parties engaged in the activity are motivated to take actions only to cover their private expected loss values, then the actions may be inadequate to cover the social expected loss value. For example, a company may use insurance to cover on-orbit losses or be willing to incorporate some spacecraft shielding to protect its own payload, but the company may not have incentives to guard against debris generation by attaching lanyards to external components or to ensure that paint does not flake off the spacecraft. Thus regulation may be desirable to address these externalities (effects for which society is uncompensated).

Table 1. *Illustrative Estimates of the Stakes (dollar values are in millions of 1992 U.S. dollars).*

Activity	Lifetime Collision Probability	Replacement Cost ($)	Private Expected Loss Value ($)	Social Expected Loss Value ($)
Geostationary Satellite orbit during the 1990s	.001	500	0.5	0.5 + g (location)
Geostationary Satellite orbit in the year 2000	.4	500	200	200 + h (location)
Hubble Space Telescope	.01	2,000	20	20 + i (location, cascade)
Piloted Shuttle Flight, 7 crew members	x	1,000 + 7(1.6 to 8.5)	x(1,011 to 1,060)	x(1,011 to 1,060) + j (location, cascade, program delay)

Sources: U.S. Congress, Office of Technology Assessment, 1990; American Institute of Aeronautics and Astronautics, 1992; Fisher, Chestnut and Violette, 1989.

IS AN OUNCE OF PREVENTION WORTH A POUND OF CURE?

It is important to note that the types of regulatory policies that are likely to be most desirable are those which would permit the minimization of the sum of debris control and damage costs, thereby allowing the widest range of opportunities to achieve given mitigation goals. In other words, regulatory alternatives should probably not be limited to reducing debris at the source. Rather, alternatives could include a host of other measures such as recycling, process changes, and "end-of-pipe" controls. Examples of these approaches include:

– source reduction: spacecraft design and operation to reduce the potential to explode or break-up; venting of excess propellant; use of lanyards to secure external components; boosting of geostationary satellites into disposal orbits.
– recycling: spacecraft or component capture and reuse.
– changes in production or operation technology: modification of orbital parameters; spacecraft shielding; reducing cross-sectional exposure; incorporation of redundant components.
– end-of-pipe controls: active debris removal; improved and increased monitoring, modeling, and measurement of the debris environment to allow advance notification for debris avoidance.

A regulatory mandate made strictly for any one or several of these options would reduce the number of ways that governments, commercial operators, and other entities can comply to contribute to a given overall level of debris reduction. Generally speaking, a "command-and-control" approach that dictates a single strategy or technological practice is usually the most expensive strategy and typically is less effective than permitting entities to choose a strategy that attains the same end but at lower cost. The cost savings from flexible policies are likely to be larger the greater are differences in compliance costs among space activities and their managers. Some evidence that these differences are large is suggested by differences in the scale and operating parameters of small versus large payloads and launch vehicles, differences between manned and unmanned activities, and differences among U.S., French, and (former) Soviet Union launch practices.[17]

TOWARDS REGULATORY OPTIONS

To summarize discussion so far, overriding principles for debris regulation should be the acknowledgment of some liveable amount of debris (benefits from the activities which produce debris may mean that zero debris is probably prohibitively costly), and flexible policies to minimize compliance costs. How might industry and governments be encouraged to undertake cost-effective strategies? Table 2 lists several strategies each evaluated on the basis of their expected benefits and costs. Following the earlier discussion, the column headings under Benefits are defined on the basis of achieving a sustainable space environment – that is, some optimal amount of debris able to be accommodated by current and future generations of spacefarers. The headings thus indicate (summarizing earlier comments):

(1) improved opportunities for future generations or users of space to adapt to debris;

(2) greater control over the parameters of the debris popula-

Table 2. *Benefits and Costs of Regulatory Strategies*

Strategy	Benefits					Costs			
	Sustainable Space Environment	Improved Adaptation Opportunities	Greater Parametric Debris Control	Risk-Related Priority Setting	Equitable Access	Compliance	Monitoring/ Enforcement	Cascade Effects	Long-run Innovation
Voluntary Actions	+/–	–	–	–	+/–	L	H	H	L
Moral Suasion	+/–	–	–	–	+/–	L	H	H	L
Command and Control	+	+	+	+	+/–	H	H	L	H
Penalties	+	+	+	+	+/–	L	H	L	?
– Linked Compensation	+	+	+	+	+	L	H	L	?/L
Taxes/Fees									
– One-shot	+/–	+/–	+/–	+	–	L	H	H	?/L
– Deposit-Refund	+	+	+	+	+	L	L	L	?/L
Tradeable Permits	+	+/–	+/–	+	+	L	H	H	?/L
Insurance	+	+/–	+/–	+	+/–	L	L	L	L
Performance Bonds	+	+/–	+	+	+/–	L	L	L	L

Key: For benefits, the symbols "+" and "–" indicate likely (+) or unlikely (–) to contribute to the benefit heading the column.
For costs, "H" designates high cost; "L" designates low cost. Thus, a combination of "+" and "L" represent desirable strategies.

tion, including modeling, measuring, monitoring, and predicting hot spots, rates of growth, etc.;

(3) the flexibility to target debris-reducing actions to riskier types of debris; and

(4) fairness in access to the space environment – that is, not unduly compromising the opportunities of non-spacefaring nations, future generations, industry, or other parties.

Column headings under Costs include (again summarizing earlier discussion):

(1) direct costs of complying with the regulation;

(2) the costs of monitoring and enforcing the regulation;

(3) social costs of debris propagation through cascade effects; and

(4) the effects of the regulation on the pace and direction of innovation in space technologies in general and debris reduction technologies in particular.

Entries in the table include these various regulatory activities that might be taken:[18]

(1) Voluntary actions which parties may unilaterally adopt to reduce debris;[19]

(2) Moral suasion, such as exhortations from international or national government groups, industry associations, or others to reduce debris;[20]

(3) Command and control, or mandatory requirements specifying exact technologies and methods by which debris mitigation is to take place;

(4) Financial penalties for debris generation, including compensation that may not be financial but may consist of in-kind resources, such as technology transfer to nonspacefaring nations or to other parties harmed by the debris;[21]

(5) Taxes or fees levied on one stage of the activity, including deposit-refund schemes whereby deposits made, say, on launch are later refunded when components are boosted to disposal orbits, excess propellant is vented, etc.;

(6) Permits to generate up to some specified amount of debris and tradeable among companies or other entities, thus allowing companies to comply flexibly with overall debris reduction goals;

(7) Reliance on insurance markets and liability law;[22] and

(8) Bonds purchased for space activities and redeemable upon proof of compliance with overall debris reduction goals (similar to insurance but specifically linked to debris mitigation actions).

Entries in table 2 are intended merely to indicate how these strategies might operate in furthering benefits at lower cost and in being perceived as "fair." In general, the command-and-control approach is likely to attain these benefits but at a fairly high cost, as command-and- control regulation, by inflexibly mandating specific technologies or procedures, typically fails to allow regulatees to use least-cost compliance techniques in

those cases where compliance could be obtained at lower cost.[23] Voluntary actions and actions taken in response to moral suasion may foster a sustainable space environment but may not be as likely as other strategies to contribute towards other benefits. Most of the strategies are likely to be difficult to monitor and enforce, although deposit-refund, insurance, and performance bonds encourage self-policing in order to secure refunds or obtain lower premiums. In these cases, regulatees may take it upon themselves to devise ways to monitor debris they generate or otherwise prove that they are in compliance with regulatory goals.

The strategies may be perceived as fair or unfair by regulatees on the basis of compliance costs and any other effects which operate to shift the distribution of wealth, perceptions and demonstrations of technological prowess, images of prestige, etc. Thus options such as voluntary actions, moral suasion, command and control, penalties, insurance, and performance bonds are likely to be fair according to regulatees for whom compliance costs and distributional effects are small, and less fair for those who face higher costs and larger redistribution. Approaches such as linked compensation that explicitly permit some form of compensation to regulatees facing higher costs, or deposit/refund and tradeable permit schemes that seek to minimize the cost burden, may be seen as fair. Taxes may be considered unfair unless the tax revenues are redistributed to regulatees or unless fees were graduated according to some generally agreed-on bases.

By way of further example, table 3 compares the application of these regulatory strategies to several specific mitigation activities identified in American Institute of Aeronautics and Astronautics, 1992: venting residual fuel, boosting geostationary satellites into disposal orbits, de-orbiting spent hardware, and reducing operational debris (e.g., attaching lanyards to releasable items). The strategies are evaluated on the basis of some of the benefits and costs from table 2. Although none of the strategies appears to outperform the others on all bases, the economic-oriented measures, especially those which encourage self-enforcement (e.g., insurance, performance bonds) may be promising.

Three general observations apply to all of the options in tables 2 and 3. One observation is that the options are all centered on a target for a level of debris mitigation (or alternatively, a liveable amount of debris) that regulators will need to have in mind. Ascertaining such a target requires a comprehensive social benefit and cost calculus to weigh the benefits of space activity against the costs of debris production and mitigation. Such an analysis would be quite challenging given the uncertainty surrounding the amount and nature of existing debris and its propagation characteristics, and given the challenge of specifying the benefits and costs of space activities in general [although some elaboration of benefits and costs, including debris, that offers a starting point for the analysis is outlined in by the National Research Council (see National Research Council, 1991)]. Moreover, specifying a desirable

Table 3. *Comparison of Regulatory Strategies*

Mitigation Activities[a]	Regulatory Strategy						Performance	
	Voluntary Actions	Moral Suasion	Command & Control	Penalties/Linked Compensation	Taxes/Fees (including Dep./Ref.)	Tradeable Permits	Insurance	Bonds
Venting residual fuel/pressurants from discarded rocket bodies	– + – –	– + + –	+ + – ? –	+ + – + –	– + + – +	N.R.	– + + + +	– + + + +
Boosting GEO satellites into disposal orbits	– + – –	– + + –	+ + – ? –	+ + – + –	– + + – +	+ – + + –	– + + + +	– + + + +
De-orbiting spent hardware at end of operational life	– + – –	– + + –	+ + – ? –	+ + – + –	– + + – +	+ – + + –	– + + + +	– + + + +
Reducing operational debris	– + – –	– + + –	+ + – ? –	+ + – + –	– + + – +	N.R.	– + + + +	– + + + +

Key: The symbols "+," "–" are assigned for these properties of the strategies (in the following order) (1) likely to achieve sustainable growth rate; (2) minimizes compliance cost; (3) permits flexible compliance; (4) may be perceived as fair; (5) is self-enforcing. The symbol "+" indicates more likely; "–" indicates less likely.

N.R. = Not Recommended

Notes: [a]Activities as listed in AIAA (1992).

amount of debris over the longer run requires choosing a discount rate for social benefits and costs that accrues well into the future. The level of intergenerational equity implied by choice of a rate is bound to be as challenging to conceptualize and specify in the case of the space environment as it is in the case of earth's environment.

A second observation relates to differences in risk associated with various types of debris and their location. Some types of debris appear worse than other types, and some locations are more valuable, more littered, or both, than other locations. Two implications of these impact and spatial differences are that it may be desirable to (1) vary the target debris levels by orbit or debris type; and (2) set a variable schedule of tax rates, fees, and insurance or bond premia to reflect these differences. Table 4 illustrates some of the risk-related factors by which these instruments may vary.

A third observation is that the various regulatory strategies can include incentives for the adoption of mitigation techniques that serve dual duty as protection and mitigation. Such techniques as shielding, for instance, can operate to protect the shielded spacecraft from debris, and to some extent contain the debris that would be generated should the spacecraft be impacted. To encourage such activities, taxes or fees could be discounted if dual-duty techniques were incorporated by the spacecraft operator.[24] As is the case in designing regulatory approaches in general, numerous details remain to be addressed before any of the strategies discussed above might be implemented. If large information gaps and issues of judgment are deemed show-stoppers in implementing these strategies in the near term, and if debris is deemed to be worrisome enough to regulate now, then one possibility might work as follows. A small fee of, say, one dollar for every one thousand dollars worth of project cost (payload plus launch) could be charged for all launches, worldwide. Assuming thirty launches annually at an average of $200 million each (payload plus launch), the collected revenues would total $6 million. This amount is almost 20 percent of the estimated U.S. budget that experts note would be desirable for annual research and other activities related to space debris (see U.S. Congress, Office of Technology Assessment, 1990). The revenues could go towards these activities (countries from whom revenue is collected would presumably share in these activities) or they could go to a trust fund for debris compensation in the event of damage to uninsured spacecraft.

Another possibility given information gaps is to allocate responsibility between government and industry by using government to specify the target amount(s) for debris mitigation and relying heavily on industry, including industrial government contractors, to devise the compliance techniques and to enforce and monitor compliance. The advantage of relying on industry is to exploit any comparative advantage it may have in the way of technical knowledge to design cost-effective techniques for compliance and monitoring. This approach might be modeled after the industry-centered recycling program for packaging materials now underway in Germany. Under that program, industry is responsible for recycling product packaging; thus industry can choose among a mix of reducing and recycling packaging provided it meets overall targets set by government. In this way, responsibility for environmental management has been divided between government and industry to exploit informational asymmetries.

THE THREE MUSKETEERS: ALL FOR ONE AND ONE FOR ALL

Space activity in general and debris issues in particular are inherently global. In the case of debris, mitigating actions by any one company, country, or even region are unlikely to be effective. Rather, the best solution will probably require the consensus of all parties: those presently using space; those who will be in space in the future; and those who may never use space directly but who indirectly benefit from space activity.

If the record of global environmental cooperation (on earth) is any blueprint, then reaching consensus about space debris policy may require the explicit sorting out of the potential clash between space environmental protection and the development of space capability by nations not presently active in space – that is, dealing with the politics of the issues as well as with the issues themselves.[25] Achieving agreement has historically often been extremely difficult, if not impossible, due to a lack of, or argument over, the specification and sharing of property rights (e.g., the Law of the Sea agreement). Frequently, agreement is reached only with the promise of financial aid and transfers of technology (e.g., as stipulated in the Montreal Protocol on chlorofluorocarbon reduction). More recently, new forms of economic incentives have emerged (e.g., debt-for-nature swaps, contracts to commercialize tropical genetic resources).

Analogous difficulties may prevail in the case of space debris. For example, the muddled specification of rights in space is bound to complicate debris policy. Assigning property rights may be seen as contrary to international law (when some equatorial countries claimed to have property rights to the geostationary orbit, nonequatorial countries protested vociferously). However, assigning "responsibility" for various orbital locations in ensuring that debris is minimized could conceivably be tried. For instance, various geostationary orbital locations are naturally of interest to countries geographically positioned to best use them; thus, these countries already have incentives to mitigate debris and also to boost spent satellites to "graveyard" orbits in order to make room for their own next-generation spacecraft. Countries or regions might also be assigned responsibility for tracking and monitoring debris generation and enforcing mitigation compliance in various lower orbits in exchange for assistance with the development of indigenous tracking and monitoring technology.

This paper has sought to illustrate a broad variety of economic issues associated with space debris, with several key

Table 4. *Possible Risk-Related Factors by Which Standards, Taxes, or Premiums May Vary*

Risk-related Factor	Example of Nature of Variation in Standard, Tax, or Premium
Altitude	Higher for LEO, operation in orbits with manned activity
Inclination	Higher for higher inclination
"All-Else-Equal" Fragmentation Likelihood	(Function of spacecraft cross-section, etc. and characteristics of existing debris population) Higher for high likelihood, likely large number of fragments, likely large mass of fragments
In-space Lifetime	Higher for longer lifetime
Interaction with Natural Environmental Effects	Higher if more susceptible to debris-generating effects of thermal cycling, atomic oxygen; higher if less likely to mitigate debris by exploiting atmospheric drag
Ease of Monitoring	Higher if fragmentation likelihood and/or fragmentation results are hard to monitor — e.g., fragmentation likely to result in large number, small mass
"Driving Record" of Operator	Higher if actuarially risky as a debris generator
Nature of Mitigation Technique	Lower (discounted) if "dual duty" in reducing risk to self and to others

lessons provided by analogies with the economics of sustainable development. The relatively "clean slate" of policy towards space debris invites thoughtful scavenging of these lessons.

ENDNOTES

1. See U.S. Congress, 1990.
2. For discussion of marketable pollution permits, see Tietenberg, 1985; for discussion of debt-for-nature exchanges, see U.S. General Accounting Office, 1991; for discussion of commercial contracts dealing with biodiversity, see Sedjo, 1992, and Simpson, 1992.
3. This definition is from *Our Common Future*, the 1987 report from the World Commission on Environment and Development, known popularly as the Brundtland Commission. For discussion, see Toman, 1992.
4. Some of the discussion of and references to environmental pollution in this section are from Helfand, 1992.
5. In fact, the FY 1991 National Aeronautics and Space Administration (NASA) Authorization Act included the following recommendation:
   (b) Sense of Congress: It is the sense of Congress that the goal of the United States policy should be that
      (1) the space related activities of the United States should be conducted in a manner that does not increase the amount of orbital debris; and

   (2) the United States should engage other spacefaring Nations to develop an agreement on the conduct of space activities that ensures that the amount of orbital debris is not increased.
6. Micrometeroids were the sole debris-related concern in the early days of space programs. See U.S. General Accounting Office, 1990 and Olsson-Steel, 1989. Manmade debris was less of a concern because there was less of it and over time, by way of natural orbital decay, the space environment was thought to be self-cleaning. Now the manmade debris population is such that according to one expert, even if all space activities stopped now, more than half the objects in various highly useful, low-earth orbits could still be there in 50 years, and after 100 years, 85 percent of the remaining objects would still be in orbit. See Wood-Kaczmar, 1990.
7. This mutual harm can result from pollution on earth, too, but seems to be a rarer situation.
8. For example, see the "Presidential Directive on National Space Policy," November 1989; the Report on Orbital Debris by the Interagency Group (Space) for the National Security Council, February 1989; and a report by the European Space Agency's Space Debris Working Group, Space Debris, November 1988. The U.S. Congress, however, has called for the cessation of debris generation (see earlier footnote).
9. The analogy here is with hypotheses, and some historical evidence, that man can adapt to climate change if it is gradual enough. See Schelling, 1983.

10. For example, even collisions with debris of 1 centimeter in diameter could be fatal for the shuttle and its crew. See U.S. General Accounting Office, 1990.

11. See discussion in U.S. Congress, Office of Technology Assessment, 1990; American Institute of Aeronautics and Astronautics, 1992; and McKnight, 1989.

12. See discussion of options and priorities in U.S. Congress, Office of Technology Assessment, 1990; Petro and Loftus, 1989; and U.S. National Security Council, 1989. Petro and Loftus offer cost estimates. Increasing funding for debris modeling and other research is another possible step. Which of these is most cost effective in attaining a sustainable space environment is not clear in the technical literature but could be a topic for future research.

13. These costs are frequently overlooked, at least in the case of environmental regulation. For an example of the discussion, see U.S. General Accounting Office, 1992.

14. See also Primack, 1992, for discussion of other social costs of debris, including light that may be reflected off of debris particles and interfere with astronomical measurements on space astronomy.

15. See discussion in U.S. Congress, Office of Technology Assessment, 1990 and American Institute of Aeronautics and Astronautics, 1992. The probability estimates are near the upper bound in the current literature (for instance, see U.S. Congress, Office of Technology Assessment, 1990).

16. Although debris has never completely penetrated a shuttle surface, windows and the surfaces of thermal tiles have displayed evidence of being hit. The risk to the shuttle from debris is expected to increase with increases in debris in shuttle orbits and with longer shuttle missions anticipated for space station construction. See U.S. General Accounting Office, 1990.

17. Petro and Loftus, 1989, offer figure-of-merit estimates of the costs of various debris-control strategies. The estimates vary markedly across options, indicating the desirability of a flexible regulatory approach if cost-effectiveness is a goal.

18. Scheraga, 1986, briefly discusses taxes as an economic-based approach to space debris, and Olmstead, 1985, discusses permits, command and control, taxes, and moral suasion as strategies for debris reduction in the geostationary orbit.

19. See American Institute of Aeronautics and Astronautics, 1992, for a survey of some actions that have been voluntarily undertaken.

20. The U.S. Environmental Protection Agency (EPA) has recently implemented a voluntary pollution reduction program in which the EPA Administrator asked more than 600 U.S. companies to voluntarily reduce pollution caused by 17 toxic chemicals. It is instructive to note that advantages to industry in participation have been explicitly cited to include "positive publicity." However, other mandatory environmental regulation that awards credit for pollution reduction, under the mandatory statute rather than the voluntary program, to obtain the credits. See Ember, 1991.

21. See Burtraw, 1991, for discussion of linked compensation in the case of environmental policy.

22. Debris-related law is presently unspecified; for example, see discussion in U.S. Congress, Office of Technology Assessment, 1990 and American Institute of Aeronautics and Astronautics, 1992.

23. The expense of command and control in regulating the environment is well-documented; a useful compilation of the results is in Crandall, 1992.

24. The discount could be set at a level to equal the reduction in expected social loss (e.g., to equal the social benefit) from the dual-duty strategy.

25. See Landsberg, 1992, for a retrospective look at the issues which characterized the first international environmental conference organized by the United Nations and which continue to challenge negotiations today. A report by the National Security Council discusses international implications of space debris (see U.S. National Security Council, 1989).

REFERENCES

American Institute of Aeronautics and Astronautics (AIAA). 1992. Orbital Debris Mitigation Techniques: Technical, Legal, and Economic Aspects (Washington, DC, AIAA, SP-016-1992).

Burtraw, Dallas. 1991. "Compensating Losers When Cost-Effective Environmental Policies Are Adopted," Resources, no. 104, Summer, pp. 1–5.

Crandall, Robert. 1992. Why Is the Cost of Environmental Regulation So High (St. Louis, MO, Washington University, Center for the Study of American Business, Policy Study Number 110, February).

Ember, Lois R. 1991. "Strategies for Reducing Pollution at the Source Are Gaining Ground," Chemical and Engineering News, July 8.

European Space Agency (ESA). 1988. Space Debris (ESA, Space Debris Working Group, November).

Fisher, Ann, Lauraine G. Chestnut, and Daniel M. Violette. 1989. "The Value of Reducing Risks of Death: A Note on New Evidence," Journal of Policy Analysis and Management, vol. 8, no. 1 (Winter), pp. 88–100.

Helfand, Gloria E. 1992. "The Simple Economics of Pollution Prevention." Discussion Paper 92–1 (Washington, DC, The Council on Environmental Quality).

Landsberg, Hans H. 1992. "Looking Backward: Stockholm 1972," Resources, no. 106 (Winter), pp. 2–3.

McKnight, Darren. 1989. Comments during "Joint Workshop on Space Debris and Its Policy Implications," sponsored by the U.S. Congress, Office of Technology Assessment and the U.S. Space Foundation, 4 April.

National Research Council, Space Studies Board. 1992. Setting Priorities in Space Research: Opportunities and Imperatives (Washington, DC, National Academy Press).

Olmstead, Dean. 1985. "Orbital Debris Management: International Cooperation for the Control of a Growing Safety Hazard," Earth-Oriented Applications to Space Technology, vol. 5, no. 3.

Olsson-Steel, Duncan. 1989. "Space Debris Versus Natural Meteoroids: Comparative Risks," Orbital Debris Monitor, vol. 2, no. 3 (July 1), pp. 4–6.

Petro, Andrew J. and Joseph P. Loftus. 1989. "Future Space Transportation Requirements for the Management of Orbital Space Debris. Paper presented at the 40th Congress of the International Astronautical Foundation, Malaga, Spain, IAF 89-244, October 7–12.

Primack, Joel. 1992. "Protecting the Space Environment for Astronomy," paper prepared for the Preservation of Near-Earth Space for Future Generations, Chicago, June. See map.

Schelling, Thomas C. 1983. "Climatic Change: Implications for Welfare and Policy," in National Research Council Changing Climate: Report of the Carbon Dioxide Assessment Committee (Washington, DC, National Academy Press), pp. 449–482.

Scheraga, J. D. 1986. "Pollution in Space: An Economic Perspective," Ambio, vol. 15, no. 6, pp. 358–360.

Sedjo, Roger A. 1992. "Property Rights, Genetic Resources, and Biotechnological Change," Journal of Law and Economics, vol. 35, no. 1, April.

Simpson, R. David. 1992. "Transactional Arrangements and the Commercialization of Tropical Biodiversity." Discussion Paper ENR92-11 (Washington, DC, Resources for the Future).

Tietenberg, T. H. 1985. Emissions Trading: An Exercise in Reforming Pollution Policy (Washington, DC, Resources for the Future).

Toman, Michael A. 1992. "The Difficulty in Defining Sustainability," Resources, no. 106 (Winter), pp. 3–6.

United Nations, General Assembly, Committee on the Peaceful Uses of Outer Space. 1992a. "Provisional Agenda for the Thirty-Fifth Session." Document A/AC.105/L.193/Corr.1 (New York, NY, United Nations, 18 May).

United Nations, General Assembly, Committee on the Peaceful Uses of Outer Space. 1992b. "Report of the Legal Subcommittee on the Work of Its Thirty-First Session (23 March–10 April 1992)." Document A/AC.105/514 (New York, NY, United Nations, 20 April).

U.S. Congress, Office of Technology Assessment. 1990. Orbiting Debris: A Space Environmental Problem–Background Paper (Washington, Government Printing Office, September).

U.S. Congress. 1991. FY 1991 National Aeronautics and Space Administration Act (Washington, DC, Government Printing Office).

U.S. General Accounting Office (GAO). 1990. Space Program: Space Debris a Potential Threat to Space Station and Shuttle (Washington, DC, GAO/IMTEC-90-18, April).

U.S. General Accounting Office (GAO). 1991. Developing Country Debt (Washington, DC, GAO/NSIAD-92-14, December).

U.S. General Accounting Office (GAO). 1992. International Environment: International Agreements Are Not Well Monitored (Washington, DC, GAO/RCED-92-32, January).

U.S. National Security Council (NSC). 1989. Report on Orbital Debris (Washington, DC, NSC, Interagency Group (Space), February.

U.S. Office of the President. 1989. "Presidential Directive on National Space Policy," January.

Warren, H. R. and M. J. Yelle. 1992. "Effects of Space Debris on Commercial Spacecraft," paper prepared for the Preservation of Near-Earth Space for Future Generations, Chicago, June. Chapter 7.

Wood-Kaczmar, Barbara. 1990. "The Junkyard in the Sky," New Scientist (October 13).

World Commission on Environment and Development (WCED). 1987. Our Common Future (New York, Oxford University Press).

19: The Economics of Space Operations: Insurance Aspects

Christopher T.W. Kunstadter

Senior Vice President, United States Aviation Underwriters, Inc., New York

INTRODUCTION

Insurance is not often considered as an element of space activities, but satellite[1] insurance has been around since the launch of the Early Bird satellite in 1965. Today, the cost of insurance can exceed 25% of the lifetime cost of a communications satellite program.

Space insurance has traditionally been bought by the owners and users of geostationary communications satellites. Therefore, technical and risk analysis by space insurance companies normally concentrates on the factors that can affect those satellites.

The space insurance community does not currently view the risk posed by orbital debris as particularly significant to its own business. This is due to two factors in particular – the geostationary orbit used by the vast majority of insured satellites, and the relatively insignificant contribution of the perceived probability of damage from debris to the overall rates charged by insurers for the insurance coverages offered.

These conditions will not continue, however, as the market for space insurance expands to include the many low-earth-orbiting systems proposed for this decade and beyond. In total, these proposed systems represent several hundred satellites. While it is unlikely that all of the systems will be launched, there will certainly be a significant increase in the number of commercial low-earth-orbiting satellites in the not-too-distant future. This additional exposure for insurance companies is certain to become a matter of concern.

THE INSURED SATELLITE POPULATION

In the meantime, geostationary satellites comprise the vast majority of insured risks. Of the 215 satellites which have ever been insured, only ten have been low-earth-orbiting. Of those ten, six were small satellites weighing less than 50 kg.

The collision risk for geostationary satellites is primarily from three sources – other active satellites, satellites which have exceeded their life and are no longer maintained, and debris jettisoned as part of normal satellite operations.

The geosynchronous and near-geosynchronous orbits are getting more crowded. There are currently 100 active insured

satellites in geostationary orbit, accounting for between 10% and 40% of the geostationary population (variously estimated at between 250 and 1,000 objects[2]). Of these 100 satellites, 80 are concentrated in the half of the orbital arc over the Americas, the Atlantic Ocean and Europe and Africa (140° west longitude eastward to 40° east longitude). This concentration of active satellites has resulted in numerous studies of the collision risk in geostationary orbit, particularly by European agencies and operators.

CO-LOCATION OF ACTIVE SATELLITES AND THE RISK OF COLLISION

The risk of collision in geostationary orbit does not appear to be with debris as much as with other satellites. Advances in communications satellite technology have allowed multiple satellites to be co-located. The separation between satellites using the same frequencies and polarizations along the arc has decreased as antenna technology has improved.

More significantly, though subject to radio frequency interference constraints, a satellite operator would prefer to have its satellites co-located with others it owns, or with others which are viewed by a large number of customers. This would increase the number of channels visible to a customer on the ground with a single antenna. Thus arises the concept of the "hot bird" – a satellite on which a large number of "premium" cable TV channels are located. This makes particular orbital slots more desirable. Frequency re-use through cross-polarization and use of different frequency bands have resulted in as many as four active satellites in use at the same time at the same nominal longitude in the geostationary arc.

As home satellite dishes get smaller, their receive beamwidth increases. In other words, they become less directional, and satellite operators must maintain their satellites in tighter windows to avoid interfering with signals from adjacent satellites. Equally important is the fact that the narrow beamwidth of larger earth station dishes (such as those used by programmers to uplink their programs, as well as by cable television systems to receive those programs) requires tight stationkeeping of the satellites in order to maintain high-quality audio and video signals for viewers on the ground.

Furthermore, and particularly in the crowded portion of the orbital arc serving Europe, individual countries are assigned just one or two specific orbital slots. As a result, European satellite operators have had more extensive experience with the co-location of satellites.

Barring co-ordinated strategies, the risk of a "near-miss" or even a collision between co-located satellites of two operators is significant. Close co-operation between European governments and satellite operating agencies such as the European Space Agency, DLR in Germany, and CNES in France, though often representing competing interests, is an example of the type of co-ordination which allows such co-location.

Co-location strategies vary, but tight tolerances are required. The two most common strategies for co-locating satellites are longitudinal separation of the satellites in adjacent station-keeping windows with a "dead band" in between, and separation of the satellites at the same longitude by using different eccentricity and inclination. Both methods are robust to failures if implemented properly. Longitudinal separation is easier operationally, but can result in degraded performance through the introduction of antenna pointing offsets. Eccentricity and inclination separation is better for more than two satellites, but requires more fuel usage and complex software for satellite orbit control.

COLLISION WITH INACTIVE SATELLITES

A potentially more serious threat comes from satellites which are no longer maintained within tight tolerances, and from satellites which have ceased to be used at all. The former may have run out of sufficient propellant to perform north-south stationkeeping, while the latter may have run out of propellant entirely, or suffered anomalies rendering them inoperable.

Operators may be reluctant to boost an old satellite to a super-synchronous orbit, as this maneuver can use up several months of propellant life. With the uncertainty of launch schedules, the operator may be concerned about the ability to ensure that its replacement satellite is on-station and checked out before the retiring satellite runs out of fuel. Thus, they may hesitate to use budgeted propellant for a de-orbit maneuver. In addition, propellant budgets often have a margin of error of as much as 3%, representing an additional 3 months of propellant over a typical ten year life which may not be available for such a maneuver.

Two factors which must be addressed in considering disposal to a "graveyard" orbit are the explosion risk and the magnitude of the disposal orbit. Disposal is useless if the satellite eventually explodes, as debris may be re-injected into the geosynchronous orbit. Even if it does not explode, the satellite may continue to cross the geosynchronous orbit periodically, lessening the effect of the disposal. Unfortunately, the orbit which may be required in order to significantly reduce the risk of collision may be significantly greater (measured in the pro-

pellant required), imposing an unacceptable commercial burden on the satellite operator.

COLLISION WITH JETTISONED DEBRIS

Debris from normal operations also poses a threat to geostationary satellites. As an example, wrap cables, which are used to restrain antennas and solar arrays during launch and orbit-raising, are typically jettisoned once the satellite has achieved synchronous or near-synchronous orbit. These stainless steel cables can be over 15 meters long, posing a serious hazard to other satellites. Alternatives exist, such as center-pull cables, which produce no external debris.

INSURANCE ASPECTS – PHYSICAL DAMAGE LOSSES

Space insurance provides coverage for the risk that a satellite or launch vehicle will fail to achieve orbit, or, having reached orbit, will fail to work properly. The coverage typically extends from ignition of the launch vehicle through deployment of the satellite, station acquisition, in-orbit checkout, and on-orbit lifetime. This would usually include the risk of accidental collision with other satellites or debris.

As there have been no recorded instances of debris collisions involving insured payloads, this risk has not been a factor in the pricing of space insurance. Looking at the collision risk issue from the point of view of the insurer, a very non-scientific approach shows that the collision rate is small. Collisions have probably disabled only a handful of the approximately 4,000 payloads launched to orbit in the past 35 years. Assuming that the rate is one collision per 1,000 payloads, and even assuming that all the recorded collisions have happened in the past ten years, yields a rate of one collision per 10,000 payloads per year (or one collision per 100 insured satellites per 100 years). Even if these admittedly unscientific calculations are optimistic by an order of magnitude or two, the impact is still negligible[3]. Taking only active payloads into account further reduces this figure.

Based on these calculations, the price which might be charged to account for the debris risk is so small when compared to the overall cost of insurance for a satellite that it would not be commercially sensible for insurers to add this cost in at this point in time. Nonetheless, the results of a collision could be extremely costly. It is not difficult to show that the value of a commercial communications satellite can be as much as $500,000,000, combining the replacement cost of the satellite and the business interruption losses (e.g., loss of revenue). When insurers finally realize the catastrophe potential (which may be too late), the price will be a more significant factor in the cost of satellite insurance.

This admittedly short-sighted view by insurers is hardly the type of pro-active stance which an organization in this field should take. Unfortunately, it would take a catastrophic loss to prove to the insurance community that the threat is real.

THIRD-PARTY LIABILITY LOSSES

A potentially larger financial exposure stems from the litigation that might stem from a collision between a satellite belonging to one party with debris or even a satellite belonging to another party. Determining fault would be difficult, but might not be impossible. The potential exposure could even exceed the $500,000,000 value of the satellite.

Insurance is available to cover this third-party liability risk, and, in fact, such coverage is required of launch vehicle operators for their commercial launch operations. Currently only one satellite operator, representing two satellites out of the 100 active insured satellite population, buys this type of insurance for its in-orbit satellites. This is surprising considering the low premiums currently charged for such coverage – typically $35,000 per year per satellite for $100,000,000 of coverage. This is even more of a bargain considering that, when a loss does occur, the price will certainly skyrocket.

SUMMARY

Insurance is not a solution to the orbital debris problem. It is, however, an important element in a complete risk management program.

The space insurance community would like to see the following steps taken to mitigate the problem of on-orbit collisions, particularly for geostationary satellites:

1. Operators of geostationary satellites should increase coordination among themselves to avoid collision risks for co-located satellites.

2. Manufacturers and operators of geostationary satellites should budget sufficient propellant to allow for a safe, commercially reasonable disposal of their satellites at the end of their lifetime, and should investigate "explosion-safing" of on-board propulsion systems.

3. Manufacturers of all satellites and launch vehicles should avoid, where practicable, the need for any jettisoned debris by using alternative hardware configurations.

The insurance community offers a service which is vital to the continued commercial development of space. Working with satellite and launch vehicle owners and operators, we continually try to improve space system reliability. By focusing on the preservation of near-earth space for future generations, we can attempt to mitigate those external factors over which the space community has direct control.

ENDNOTES

1. As used throughout this paper, "satellite" means a payload placed into orbit to perform a function, and not simply an object (e.g., spent rocket stage, debris, etc.) in orbit.

2. References 6, 7, 11, 14 & 15.

3. Kessler (Ref. 14) cites Hechler's (Ref. 4) estimate of a probability of collision of 0.16 over a 20 year period.

REFERENCES

1. "A Review of Orbital Debris Monitoring in Europe", D. Rex, J. Zhang & P. Eichler, AIAA 90-1354, presented at the AIAA/NASA/DOD Orbital Debris Conference: Technical Issues and Future Considerations, Baltimore, Maryland, April 16–19, 1990.

2. "Breakups and Their Effect on the Catalog Population", D.S. McKnight & N.L. Johnson, AIAA 90-1358, presented at the AIAA/NASA/DOD Orbital Debris Conference: Technical Issues and Future Considerations, Baltimore, Maryland, April 16–19, 1990.

3. "Collision Probability and Spacecraft Disposition in the Geostationary Orbit", W. Flury, Adv. Space Res., Vol. 11, No. 12, 1991.

4. "Collision Probability at Geosynchronous Altitudes", M. Hechler, Adv. Space Res., Vol. 5, No. 2, 1985.

5. "Debris Chain Reactions", P. Eichler & D. Rex, AIAA 90-1365, presented at the AIAA/NASA/DOD Orbital Debris Conference: Technical Issues and Future Considerations, Baltimore, Maryland, April 16–19, 1990.

6. "ESA Concepts for Space Debris Mitigation and Risk Reduction", H. Klinkrad, presented at The Preservation of Near-Earth Space for Future Generations Symposium, Chicago, Illinois, 1992. See chap. 11.

7. "Orbiting Debris: A Space Environmental Problem–Background Paper", U.S. Congress, Office of Technology Assessment, OTA-BP-ISC-72, September 1990.

8. "Removal of Debris from Orbit", P. Eichler & A. Bade, AIAA 90-1366, presented at the AIAA/NASA/DOD Orbital Debris Conference: Technical Issues and Future Considerations, Baltimore, Maryland, April 16–19, 1990.

9. "Results in Orbital Evolution of Objects in the Geosynchronous Region", L.G. Friesen, A. Jackson, H. Zook & D. Kessler, AIAA 90-1362, presented at the AIAA/NASA/DOD Orbital Debris Conference: Technical Issues and Future Considerations, Baltimore, Maryland, April 16–19, 1990.

10. "Space Debris: How France Handles Mitigation and Adaptation", J-L. Marcé, presented at The Preservation of Near-Earth Space for Future Generations Symposium, Chicago, Illinois, 1992. See chap. 12.

11. "Space Debris: The Report of the ESA Space Debris Working Group", European Space Agency, ESA SP-1109, November 1988.

12. "Special Considerations for GEO – ESA", A.G. Bird, AIAA 90-1361, presented at the AIAA/NASA/DOD Orbital Debris Conference: Technical Issues and Future Considerations, Baltimore, Maryland, April 16–19, 1990.

13. "Techniques for Debris Control", A. Petro, AIAA 90-1364, presented at the AIAA/NASA/DOD Orbital Debris Conference: Technical Issues and Future Considerations, Baltimore, Maryland, April 16–19, 1990.

14. "The Current and Future Environment: An Overall Assessment", D.J. Kessler, presented at The Preservation of Near-Earth Space for Future Generations Symposium, Chicago, Illinois, 1992. See chap. 3.

15. "The Earth Satellite Population: Official Growth and Constituents", N.L. Johnson, presented at The Preservation of Near-Earth Space for Future Generations Symposium, Chicago, Illinois, 1992. See chap. 2.

V. Legal Issues

20: Environmental Treatymaking: Lessons Learned for Controlling Pollution of Outer Space

Winfried Lang

University of Vienna

"Oh, and tell them to spy on the ozone layer, will you, Ned. Its dreadfully hot in St. Agnes for the time of the year."

George Smiley in John Le Carre, The Secret Pilgrim.

A. ENVIRONMENTAL TREATYMAKING

In this chapter a limited number of environmental treaties will be scrutinized. The main purpose of this exercise is to highlight the specific features of these treaties not only as regards substance but also as regards legal techniques and institutional issues.

1 Marine pollution

As the legal situation of the high seas has sometimes been compared with that of outer space – see only the concept of "Common heritage of mankind" or the issue of space debris in relation to the salvage of ship wrecks (Christol 1990, 268–276; Cocca 1990, 71–76; Reijnen – de Graff 1989, 45–48) – it is appropriate to look into treatymaking endeavors in respect of marine pollution. Since there are over 70 multilateral instruments of one kind or the other which deal with problems of marine pollution one has to pick the most obvious of these instruments, which means a treaty which is global in character and general in scope, yet specific enough in order to be operational: *International Convention for the Prevention of Pollution from Ships 1973* (MARPOL 73').

Here we are faced with a situation that we shall encounter in many other areas of environmental treatymaking: the first instrument generates a number of "satellites" in the form of protocols, amendments, annexes or other types of arrangements, which are separate in form but dependent in function (Johnston 1988, 200). Another preliminary observation would be, that it sometimes proved difficult to induce a substantial majority of states to sign or ratify these global instruments simply because economic and technological assumptions underlying the text were open to question, if not clearly invalid, at the time of their conclusion or later (Johnston 1988, 203). Thus it took ten years for MARPOL to enter into force and this mainly was made possible because a 1978 protocol, generally known as the "Tanker Safety and Pollution" Protocol, allowed states to become parties to the Convention while only accepting its Annex I (Oil Pollution Regulations) whereas in the original version also Annex II (Noxious Liquid Substances) had to be accepted in order to become a full party to this treaty (Gold, 1985, 39).

As oil pollution from vessels seems to have abated in the last eighteen years it has to be acknowledged that this was mainly due to technological innovation supplied by industry and to the upvaluing of oil in the market place since 1973. Thus it was clearly demonstrated that treatymakers (mainly lawyers) are highly dependent on the interaction of the relevant technical elites and on their ability to depoliticize the negotiating process (Johnston 1988, 204/205). This priority of technical considerations can be achieved because these elites operate on transnational circuits or networks including various kinds of non-governmental associations of a scientific or technical character. Treatymaking in the legal and diplomatic sense has to be preceded or at least accompanied by a continuous dialogue between these technical and scientific experts which have to arrive at a common understanding as regards the feasibility of the various prescriptions. These networks or circuits have become known as epistemic communities (Haas 1992, 189).

The original 1973 MARPOL Convention attempted to eliminate ship source pollution almost at one stroke. Unfortunately a number of very complex technical problems, primarily related to measures designed to prevent pollution by chemicals, were such that many states could neither accept nor implement the new convention in the very ambitious timeframe originally planned. However, a series of very grave tanker accidents which took place in 1977/78 caused pressure to the effect that MARPOL was put into effect without its Annex II concerning pollution by chemicals (Gold 1985, 58). The other Annexes (III on packaged harmful substances, IV on sewage from ships, V on garbage from ships) are only of an optional character and therefore not a condition for full membership in the MARPOL-regime. The main lesson to be learned from this early exercise is, that the more comprehensive an instrument is conceived the more likely are delays as regards ratifications and thus its entry into force. A step-by-step approach, a piecemeal approach looks much less ambitious and satisfactory from an environmentalist perspective. It takes, however, into account various obstacles related to technological/economic feasibility and scientific uncertainty, and should therefore be preferred to the previous one.

Turning now to a closer examination of the treaty itself one realizes immediately that one is confronted with the framework type of convention. The treaty itself does not contain more than 20 articles; it is, however, surrounded by a host of other instruments, the afore-mentioned annexes and still two protocols, one on the reporting of incidents involving harmful substances and the second on arbitration. Of special importance are the following provisions:

– *Article 1:* It sets out the overall objective, namely to prevent the pollution of the marine environment by the discharge of harmful substances or effluents; this objective is to be attained by giving effect to the provisions of the Convention and those Annexes a state is a party to.
– *Article 2 and 3* deal with definitions and application: War-ships as well as other ships owned or operated by a state and used for non-commercial service are excluded from the purview of the treaty (Would this exclusion also apply to some types of satellites?)
– *Articles 4 to 7* deal with sanctions in case of violations (proceedings against ship operators), the issuing of certificates and the inspection of ships.
– *Article 8* (in conjunction with Protocol I) establishes obligations as well as machinery for the reporting of incidents (some limited role for the International Maritime Consultative Organization IMCO; nowadays IMO).
– *Article 11* follows up on Article 8 and creates a broad communication network, the focal point of which is to be the aforementioned international organization.
– *Article 12* deals with the investigation of casualties, if such casualty has produced a major deleterious effect upon the marine environment.
– *Article 13–16* constitute a central element as they try to pull together the various side-instruments and the main convention; these provisions for entry into force and amending the various texts served as an important precedent for later treaty-making in the field of the environment:

A double majority – minimum number of ratifications (15) which reflects a certain percentage of the gross tonnage of the world's merchant shipping (50%) – is required for the entry in force of the Convention and for certain amendments. This should avoid too important discrepancies as regards the working conditions (costs) for the main commercial shipping fleets; this should avoid a disadvantage in competition for those shipping industries the governments of which were adhering to the Convention.

A simple no-opposition procedure (no objection of at least one third of the Parties) was established in respect of amend-

ing the highly technical appendices to the various Annexes. Thus flexibility was assured as regards the adaptation of these texts to the evolution of science and technology.

These legal techniques, which aim at attaining the broadest possible participation as well as a minimum of flexibility, will be encountered in numerous other environmental treaties.

– *Article 17* provides for technical cooperation or technical assistance; this does not necessarily refer to the highly sensitive issue of transfer of technology, including access to and protection of intellectual property (patents), an issue that has embittered many negotiations which tried to bridge the North-South gap.

In concluding this brief summary one should bear in mind that treaties are not likely to operate by themselves or by the direct cooperation of their parties. Treaties of the multilateral kind, in order to become and remain operational, need the support of an international organization. Powers conferred by this Convention to IMCO/IMO are not impressive (as compared to later environmental treaties), they are however sufficient to convey to this organization the role of a guardian which does not only look after the full implementation of the provisions of the treaty but is also concerned with the progressive development of these provisions. Thus, international law does not reflect only some "status quo" but becomes something of a "living organism" serving the changing needs of the society.

2 Air pollution

Many years before air pollution became a subject of multilateral treatymaking it had already been a bone of contention between the United States and Canada. This so-called Trail-Smelter dispute was settled by arbitration proceedings, the outcome of which has had a strong impact on the evolution of international environmental law (Kiss-Shelton 1991, 122–126). Since 1896 US-farmers had suffered damage due to emissions of sulphur dioxide which resulted from the activities of a Canadian smelter of zinc and lead ores, located in Trail, British Columbia. The final arbitral decision, issued in 1941, settled not only the matter in preventing future damage but contained also a famous dictum of law, which should be quoted here because it is to some extent also applicable to environmental protection in outer space: "No State has the right to use or permit the use of its territory in such a manner as to cause injury in or to the territory of another state or to the properties or persons therein, when the case is of serious consequence and the injury is established by clear and convincing evidence".

The basic thinking contained in this arbitral decision re-emerged in 1972, when the Stockholm Conference on the Human Environment drafted and approved a catalogue of principles which should guide international action in the field of environment. The most famous of these principles, also applicable to outer space, namely Principle 21 should be quoted here at some length: "States have ... the sovereign right to exploit their own resources pursuant to their own environmental policies, and the responsibility to ensure that the activities within their jurisdiction or control do not cause damage to the environment of other states or of areas beyond the limits of national jurisdiction". This principle quoted or referred to in many other instances since 1972 is considered by legal opinion at large as a basic norm of customary international environmental law.

One of those instances was the *Convention on Long-Range Transboundary Air Pollution (LRTAP)* signed on 13 November 1979. There principle 21 is quoted in the preamble and qualified as "expressing common conviction". A brief glance at the history of this instrument (Brunnée 1988, 175; Chossudovsky 1989, 39–110; Fraenkel 1989, 447–476; Gündling 1986, 19–31; Rosencrancz 1981, 975–982; Lang 1989, 22–34) tells us that environmental concerns – acidification of inland waters in Scandinavia as a consequence of sulphur deposition coming from abroad – do not suffice to trigger environmental treatymaking. During most of the 1970's Scandinavian diplomacy had tried to put this issue on the international negotiating agenda. Only when this drive met with the wish of the East European states to give more concrete content to "detente", in the aftermath of the Helsinki Conference on Security and Cooperation in Europe 1975, and to organize high-level-meetings on environment, transport and energy – also an effort to divert attention from the human rights issue – the problem of air pollution attained sufficient visibility in order to be treated at the political level. As any regulation that dealt with sulphur emissions or their transboundary fluxes, which emanated from the industrial belt of Europe (United Kingdom, Germany, Poland etc.), was likely to affect industrial activities in these countries, the issue of economic feasibility of restrictions and reductions ranked high among the questions to be taken into account. Those countries to be mainly affected by such regulation resisted for a long time to any kind of firm legal commitment, in particular many member countries of the European Communities. Only highest-level persuasion between Sweden and the FRG achieved a first break-through, namely that a modest political declaration be replaced by an equally modest but at least legally binding treaty. This treaty, very soon – also as a consequence of the "greening" of politics in Germany – was considered as constituting only a framework-convention which had to be supplemented by protocols which contain more specific commitments related to specific substances (SO_2, NOx etc.).

The main contents of the framework-treaty can be summarized as follows:

- *Articles 2–4* reflect some fundamental principles: Parties endeavor to limit and "as far as possible" gradually reduce and prevent air pollution including long-range transboundary air pollution. By means of exchanges of information, consultation, research and monitoring. Parties shall develop "without undue delay" appropriate policies and strategies "taking into account efforts already made". They shall, furthermore, exchange information on and review their policies, scientific activities and technical means etc.

- *Article 5* introduces the concept of prior consultation into international environmental law, a principle already well known in the law of outer space (Article IX of the Outer Space Treaty 1967). Consultations shall be held between parties actually affected by or exposed to a "significant" risk of LRTAP and the party within which and subject to whose jurisdiction a "significant" contribution to LRTAP originates or could originate. Here again one recognizes the concept of a threshold, already contained in the above-mentioned dictum of the Trail-Smelter-decision, namely that not all pollution should be prevented or combatted but only pollution beyond a certain level of nuisance or damage. This raises the question: Where to fix that level? By whom should it be fixed?

- *Article 6* on air quality management is the first more concrete provision: Parties undertake to develop best policies and strategies and control measures "compatible with balanced development, in particular by using the best available technology which is economically feasible". Again we are faced with relatively "soft obligations"; the same question is asked again: Who determines the availability of pollution abatement technologies? Who determines their economic feasibility? Who determines the compatibility of control measures with balanced economic development? Are there any objective yardsticks at our disposal? Is the state that accepts certain obligations the sole master to determine the scope of its proper obligations? An affirmative answer to this last question assumes that really effective procedures of disputes settlement are available and that states are ready to fulfill their obligations in good faith.

- The remaining articles are of a technical and institutional nature. This again confirms the view that legal obligations have little bearing on real life unless they are made operational by an appropriate machinery. This set-up includes also a monitoring network to check the long-range transmission of air pollutants (Article 9). Although this convention is considered as a model of the "framework convention cum protocols"-structure, nowhere in its text there is made any provision for adopting such protocols. These protocols came only about as a consequence of later insight mainly as a consequence of public pressure.

Among these protocols one should mention the *Protocol on Sulphur Emissions or their Transboundary Fluxes by at least 30 per cent*, which was signed in Helsinki in July 1985. This instrument represents a real step from pious wishes and vague commitments to obligations that can be measured against a time-scale and in the light of quantifiable steps of reduction. Certainly, some leeway is still given to parties as regards actual emissions or only transboundary fluxes – the latter was the maximum some East European states were ready to go – but for the first time a threshold had been fixed, a threshold beyond which emissions are to be considered as illegal. Such across-the-board reductions may well be regarded as relatively crude measures not taking into account the specific situation of some countries; they are, however, a first step towards even more stringent obligations. Parties are about to reconsider their endeavor as they approach their 1993-target. Especially the position of economies in transition has to be reevaluated: To what extent should their drop in industrial production and consequently their levelling off in respect of certain emissions be considered as a fulfillment of their obligations under this protocol?

Turning now to the 1988 *Protocol concerning the Control of Emissions of Nitrogen Oxides or their transboundary Fluxes* one becomes aware of a new technique in environmental treatymaking:

- Instead of imposing strict obligations of reduction treatymakers are already satisfied when contracting parties are ready to freeze their emissions by a certain date (31 December 1994) at some previous level (1987), which in real terms may well amount to some kind of reduction.
- Instead of imposing reduction obligations parties are only expected to commence negotiations on a specific subject at an early date ("pactum de negotiando"), without any cut-off date as regards the outcome of these negotiations.
- The across-the-board approach of the previous protocol is replaced by the "critical loads"-approach, which necessitates individual treatment of specific regions, especially in respect of their vulnerability to emissions of the NOx-type.

A brief glance at the various provisions of this Protocol tells us that "best available technologies which are economically feasible" are the cornerstone of this agreement be it new stationary or new mobile sources of pollution which are concerned. Again, procedural duties related to exchange of technology, to research and monitoring in respect of the "critical loads"-approach are most prominent among the overall set-up of commitments to be carried out by the contracting parties. Furthermore, parties are requested to make unleaded fuel available, as a minimum along main international transit routes.

In view of the limited membership of this Protocol (Europe – North America) there do not exist major exceptions for developing countries. Amendments to the Protocol require consensus for adoption and an unqualified two-thirds majority for their entry into force. Amendments to the Technical Annex also require to be adopted by consensus, which means quasi-unanimity; they become effective thirty days after they have been communicated to the parties.

Looking back on the achievements of treatymaking in the field of air pollution one becomes aware that the "duty of states to inform and consult each other is widely recognized as a principle of international environmental law" (Gündling, 1986, 23). This duty is to some extent already part of existing outer space law: Article II of the Convention on Registration of Space Objects launched into Outer Space imposes on the launching state the duty to inform the Secretary General of the United Nations (an information obligation comparable to those under the IAEA-treaties on nuclear accidents). Article IX of the Outer Space Treaty imposes on the state of origin of some space activity, which has reason to believe that such activity would cause potentially harmful interference with activities of other parties (language comparable to the Basel convention on wastes), the duty to undertake appropriate international consultations before proceeding with any such activity or experiment. Thus one may recognize some similarity between environmental and outer space law as regards procedural obligations. However, as far as obligations of substance are concerned environmental law appears to be much more advanced and concrete.

Finally, attention should be drawn to the institutional aspect. Regulations on air pollution were drafted in the context of the Economic Commission for Europe, one of the regional bodies of the United Nations, a body that has bridged the East-West-gap during the last forty-five years. The machinery established to administer the air pollution-regime remains closely associated with this Commission. Thus, institutions in charge of a specific negotiation are likely to be in charge of executing the outcome, of assisting states in implementing these instruments, of monitoring states behavior in respect of compliance with their duties.

3 Nuclear accidents

For many years nuclear accidents were the subject of treatymaking between neighboring states; a basic machinery of ecological crisis-management was established in order to prevent the transboundary spill-over of radioactive releases resulting from such accidents; sometimes coordinated emergency and mutual assistance arrangements were agreed upon (Bruha 1984, 1–63).

Treatymaking at a broader multilateral scale was only undertaken after the Chernobyl accident (1986) had amply demonstrated that the effects of major accidents do not only affect the state of origin and its immediate vicinity but could also cause damage to far away countries. The International Atomic Energy Agency (IAEA) had already issued guidelines on emergency assistance, reporting etc. in early 1985. These guidelines had, however, no legally binding force; at most could they be considered as "soft law"; evidently there was not sufficient political will to turn them into legally binding commitments. The excitement of public opinion, at least in Europe, in the immediate aftermath of this accident, generated that political will. During a few weeks in summer 1986 two conventions were finalized, one on early notification, the other on assistance. Both instruments do not constitute framework treaties in the more traditional sense, which means that they have to be supplemented by protocols in order to have at least some impact on real life. However, as both instruments, especially the latter one on assistance, leave many questions open, countries concerned by any such threat are well advised to enter into more detailed supplementary agreements in order to have a fully working system of ecological crisis management.

The history of these negotiations was relatively short (Adede 1987; Gorkom 1989, 149-163; Hafner 1988, 19-39; Handl 1988, 203-248; Heller 1987, 651-664; Lang 1988a, 9-16; Moser 1989, 10-23; Sands 1988; Zehetner 1987, 118-140). Due to the swiftness of action no major confrontation of the traditional East-West or North-South kind could develop. Certainly, countries highly dependent on nuclear energy for the energy needs of their economy were reluctant to impose too strict and comprehensive obligations on their industry. Sometimes the argument of sovereignty was used in these negotiations in order to reject too intrusive regulations, whereas the real reasoning behind such arguments was to preserve a maximum room of maneuver for the respective nuclear industry.

Casting a brief glance on the *Convention on Early Notification of a Nuclear Accident* it should be noted that it also applies to outer space insofar as facilities and activities to be covered include "any nuclear reactor wherever located". The convention applies in the event of any accident from which a release of radioactive material occurs or is likely to occur and which has resulted or may result in an international transboundary release that could be of radiological safety significance for another State. In the event of such accident the state of origin has to comply with a number of obligations (notification, information, consultation) vis-a-vis other states which are or may be physically affected. However, in view of the above-mentioned criteria of a reportable accident, these duties of notification etc. depend to a large extent on self-determination by the state of origin, which may consider at some length if such accident is or is not within the scope of the convention before it triggers the international alarm. Another weakness of this convention is the regulation related to accidents caused by nuclear weapons; in such instances the state of origin "may notify" but is not obliged to do so. Thus nuclear weapon states have obtained a maximum of liberty, although during the conference they issued unilateral declarations to the effect that they will report also all weapons-related accidents. It has to be seen if this unilateral, political, but not legal commitment is fully honored by all nuclear weapons states.

In both conventions major tasks are entrusted to the IAEA; these tasks include information and transmission functions as well as assistance (preparing emergency plans, training programmes, investigating feasibility of monitoring systems etc.). Again, we are faced with the fact that environmental treaties, in order to be fully operational, need the support of an international organization. The IAEA, which to some extent had

sponsored the negotiations leading to these treaties, became by means of these same treaties their proper guardian.

The *Convention on Assistance in the Case of Nuclear Accident or Radiological Emergency* does not contain any firm obligation of assistance. The requested state has only the obligation to "promptly decide and notify the requesting state, whether it is in a position to render assistance" (Article 2). On a broader scale "parties shall cooperate ... to facilitate prompt assistance"; and this cooperation may be realized by means of "bilateral or multilateral arrangements ... for preventing or minimizing injury or damage" (Article 1). Especially in the light of this last quotation the assistance-convention may be assimilated to a framework-convention. The remainder of its text is devoted to the modalities of assistance, which range from the direction and control of assistance operations to delicate issues such as the reimbursement of costs, the privileges, immunities and facilities for the assistance personnel or the problem of claims and compensation in case of damage that occurred during the assistance operations. Here one is faced with another legal technique ("opting out"), which is frequently used by negotiators, if they are unable to achieve full consensus: Parties which are not ready to accept treaty provisions related to privileges and immunities or to claims proceedings may dissociate themselves from these provisions by a specific declaration to that effect when becoming a party to the convention; the legal effect is the same as if they had entered a reservation in respect of these provisions. Real life consequences of such action are evident: parties that are not ready to grant special treatment to assistance personnel, risk to be left without assistance at all. Here again subsidiary arrangements of a bilateral or regional nature may fill the gap, provided they are agreed upon well in advance and not only when the accident has already occurred.

It should be noted that efforts aimed at including a reference to Principle 21 of the Stockholm Declaration into the preambles of these two conventions had failed. Evidently, those negotiators, who had in mind a maximum freedom of manoeuvre for their nuclear industries did not appreciate any language related to "the responsibility to ensure that activities within their jurisdiction or control do not cause damage to the environment of other states or of areas beyond the limits of national jurisdiction". Looking back to the LRTAP-Convention, where such language had been included in the preamble, it should be recalled, that in that other context efforts aimed at including nuclear radiation into the concept of long-range transboundary air pollution had equally failed. Nuclear issues are sometimes shelved in environmental negotiations by the argument that they are something special that should receive special treatment. This argument was used again when during negotiations leading to the Basel Convention on wastes, to be presented in a later section, the question of managing and transporting nuclear wastes was raised. It has to be expected that a similar line of argument will be used in the process of treatymaking related to space environment.

These two conventions on nuclear accidents break new ground. In spite of their gaps and deficiencies, amply demonstrated above, they are setting rules in an area, where no rules had been in existence before. Confronted with the choice between a weak convention or no convention at all even ambitious negotiators favored the former option, simply because they believed that future endeavors would be able to build on the groundwork already achieved. A similar reasoning was applied during recent negotiations on global climate change (see below). Treatymaking in the field of environment has to be considered as an ongoing process, an area where legal rules are not fixed once and for all, but as an area where legal rules are subject to continuous adaptation in the light of changing circumstances (scientific and technical progress, evolution of economic interests etc.).

4 Depletion of the ozone layer

The thinning out of the stratospheric ozone layer as a consequence of the increasing impact of chlorides on that protective shield has not only caught public attention (ozone-hole) but has also generated a sustained diplomatic activity. Legal instruments at present available to create and coordinate international action in favour of the ozone layer may well be considered a full fledged "international regime" (Keohane – Nye 1989, 19; Krasner 1983; Young 1989, 22), which is evolving in order to adapt itself to changing circumstances (new insights of science, technological developments). These legal instruments also represent a typical case of the structure becoming more and more common in environmental treaty-making: a framework-convention supplemented by protocols or a single protocol. The negotiating history of this regime goes back to the early eighties and has to be considered, at least for the time being, as open-ended because it remains subject to renewed negotiations either on accelerating existing reduction schedules or on including new substances into the existing schedules (Benedick 1991, Brunnée 1988, Lammers 1988, Lang 1988b, Szell 1991).

In 1985 the *Vienna Convention on the Protection of the Ozone Layer* was signed. It served as the initial stepping stone for all international endeavors undertaken hence in this area. However, apart from the institutional structure (Conference of the Parties, Secretariat) this treaty contains only procedural duties, which may be summarized as follows:

–   to take appropriate measures to protect human health and environment against adverse effects resulting or likely to result from human activities which modify or are likely to modify the ozone layer;

–   to cooperate by means of systematic observations, research and information exchange in order to better understand the modification of the ozone layer;

–   to adopt legislative or administrative measures and to harmonize policies aimed at controlling, limiting etc. human activities detrimental to the ozone layer;

–   to cooperate in the formulation of agreed measures etc., with a view to the adoption of protocols.

Especially this last mentioned duty, to be qualified in legal terms as a "pactum de negotiando", has become the basis of further negotiations leading to more concrete and substantial obligations. An ambitious group of states (USA, Canada, Scandinavian countries) had wished that already in the Vienna Convention were included certain reduction obligations in respect of substances with an high ozone depletion capacity. As, however, the entire process was built upon consensus this goal could not be attained in the light of considerable resistance coming from the European Communities and Japan. Less than 18 months later, the Convention had not yet entered into force, this resistance had weakened to the extent that negotiations on a protocol covering in the first instance only chlorofluorocarbons could take off.

In order to obtain a sufficient amount of flexibility that would be built into these protocols, the Convention provided for a less stringent amendment procedure in respect of protocols than in respect of the Convention itself (two-thirds vote instead of three-fourth majority). As regards changes in respect of technical and other annexes even a non-objection procedure was provided for in order to facilitate quick adaptation to new circumstances. A similar graduation applies in respect of the entry into force: 20 instruments of ratification are required for the Convention, whereas only 11 such instruments are required for a protocol, except otherwise provided in such protocol (this exception was then used when the Montreal Protocol was adopted).

After less than ten months of negotiations the *Montreal Protocol on Substances that Deplete the Ozone Layer* was adopted in September 1987. By not waiting for the formal entry into force of the frame-work convention this protocol started very early to emancipate itself from the convention. In addition to procedural duties (assessment and review of control measures, reporting of data, control of compliance, cooperation in areas such as research, development, public awareness and exchange of information, promotion of technical assistance to developing countries) and the establishment of appropriate organs (Meeting of Parties, Secretariat), which are distinct from the institutions under the framework convention, the treaty establishes some important substantial obligations:

–   The consumption of certain chlorofluorocarbons has to be reduced by 20% until 1993/1994;

–   The consumption of the same substances has to be reduced by another 30% until 1998/1999 unless a two-thirds majority representing two thirds of the total consumption of parties decides otherwise;

–   The consumption of halons has to be frozen at the 1986 level at the latest three years after entry into force.

This set of obligations is surrounded by a host of regulations, which contain exceptions and special treatments for various groups of countries (EEC, planned economies, developing countries). These exceptions had to be consented to in order to have a membership as broad as possible. Furthermore, in order to avoid violations of the protocol by transferring the production of controlled substances into the territory of non-parties, a relatively tight system of controlling trade with such non-parties had to be added. The establishment of mechanisms for controlling the compliance of states with their duties was only mentioned; firm action was postponed until after the entry into force.

This entry into force was subjected, as mentioned above in respect of other treaties (MARPOL), to a double majority; the eleven instruments to be deposited for that purpose would have to represent at least two thirds of 1986 estimated global consumption. By means of this double majority the United States and their chemical industry were assured that the Protocol would not enter into force unless ratified by the European Communities, which again would imply that all major competitors could act on a level play ground.

An elaborate system of adaptation to changing circumstances was devised to give the protocol the necessary flexibility. So-called "adjustments", which may render reduction schedules in respect of substances already included much steeper, could be decided upon by a two thirds majority representing 50% of the total consumption; such decisions enter into force within 6 months and do not need any further intervention by governments or parliaments. Only "amendments", which would cover substances to be added to the existing schedules or removed from them, would be subject to the traditional amendment procedures (adoption by two-thirds and ratification by two thirds, unless otherwise provided for in the protocol). Only three years later the political importance of these seemingly technical provisions was fully recognized.

In 1990 a complete overhaul of the Montreal Protocol, less than two years after its entry into force, took place:

- Under the heading "Adjustments" a total phase-out of CFC's was fixed for the year 2000; as far as halons were concerned they also became part of the reduction scheme and should be phased out by 2000.
- Under the heading "Amendment" the existing coverage was extended to new substances such as carbon tetrachloride or methyl-chloroform.
- Solidarity with developing countries was widely extended; in addition to a ten-year-grace-period for complying with their duties, already granted in the Montreal Protocol itself, these countries should also benefit from a Multilateral Fund, which is supposed to facilitate the switch to less or not polluting substances by means of an ongoing transfer of technology; these countries may even justify their non-compliance by reference to insufficient support received in this context.
- Trade barriers in respect of non-parties were further developed, i.e. trade should not only be restricted in respect of controlled substances but also in respect of downstream products.
- A relatively elaborate complaints machinery was established in order to scrutinize the actual behavior of parties (Implementation Committee).

Developments since 1990 do not justify any extreme kind of optimism: parties are lagging behind as regards the reporting of data; financial contributions to the Multilateral Fund have not been paid according to schedule; efforts persist to water down the role of the Implementation Committee, in order to reduce it to a simple advisory body (Lang 1992).

Which preliminary lessons are to be learned from the ozone depletion regime for treaty-making in outer space?

- Any kind of regime supposed to cover the global commons should be as comprehensive as possible; all stakeholders should be on board.
- If parties are unevenly affected by the treaty, if the burden of treaty-compliance is unequally distributed, it is necessary to provide for special treatment of those who may be tempted to stay outside the regime ("hold-outs"); by means of exceptions and other advantages they should be enticed to join the regime.
- To the extent that the regime is affected by uncertainties in respect of scientific and technological knowledge as well as economic feasibility it should be constructed as flexible as possible in order to follow quickly the respective changes.
- Special attention should be given to regime-maintenance (control of treaty-compliance); sensitive issues such as sovereignty or the exclusiveness of internal matters arise mainly during the implementation phase.

5 Management and movement of hazardous wastes

Space debris has become a source of concern for all countries interested in the various uses of outer space. Although such debris cannot be assimilated to wastes and hazardous wastes on earth a brief glance at treatymaking in this particular field may be helpful.

International action in this field became urgent as the increase of transboundary transfers of hazardous wastes seemed to escape the control of national authorities. Public attention was aroused when some cases of illegal shipments to developing countries became known. Different levels of economic development are frequently at the origin of these transfers. As disposal within industrialized states becomes more and more difficult and expensive – a consequence of public awareness and the ensuing environmental legislation – the generators of waste try to transfer it abroad. There disposal may be less expensive because of a lower level of sensitivity in public opinion, less stringent laws, less efficient administrative agencies, less independent law courts,underpopulated areas serving as disposal sites, the need to earn foreign exchange in order to reduce foreign debt etc. Thus developing countries – including in the past the Communist countries of Central and Eastern Europe – were the most likely candidates for importing hazardous wastes.

Wastes, their disposal and transport, were the subject of several rule-making efforts. However, most of them were either restricted to the regional level (EC-directives) or had only soft-law quality (UNEP-Cairo-Guidelines, various OECD recommendations and the IAEA-Code of Practice) or were limited to a single medium (London Dumping Convention).

Thus the *Basel Convention on the Control of Transboundary Movements of Hazardous Wastes and their Disposal* of 1989 represents the most comprehensive treaty-making effort in this area (Lang 1991b, Cusack 1990, Hackett 1990, Handl 1989, Handl-Lutz 1989). It has been subject of considerable criticism as regards lacunae, vague formulations, issues set aside for further consideration (e.g. responsibility) etc. However, a full evaluation of a treaty can only take place after it has stood the test of reality at least for a number of years. Its main provisions should be briefly spelled out as follows:

– Parties shall reduce the generation of waste; this duty is not an absolute one because social, technological and economic aspects may be taken into account.
– Adequate disposal facilities should be made, if possible, available in the country of origin.
– Transboundary movements of waste shall be reduced to the minimum, if this is consistent with the sound and efficient management of waste.
– Exports are not to be allowed to states were such imports are prohibited or where the government of the exporting country has reason to believe that the wastes will not be managed in an environmentally sound manner; this same condition applies to the government of the importing country.
– Illegal traffic, as for instance traffic without appropriate notifications or without consent (importing state) has to be prevented or punished.
– Trade restrictions are to be imposed in respect of non-parties.
– The state of export has to ensure that the wastes are taken back by the exporter, when the transboundary movement cannot be appropriately completed.

These duties to be fulfilled by the states parties rest upon a complex network of relations between various governments (exporting, transit, importing state) and the individual generator, exporter and importer of wastes. Issues of administrative law, trade law and penal law are closely intertwined. To the extent that the generators of space debris are to be private entities, experience gained during the implementation of the Basel Convention may be of some interest to those engaged in rule-making for outer space and the application of these rules.

In addition to this substantive regulations the Convention contains a set of procedural obligations at the intergovernmental level (cooperation in respect of information exchange etc.) and an equally impressive institutional structure. This machinery will be badly needed, especially in order to clarify whatever uncertainties were written into the text by its original drafters. This may apply inter alia to the procedure of compliance control (verification), which does not attain the level of concreteness realized by the Montreal Protocol and its follow-up.

As regards specific legal techniques for the coming into life of the treaty or its further evolution and adaptation to changing circumstances it should be mentioned that the possibility of protocols is provided for and that amendment conditions (convention, protocols, annexes) follow closely the graduation-rule contained in the Vienna Convention on the Protection of the Ozone Layer. There do not exist any double majority requirements (entry into force, amendments) as established in the Montreal Protocol.

6 Global climate change

Except for the Law of the Sea negotiations rarely a negotiating process has generated so many confrontations among a broad variety of actors. Regime-building has rarely been as arduous as in this case: high economic stakes were involved – some developing countries depend for much of their income on fossil fuels; one industrialized state refused to change its life-style highly dependent on the consumption of fossil fuels; most industrialized states were ready to accept quantified cuts in respect of greenhouse gas emissions; low-lying island states were afraid that the rise of ocean levels may threaten their very survival; many developing countries believed that the climate issue would give them a better leverage in order to create a global economic system which reflects their interests to a much greater degree. Against this background of economic and political tensions exacerbated by the media and an increased involvement of the private sector (business, green movements etc.), it should be considered no minor achievement that negotiators were after all able to agree on a text. Although considered as totally insufficient by some governments and many interested groups located in the North, this treaty may well serve as the basis for further action in the years to come.

The *UN-Framework Convention on Climate Change* signed at the Rio Summit (UNCED) by 154 governments confirms an already well-known trend in environmental treaty-making: a step-by-step approach, that is based on an umbrella-treaty with relatively broad and indeterminate obligations, a treaty to be supplemented by so-called "protocols" which are supposed to contain time-targeted and quantitatively specified obligations. A similar approach had been followed in the above-mentioned instruments in respect of air pollution and ozone depletion. As this Climate Convention has frequently been compared with the various steps of the ozone depletion regime a preliminary assessment should be added: the Climate Convention goes well beyond the content of the Vienna Convention (especially as regards targets to be attained) and to some extent it even anticipates the London Amendment to the Montreal Protocol (institutions to control implementation and compliance, link between compliance by developing countries and the flow of funds and the transfer of technology).

Having thus situated the exact place of the Climate Convention in the broader process of environmental treaty-making, its specific features should hereinafter be highlighted:

- The overall objective of the Convention is to stabilize greenhouse gas concentrations in the atmosphere at levels that would prevent dangerous anthropogenic interference with the climate system. Whatever caveats were added to this objective – natural adaptation of ecosystems to climate change, security of food production, sustainable development – a basic albeit very broad obligation has been determined.
- Commitments under the Convention comprise two categories, those applying to all parties and those applying to industrialized states only. This distinction and differentiation may well collide with longstanding principles such as the sovereign equality of states. It does, however, reflect a real life situation, namely that industrialized states contribute to a much larger extent to global warming than developing countries. It should also be recalled that the principle of differentiation and graduation is not only well known in international trade law (GATT) but also in the Law of the Sea Convention.
- Commitments of all parties include obligations to report on the status quo (inventories of emissions by sources and removals by sinks) and to "formulate, implement, publish and regularly update ...programmes containing measures to mitigate climate change...". In spite of the requirements of reporting and "peer review" contained in other provisions of the Convention, the "soft" nature of these general obligations cannot be denied.
- To this "soft" nature of general commitments corresponds the equally weak nature of specific commitments industrialized countries are supposed to fulfill. These states recognize that "the return by the end of the present decade to earlier levels of emissions" would contribute to a modification of present trends without, however, formally committing themselves to such a return. Indirectly, namely by referring to duties of information on policies and measures, they announce their intention to return individually or jointly to their 1990 emission levels, without, however, committing themselves in clear legal terms to this target. Again "soft law" provisions prevail over "hard law" obligations. This vagueness of obligations amply demonstrates that the industrialized states, even if they had been ready to enter into more stringent obligations, preferred a weak convention to a convention without the participation of the United States, which from the outset were opposed to firm commitments.
- As regards so-called "general" commitments (reporting etc.), the only ones also incumbent upon developing countries, it should be noted that these obligations are to be met by developing countries only to the extent that industrialized states are ready to assist them by transferring financial resources and technology. Thus a new

"conditionality", mentioned for the first time in the environmental field in the London Amendment to the Montreal Protocol, has gained broader recognition in international law.

- Furthermore a host of special situations may be invoked by states and especially developing countries in order not to abide by their obligations under the Convention; these exceptions range from small island countries to rich petroleum producing countries. Almost all countries are eligible for special treatment provided they encounter serious difficulties in switching to alternative sources of energy (i.e. other than fossil fuels).
- A broad body of institutional provisions tries to make up for the lack of more substantive and stringent obligations related to concrete measures. Whereas the Conference of the Parties can rely on a subsidiary body in charge of assessing the implementation of the Convention, an achievement of some importance, no definite solution could be found for a financial mechanism. Distrust between developing countries and the Bretton Woods institutions persisted and continued to cast a shadow on these provisions. The Global Environment Facility even in its revised version was only accepted on an interim-basis.

Turning to specific legal techniques for flexibility and adaptation of the treaty to changing circumstances no new devices have been developed; amendments need a three-fourth majority for their adoption and entry into force, whereas annexes after their adoption along the same lines enter into force only for those parties that did not object within a certain period. As regards protocols to be approved at some later date specific provisions have still to be adopted in order to facilitate their adaptation to scientific and technological progress.

This Convention is by legal standards certainly not the most advanced achievement in environmental treaty-making. In view of the aforementioned political and economic difficulties it represents, however, the maximum of progress to be attained in 1992.

B. ENVIRONMENTAL NEGOTIATIONS

In this chapter shall be summarized the attributes of environmental negotiations as they are reflected in recent accounts of negotiating experience and related research (Lang 1991a, Sjöstedt 1993, Lang 1993). The focus of this chapter will not be on substance but mainly on procedure.

1 Number of actors

Except for problems occurring in the immediate vicinity, which are to be settled in a bilateral way between governments directly concerned, most environmental negotiations take place among a larger number of actors; the context may be a regional one (see LRTAP-negotiations) or a global one (see the other above mentioned negotiations). Thus the specifics of multilateral negotiations have to be taken into account; these include issues such as coalition-building, leadership in a variety of forms (governments, individuals etc.), the role of international organizations or the performance of regional and other groups.

In most recent writings on negotiations the term "actor" has a double meaning; it covers the participating governments as well as the individuals sitting at the conference-table. There does exist a strong link between these two "actors": governments give instructions to the individual negotiator, who is reporting back requesting new instructions, suggesting a slight modification of the initial position or proposal etc. The individual negotiator in many instances may be much more than a mere "puppet on the strings"; he or she may enjoy a certain leeway within the instructions they have received; he or she are supposed to use their personal skills to persuade the other side, extract concessions from the other side etc. Thus the personal factor, the professional and cultural background of a negotiator have their impact on the course of negotiations.

The term "actor" goes also well beyond national governments. Especially in recent years a new actor has emerged, the so called regional integration organization, also known as European Economic Community. This "collective actor" made its first appearance in international trade negotiations. As it was able to acquire more and more competences, including those in the field of the environment, it became an active participant in more and more environmental negotiations. Even prior to the amendment of its basic document (Single European Act) which conferred major functions in the field of the environment on the Community, its representatives insisted on being treated as full-fledged actors in these negotiations and even requested that the Community be considered a party to the legal instruments resulting from these negotiations. Whereas in 1979 this status could only be obtained with some difficulties (LRTAP-negotiations were delayed for several months) the Community was already a key-player in the negotiations on the protection of the ozone layer (1985, 1987); in the latter case this role accrued to the Community almost naturally as it was a major stakeholder in the entire exercise, producing and exporting a great part of those substances (chlorofluorocarbons) that were to be controlled and prohibited by the agreement.

The Community is not an actor easily to be dealt with, because its positions, views and proposals have to be agreed upon among its constituent members before these positions etc. are presented to the negotiating conference as a whole. Thus, from an EC-members perspective global environmental negotiations present themselves as a two-tier endeavor; first agreement has to be sought within the Community; second, this "common view" is introduced into the global forum. Such cumbersome proceedings have to be repeated every time the Community has to adjust its position. Any position taken by negotiators acting on behalf of the Community carries much weight because it reflects the impressive political and economic power wielded by its twelve member-states.

2 Role of international organizations

International organizations as fora as well as participants (and therefore also some kind of "actor") are a well known feature of multilateral negotiations. Their visibility was high in most treatymaking endeavors presented in the previous chapter. The International Maritime Organization, the Economic Commission for Europe, the International Atomic Energy Agency, the United Nations Environment Programme were among the institutions which performed a variety of functions:

First and foremost they were the meeting ground; second, they provided for objective expertise of the scientific and technical as well as of the legal kind; third, their chief executive officers or legal advisers were helpful in pushing ahead stalled negotiations or devising compromise formulae that were all the more acceptable to governments as these formulae did not come from another self-interested participant but from a seemingly impartial institution; fourth, as almost all legal instruments, in order to be fully operational need some executive agency, these functions normally accrue to the organization already in charge of the preceding negotiations.

Since international relations remain highly compartimentalized there does not exist a single international organization, which would be responsible for environmental treatymaking as a whole. Although such effort at centralization, or at least improved coordination, could well be justified under the principle of cost-effectiveness, it is most unlikely to occur. Lack of coordination in international affairs reflects to a large extent lack of coordination within national governments. As an example one may refer to the treatment of environmental issues in the context of the General Agreement on Tariffs and Trade: only in 1991/1992 was the broad membership of GATT ready to acknowledge a challenge, namely that certain environmental treaties (Montreal Protocol, Washington Convention on Endangered Species -CITES) contained regulations on trade which may or may not be fully compatible with the rules of international trade law.

International organizations are an indispensable element of environmental negotiations; their role is certainly not superior to that of governments, but governments would be at a loss if they had not international/intergovernmental organizations available in order to perform those multiple tasks referred to above.

3 Role of science

Science and its various roles to be played before, during and after negotiations constitutes certainly one of the most outstanding features of environmental negotiations. Science, being understood at present as the natural sciences, intervenes at various moments in the rule-making or treatymaking process. At first, science is supposed to perceive any new, or yet undiscovered threat to the environment. Second, once scientific insight has generated by means of public opinion a sufficient amount of political will, which triggers national and international action, it accompanies negotiations conducted by lawyers, diplomats and others supporting their arguments with scientific evidence. Third, science will be requested to participate in most endeavors required to monitor the actual compliance of states with their obligations entered into under the various treaties.

Among the above-mentioned case-studies science was most important in negotiations on ozone depletion, because only scientists could decide on the basis of certain specific criteria which substances had the highest ozone depleting capacity and therefore had to be eliminated on a priority basis. In case of the MARPOL-Convention it were the various informal networks linking scientists across national boundaries, which largely facilitated the conclusion, and later on the coming-into-force of the agreement.

Scientific uncertainty remains a major problem in environmental negotiations. As long as there is no formal and hard proof of certain threats governments continue to be reluctant in accepting commitments which generate costs, at least short term costs, to be borne by the respective economy. The so-called "precautionary principle", which would authorize action even in the absence of firm evidence, has recently been included in several political declarations but cannot yet be considered a principle of international law.

One may even qualify "science" as an "actor" in its own right. This view could, however, be challenged, because actors are supposed to represent political or economic interests in the more traditional meaning of the term. On the other hand science may be deemed to act on behalf of the "environment" or the "biosphere" which do not easily find strong entities defending them against the encroachment of various human activities.

4 Role of non-governmental organizations

Non-governmental organizations (NGO's) do appear not only in environmental negotiations but also in other international activities such as the protection of human rights. In the specific context of the environment they represent either business interests or ecological interests. In order to achieve an optimal defense of their interests these organizations closely associate themselves with science and media, which frequently are their most trusted allies and/or instruments. NGO's approach environmental negotiations at two different levels. In the context of national decision-making they aim at influencing the competent governmental authority that issues instructions to negotiators at the conference table. At the international level NGO's try to influence negotiators either directly at the conference-site by means of persuasion or indirectly by appealing to governments back home to issue new and/or different instructions to the negotiator.

The impact of NGO's on the negotiating process cannot be determined with any kind of precision. They represent social forces which cut across national boundaries and they are subject to one major criticism: To whom are they accountable? Whom do they really represent? How democratic are their internal mechanisms of decision-making? As long as these questions have not received adequate and satisfactory answers, most governments refrain from giving NGO's a major say in environmental or other negotiations. Governments, at present, are also likely to limit the role of NGO's, because NGO's although having a high capacity to articulate views and to postulate certain outcomes, remain unable to implement commitments. This situation could be somewhat remedied by including NGO's in processes of compliance control and monitoring, thus raising their sense of responsibility and commitment.

5 Role of the media

Problems of environmental protection concern everybody. Everybody feels threatened when a major accident with radioactive or chemical release occurs in a neighboring country; many people feel threatened by the perspective of ultraviolet-radiation, which is likely to grow as a result of any further depletion of the ozone layer. Thus environmental negotiations score higher in many media than most other types of international negotiations. This media impact, however, remains rather diffuse, unless it is focussed on specific issues either by NGO's or by certain national delegations. During the negotiations on the ozone layer it was the US-delegation, which by means of continuous contacts with the media tried to build-up a climate of public expectations which should induce still reluctant delegations (mainly those with EC-membership) to agree to substantial reductions of emissions. Further research will tell us, whether the relatively flexible stance finally adopted by the European Community was brought about by this manipulation of public opinion from the outside or rather by an internal process of rethinking threats and options. Finally it should be acknowledged that due to the increasing interdependence of politics and economics as well as a consequence of the so-called media-revolution ("global village") the interest of the broader public for foreign policy issues has grown during the last decades much faster than in any earlier period. This also explains the heightened interest of the public for environmental negotiations.

6 Other factors

Environmental negotiations are influenced by a number of factors one may find to a lesser extent in other negotiations:

- The *time factor* plays an important role insofar as frequently short term costs have to be exchanged against long term benefits. It is crucial to persuade those supposed to bear the short term costs (emission reductions, disadvantages in competition etc.) that these costs will be more than compensated by long term benefits (e.g. income from licensing of new technology). If such expectations don't come true within a reasonable period of time reluctance to swallow other restrictions and ensuing costs will grow. This time factor should also remind us, that future generations, the real victims of our present ill-behavior, do not have fully recognized defenders of their interests, that the question of "intergenerational equity" has not yet received a satisfactory answer.

– Some regions or countries are less exposed or more resistant to environmental threats or damage than other regions or countries. This *variation of exposure and/or vulnerability* leads to different reactions when confronted with the issue of burden-sharing. Countries close to Antarctica feel much more threatened by ozone depletion than countries in more moderate regions; low lying island countries feel much more threatened by global warming and the ensuing rise of ocean levels than mountain regions. The so-called "critical-loads"-approach in the LRTAP-context tries to differentiate along the lines of this variation of vulnerability or absorption capacity of different regions.

– Because environmental threats continue to exist even beyond the formal conclusion of negotiations and the implementation of their outcome, most of these negotiations have to be considered as an *open-ended* exercise. New substances may have to be included into the regime; initial CFC-reduction schedules were too slow in the light of the rapid thinning-out of the ozone layer; new techniques in ship-building required the modification of provisions related to marine pollution. As environmental treaties have to be closely associated with the progress of science and technology they have to be construed as highly flexible mechanisms. This implies that these negotiations rarely know a final point.

– Traditional notions such as *national sovereignty* or *national borders* are not likely to retain their importance. Paling against the size of various threats these concepts, although formally adhered to, lose much of their content, if they come to be considered as obstacles to international cooperation. The relevance of sovereignty also remains an important issue in the North-South dialogue on environment and development. Third World countries, many of which acceded only a few decades ago to formal statehood, and which suffer from the weaknesses of their respective governments, cling to sovereignty almost as the main remedy for some of their problems. Sometimes sovereignty is also used as a shield to protect concrete economic interests; some actors in the negotiations on nuclear accidents used this argument in order to protect their nuclear industries against too intrusive obligations of notification and information.

– *Linkages* do exist between environmental negotiations and a number of *vested interests*. Most prominent among those interests is the economic feasibility of restrictions, reductions and other control measures. As it has been demonstrated in the above case-studies (LRTAP and others) these interests are practically never neglected by governments. No government can afford to put its industry at a disadvantage in international competition by imposing extra costs. Some environmental negotiations may even aim at preparing a level ground for industries competing on world markets as it happened in the context of ozone depletion negotiations. These same negotiations should teach us caution in respect of using economic instruments such as trade barriers to achieve environmental goals. Trade distortions caused by barriers established in order to create tightly controlled environmental regimes should be kept to the absolute minimum. Otherwise it could well happen that the basis of these regimes is eroded by a backlash of economic interests. Economy and ecology have to attain a situation of equilibrium, if damage is to be avoided for both.

7 Evolution of the negotiating situation

In the early part of the nineties the overall political situation has changed. The traditional East-West-confrontation of two military blocks, of two incompatible political systems has ceased; market economy proved to be superior to planned economy; stock-taking in previously Communist countries reveals an extremely sad state of the environment.

In the diplomatic arena environmental issues rarely were a bone of contention between East and West. Quite to the contrary, environmental issues were frequently used to highlight things in common, to bridge the political gap. Scandinavian countries striving for international action against long-range transboundary air pollution, especially transboundary fluxes of sulphurdioxide, were greatly helped by the Soviet wish to strengthen détente, to weaken the human rights issue within the CSCE context, by diverting attention from human rights to environment. This ad-hoc coalition of East and North was the point of departure for the LRTAP-convention.

The evolution of internal politics of leading countries may also affect the overall negotiating situation. The "greening" of German politics changed to a considerable extent the orientation of German environmental diplomacy. Since the early eighties the FRG-stance on air pollution, for instance, has changed from reluctance towards international action to a very positive attitude. Thus, it should be recalled that environmental negotiations are to a great extent a prolongation of domestic environmental policies. Environmental negotiations may even be considered an instrument of domestic policies insofar as the responsible decision-makers realize that purely domestic measures do not suffice to combat a particular threat such as ozone depletion or global warming.

This close relationship between domestic politics and environmental negotiations is also one of the reasons behind the newly emerging North-South conflict. Leaders in the South are primarily pre-occupied with development, with building their society and nations upon firm ground. They espouse environmental goals only to the extent that these goals do not collide with their development objectives. For many of them environmental negotiations and environmental agreements do have one primary purpose: to obtain new financial assistance from the North, to strengthen technological cooperation, i.e. to remove obstacles to the transfer of technology. Any kind of global environmental negotiations will be deeply affected by these mainly financial issues.

C. LESSONS

1992 has been proclaimed the International Space Year, a year in which also international action in favour of the environment attained its strongest momentum so far. It therefore seems quite natural to merge these two endeavors: to mobilize support of space activities for the protection of the environment on earth and to move ahead as regards the protection of the space environment itself.

A review of legal opinions on the state of space law reveals a broad consensus to the effect that rulemaking or treatymaking in favour of the space environment should be accelerated, that environment-related provisions in the existing space treaties are far from satisfactory. Most papers delivered during various colloquia of the International Institute of Space Law (1987, Brighton; 1989, Torremolinos-Malaga; 1990, Dresden) reflect this emerging consensus; the same is true when reading papers submitted to the Cologne Colloquium in 1988 (Böckstiegel 1990). This quest for a new international regime has been supported by the most prominent authorities in space law (Christol 1991, 261). It has also found expression in the conclusions and recommendations adopted by the International Law Association during its Cairo Meeting in April 1992.

As the views of academia are sometimes much ahead of political realities one should enquire whether diplomatic-legal activities have kept pace with the aforementioned ambitious goal-setting (Malanczuk 1990–1992). As a matter of fact it took the Legal Subcommittee of the UN-Committee on the Peaceful Uses of Outer Space (COPUOS) more than ten years to arrive at the penultimate phase of drafting its "Principles relevant to the Use of Nuclear Power Sources in Outer Space" (Kopal 1991, 103-122). These principles represent basic political commitments but are still far away from legally binding obligations. Even less successful were efforts to put the issue of space debris on the agenda of the Committee.

Turning now to the very subject of this paper, namely lessons to be learned from environmental treatymaking one is confronted with a legitimate question: Is there a need for space law to learn lessons from environmental law and environmental treatymaking? In some instances space law has been for some time well ahead of environmental law. This applies in particular to the duty of consultation which was hailed as a major achievement when it was included into the LRTAP-Convention in 1979, whereas such obligation was already contained in the Outer Space Treaty of 1967 (Article IX). A similar advance of space law can be recognized in respect of responsibility and liability of states. Whereas there does not

exist any major environmental treaty expressly stipulating the responsibility of states, the 1972 Convention on International Liability for Damage caused by Space Objects does contain very specific and stringent obligations to that effect. A brief glance at the LRTAP-Convention teaches us even the opposite; the question of responsibility was excluded by means of a footnote from the purview of this legal instrument. Still, the question asked at the beginning of this paragraph should be answered in the affirmative. Why? Simply because only little progress has been achieved in respect of space environment. Since the Moon Treaty in 1979 no major legal instrument was adopted in respect of outer space. In the meantime treatymaking and rulemaking in respect of the environment on earth not only has caught up; it even went well beyond the achievements of space law. This lagging behind of space law most probably was caused by political considerations and the impact of security perceptions related to the concept of "Star Wars", which dominated the thinking of superpowers during most of the eighties. Now, the superpower rivalry having passed away there should at least be a "window of opportunity" to push ahead with treatymaking in respect of outer space and in particular its environment.

Such treatymaking could well draw some lessons accumulated during negotiations on the environment:

– The first lesson would be that *all stakeholders should be part of the consensus* upon which the new legal instrument is to be built. In this respect there does exist a major difference between space and environment in general. Only a very limited number of states are able to engage in space activities which may pollute space environment, whereas almost all states may be considered as polluters of environment on earth. Thus, one may be tempted to restrict space treaties to these space-faring nations. However, as space is deemed to be part of the "common heritage of mankind" and as every nation may feel adversely affected by pollution resulting from space activities, it is unlikely that an agreement on various elements of the space environment would meet with the approval of the community of states as a whole if its membership were restricted to space-faring nations only.

– The second lesson, following from the first one, reads as follows: As not every contracting party will bear the same burden, as not everyone will be affected by treaty obligations in the same way, *such treaties are likely to be assymetrical.* This lesson has already been learned in respect of the Montreal Protocol and its follow-up; this lesson also emerged during negotiations on global climate change which means that industrialized states have to assume much more important obligations than developing countries. These latter countries request a special treatment and are likely to obtain it. However, as their impact on the space environment will be moderate, at least during the foreseeable future, their leverage in this context will be much weaker than in the climate change negotiations.

- The third lesson relates to *equity*. Space treaties shall enter into force only if all or most space-faring nations have joined the treaty; this implies that a *double majority requirement* should be included, especially in order to maintain a minimum of balance in respect of costs and burden.

- The fourth lesson relates to *flexibility*. Space treaties, to no lesser extent than environmental treaties, are likely to be subject to rapid change of scientific and technological know-how. Thus, they need some in-built mechanism of amendment and adaptation. Scientific and technical issues should only be addressed to in protocols or annexes, which may be modified more easily than the main body of the treaty.

- The fifth lesson would be to leave *substantial leeway to science*. Science should be able to articulate itself in special bodies or fora, in which views can be expressed without political or other constraints. Science, most probably, will not be the master of the further evolution of the regimes; science should, however, be the adviser of decision-makers who bear the final, political and economic responsibility.

- The sixth lesson covers the *step-by-step approach* followed in numerous environmental treaties: a relatively general framework convention covering one of the elements of space environment is to be supplemented by protocols which contain more stringent rules and regulations. It may well be that this framework convention has to be preceded by "soft law"-rules (see the above mentioned Principles on Nuclear Power Sources), in order to overcome the initial resistance of some space-faring nations. It should be noted that this sequence of "soft law" and "hard law" repeatedly occurred in the field of environment law (see IAEA guidelines being transformed into conventions on nuclear accidents; see also OECD soft-law on transboundary pollution influencing the ECE-draft on LRTAP).

- The seventh lesson teaches us the *need for a sectoral approach*: Governments rarely are ready to enter into general commitments the scope of which they cannot easily evaluate. Therefore a treaty on space environment in general, even if it were a framework convention, is unlikely to succeed. One should rather try to draft several treaties on distinct issues or elements such as "geostationary orbit", "nuclear power sources", "space debris" etc.

- The eighth lesson relates to the *importance of institutions*. Since most legal instruments do not affect real life unless they are supported by some machinery, any treaty aiming at the protection of some elements of space environment should establish institutions which are sufficiently strong to facilitate the continuous evolution of treaty texts and to control compliance with treaty obligations.

- The ninth lesson draws attention to the issue of *responsi-*

bility/liability. In this respect environmental treaties have failed. Except for stipulating the civil law liability of various operators (nuclear installations, oil tankers) they have not been able to fix the issue of state responsibility. Here the provisions already contained in the Liability Convention should be applied to other elements of the outer space environment (Gehring-Jachtenfuchs 1988, Malanczuk 1991); it may even be advisable to further develop these provisions.

- The tenth lesson covers the question of *compliance control* and monitoring. In order to attain a minimum level of credibility any agreement on elements of space environment should dispose of a minimal outfit in this respect: peer review (national reporting scrutinized by other parties) appears to be the bottom-line of such outfit; third party surveillance would be the ultimate achievement, be it control by a political or a judicial body. Considerations of protecting national sovereignty are, however, likely to delay the early achievement of this task.

Whatever lessons learned during environmental negotiations are to be applied to treatymaking for space environment, there will always remain major differences, especially in respect of the strong security considerations prevailing in the evolution of space law. A continuous dialogue between experts knowledgeable in these two branches of international law would further progress in each area; whatever barriers exist between them should be torn down for the benefit of mankind as a whole.

BIBLIOGRAPHY

ADEDE A.O., The IAEO Notification and Assistance Conventions in Case of a Nuclear Accident, London 1987;

BENEDICK Richard, Ozone Diplomacy, Cambridge (Mass.) 1991;

BÖCKSTIEGEL Karl-Heinz (ed.), Environmental Aspects of Activities in Outer-Space, Proceedings of the Cologne Colloquium of the International Institute of Space Law, Köln 1990;

BRUHA Thomas, Internationale Regelungen zum Schutz vor technisch-industriellen Umweltnotfällen, Zeitschrift für ausländisches öffentliches Recht und Völkerrecht 1984, 1–63;

BRUNNEE Jutta, Acid Rain and Ozone Layer: International Law and Regulations, Dobbs Ferry (N.Y.) 1988;

CHOSSUDOVSKY Evgeny, "East-West"-Diplomacy for Environment in the United Nations: The High Level Meeting within the Framework of the ECE on the Protection of the Environment, Geneva (UNITAR) 1989;

CHRISTOL C.Q., Suggestions for Legal Measures and Instruments for Dealing with Debris, in: BÖCKSTIEGEL, Environmental Aspects of Activities of Outer Space (Cologne Colloquium 1988) Köln 1990;

CHRISTOL Carl Q., Space Law, Past, Present and Future, Deventer-Boston 1991;

COCCA Aldo Armando, Environment as a Common Heritage of Mankind, in: Proceedings of the 32nd Colloquium, Torremolinos-Malaga, International Institute of Space Law 1989, Washington 1990, 71–76

CUSACK M., International Law and the Transboundary Shipment of Hazardous Waste to the Third World: Will the Basel Convention make a Difference? American University Journal of International Law and Policy 1990, 393–423;

GEHRING Th. – JACHTENFUCHS M., Haftung und Umwelt, Frankfurt a.M. 1988, 85–110;

FRAENKEL Amy, The Convention on Long-Range Transboundary Air Pollution: Meeting the Challenge of International Cooperation, Harvard International Law Journal, Spring 1989 (vol 30/2), 447–476;

GOLD Edgar, Handbook on Marine Pollution, Arendal-Halifax 1985;

GORKOM Lodervijk, Nuclear Accidents: Two Conventions through Effective IAEA Action after Chernobyl, in: KAUFMANN (ed.), Effective Negotiation, Dordrecht 1989, 149–163;

GÜNDLING Lothar, Multilateral Cooperation of States under the ECE Convention on Long Range Transboundary Air Pollution, in: FLINTERMAN-KWIATKOWSKA-LAMMERS (ed.), Transboundary Air Pollution, Dordrecht 1986, 19–31;

HAAS Peter, Banning chlorofluorocarbons: epistemic community efforts to protect stratospheric ozone, International Organization, Winter 1992, 187–224;

HACKETT D., An Assessment of the Basel Convention on the Transboundary Movements of Hazardous Waste and their Disposal, American University Journal of International Law and Policy 1990, 291–323 ;

HAFNER Gerhard, Das Übereinkommen über Hilfeleistungen bei nuklearen Unfällen oder strahlungsbedingten Notfällen, Österreichische Zeitschrift für öffentliches Recht und Völkerrecht 1988, 19–39;

HANDL Günther,Transboundary Nuclear Accidents: The Post Chernobyl Multilateral Legislative Agenda, Ecology Law Quarterly 1988, 203–248;

HANDL Günther, The 1989 Basel Convention on the Transboundary Movement of Hazardous Waste: A Preliminary Assessment, Proceedings of the 1989 Conference of the Canadian Council on International Law, 367–377;

HANDL G. – LUTZ R. (eds.), Transferring Hazardous Technologies and Substances, The International Legal Challenge, London 1989;

HELLER Michael, Chernobyl Fallout: Recent IAEA Conventions Expand Transboundary Nuclear Pollution Law, Stanford Journal of International Law 1987, 651–664;

International Institute of Space Law, Proceedings of Colloquia, Brighton (1987), Torremolinos-Malaga (1989), Dresden (1990);

JOHNSTON Douglas M., Marine pollution agreements: successes and problems, in: CAROLL J. (ed.), International Environmental Diplomacy, Cambridge (CUP) 1988;

KEOHANE Robert – NYE Joseph, Power and Interdependence, Glenview (Ill.) 1989, Second Edition;

KISS Alexandre – SHELTON Dinah, International Environmental Law, Ardsley-on-Hudson – London 1991 (Transnational Publisher Graham and Trotman);

KOPAL Vladimir, The Use of Nuclear Power Sources in Outer Space: A New Set of United Nations Principles? Journal of Space Law 1991, 103–122;

KRASNER Stephen (ed.), International Regimes, Ithaca (N.Y.) 1989 (5th Printing);

LAMMERS Johan G., Efforts to develop a Protocol on Chlorofluorocarbons to the Vienna Convention for the Protection of the Ozone Layer, Hague Yearbook of International Law 1/1988, 225–269;

LANG Winfried, Frühwarnung bei Nuklearunfällen, Österreichische Zeitschrift für öffentliches Recht und Völkerrecht 1988, 9–18 (1988a);

LANG Winfried, Diplomatie zwischen Ökonomie und Ökologie, das Beispiel des Ozonvertrages von Montreal, Europa-Archiv 4/1988, 105–110 (1988b);

LANG Winfried, Internationaler Umweltschutz, Völkerrecht und Aussenpolitik zwischen Ökonomie und Ökologie, Wien 1989 (Orac);

LANG Winfried, Negotiations on the Environment, in: KREMENYUK V. (ed.), International Negotiation, San Francisco 1991 (1991a);

LANG Winfried, The International Waste Regime, in: LANG-NEUHOLD-ZEMANEK (eds.), Environmental Protection and International Law, London 1991, 147–161 (1991b);

LANG Winfried, Ozone Layer, Yearbook of International Environmental Law 1991, London 1992 (1992);

LANG Winfried, Specific Characteristics of Environmental Diplomacy, in: HÖLL O. (ed.), Environmental Cooperation in Europe, Boulder (Colorado) 1993 (1993);

MALANCZUK Peter, Outer Space, Yearbook of International Environmental Law, 1990–1992;

MALANCZUK Peter, Haftung, in: BÖCKSTIEGEL K.-H. (ed.), Handbuch des Weltraumrechts, Köln 1991 (Heymanns Verlag), 755–803;

MOSER Berthold, The IAEA Conventions on Early Notification of a Nuclear Accident and on Assistance in the Case of a Nuclear Accident or Radiological Emergency, Nuclear Law Bulletin, December 1989, 10–23;

REIJNEN G.C.M. – GRAFF W. de, The Pollution of Outer Space, in particular of the Geostationary Orbit, Dordrecht 1989;

ROSENCRANZ Armin, The ECE Convention of 1979 on Long Range Transboundary Air Pollution, American Journal of International Law, October 1981, 975–982;

SANDS Philippe (ed.) Chernobyl: Law and Communication, Transboundary Nuclear Air Pollution, Cambridge 1988;

SJÖSTEDT Gunnar (ed.), International Environmental Negotiation, Newbury Park (Calif.) 1993;

SZELL Patrick, Ozone Layer and Climate Change, in: LANG-NEUHOLD-ZEMANEK (eds.), Environmental Protection and International Law, London 1991, 167–178;

YOUNG Oran R., International Cooperation, Building Regimes for Natural Resources and the Environment, Ithaca 1989;

ZEHETNER Franz, Grenzüberschreitende Hilfe bei Störfällen und Unfällen, in: PELZER (ed.), Friedliche Kernenergienutzung und Staatsgrenzen in Mitteleuropa, Baden-Baden 1987, 118–140.

21: Regulation of Orbital Debris – Current Status

Howard A. Baker*

Counsel, Trade Law Division (JLT), Department of Foreign Affairs and International Trade, Government of Canada

INTRODUCTION

Currently, there is no legal regulation of space debris. There is, however, a framework of international law into which any regulatory scheme ought to fit.

Although no law is yet in place, efforts are being undertaken by various members of the global space-user community to lay the foundation for future regulation. Technical experts are developing methods for reducing the current quantity of space debris and advocating voluntary restraint rather than legal regulation.[1] States involved in the use and exploration of outer space are developing policies to address the question of space debris management.[2] International organizations are beginning to study the technical, economic, legal and policy aspects of the risk to their space activities posed by space debris.

Eventually and, in this author's opinion, hopefully sooner than later, serious consideration will be given to the implementation of an international legal regime for the regulation of space debris.

The purpose of this paper is to provide an overview of some provisions in public international law which could be applied to the regulation of space debris. Two areas of law will be canvassed: international space law and international environmental law.

In view of the descriptive nature of this paper, editorializing on the impact of current international law on the legal regulation of space debris will be restricted to the following: Any regulation of space debris afforded by current principles of public international law is by inference at best. In most instances, that inference is very weak; yet even if it were stronger, creation of substantial amounts of space debris would be permissible, due to the inadequacies of the existing legal regime.[3]

A. INTERNATIONAL SPACE LAW

Not yet 40 years old, the "space age" has spawned five international treaties relating to the exploration and use of outer space. These are the Outer Space Treaty,[4] the Return and Rescue Agreement,[5] the Liability Convention,[6] the Registration Convention,[7] and the Moon Agreement[8]. While these treaties form the bulk of international space law, there are other agreements which also should be examined for their application to space debris regulation. These include the Partial Nuclear Test Ban Treaty,[9] the Environmental Modification Convention,[10] and the ITU Convention[11].

The major provisions in international space law for protection of the outer space and Earth environments are Article IX of the Outer Space Treaty and Articles 7 and 15 of the Moon Agreement. Other relevant provisions relate to the definition of space debris, the question of State jurisdiction and control over space debris, international responsibility for space debris, the identification of space debris, liability for damage caused by space debris, the questions of nuclear activities and environmental modification, and harmful interference with satellite telecommunication.

1. Article IX of the Outer Space Treaty

Article IX of the Outer Space Treaty may be viewed as the basic provision in space law for environmental protection, in that it attempts to regulate the unfettered freedom to use and explore the outer space environment.[12] As such, space debris is a harm which can be brought within its scope.

Sentence 1 of Article IX provides that in the exploration and use of outer space, due regard should be given to the corresponding interests of all States Parties to the Outer Space Treaty. The principle of due regard would seem to require that States Parties avoid the creation of space debris and attempt to reduce and remove any space debris causing (a) harmful contamination in outer space, including the Moon and other celestial bodies, (b) adverse changes in the Earth's environment, or (c) potentially harmful interference with space activities. These corresponding interests are provided for in sentences 2, 3 and 4 of Article IX. It is not clear, however, whether "corresponding interests" can be found outside of Article IX.

Sentence 2 of Article IX provides for the avoidance of both harmful contamination in outer space, including the Moon and other celestial bodies, and adverse changes to the environment of Earth. Given that "harmful contamination" is not defined, it is unclear which types of space debris, if any, come within its scope.[13] As well, neither harmful contamination nor adverse changes are prohibited in sentence 2, but rather are to be avoided through the implementation of regulatory controls, where necessary.

Sentences 3 and 4 of Article IX apply to scientific, commercial or governmental space activities which may cause potentially harmful interference with space activities of other States.

Under sentence 3, a State Party conducting a space activity has a duty to undertake international consultation, if that State has a reasonable belief that its space activity would prevent the future use of outer space for scientific, commercial or governmental activities. While the consulting State would seem to be obliged to provide information as to the nature of the activity or experiment for which consultation was sought,[14] there is no requirement that the information be either complete or delivered in time for sufficient study prior to consultation. As well, sentence 3 does not provide for consultation procedures; it does not address the question of settling disputes which may arise during the consultation process; nor does it require that any recommendations resulting from consultation bind the parties.

In sentence 4, if a State Party has a reasonable belief, and can demonstrate, that the space activity of another State would prevent the future use of outer space for scientific, commercial or governmental activities, the former State has a right to request consultation. On accession to the request, the requesting State would seem to have a right to receive from the acceding State any additional information as to the nature of the activity for which consultation is sought,[15] although this information need be neither complete nor timely. Sentence 4 is limited in the same manner as sentence 3. Additionally, there is no obligation for the State conducting the activity to accede to the request for consultation.

States need not be space-capable to initiate a sentence 4 request, but would be required to be parties to the Outer Space Treaty. Such States, wishing to raise concerns about space debris, could request consultation if they can identify the space debris and if they can determine which space activity of which State is responsible for that debris.

2. The Moon Agreement

By virtue of its Article 1, the Moon Agreement includes within its scope the Moon, orbits around or other trajectories to or around the Moon, and other celestial bodies in our solar system.[16] It is unclear whether "orbits" and "trajectories" are to be construed as areas of space or as isolated locations in time. Should these terms be interpreted to mean areas of space, then the scope of the Agreement could include all the space in the plane of the Moon's orbit around Earth and enclosed in that orbit, given that a trajectory to the Moon may be plotted anywhere within that plane.[17]

Article 7 of the Moon Agreement[18] enhances the environmental obligations found in Article IX of the Outer Space Treaty through the expression of specific standards of conduct to be followed on the Moon and other celestial bodies.

Article 7 paragraph 1 provides that the "existing balance" of the Moon's environment is not to be disrupted and that measures are to be taken to avoid harmfully affecting Earth's environment. The non-disruption of the existing balance would appear to be a more objective standard than that of "potentially harmful interference" with space activities, found in Article IX of the Outer Space Treaty. The former, therefore, may be more conducive to scientific definition.

Article 7 paragraph 2 obliges States Parties to the Agreement to give notice of all preventive measures taken, thereby increasing the effectiveness of the duty to prevent disruption. This notice may be after the fact, except for the placement of radioactive materials, advance notice of which need only be given to the maximum extent feasible.

Article 15 paragraph 2 of the Moon Agreement provides that a State Party may request consultation either if it reasonably believes that another State Party has breached its duties under the Agreement or is interfering with the rights of the requesting State under the Agreement, or if any activity by another State Party causing potentially harmful interference also disrupts the existing balance of the Moon's environment.[19] Any State Party receiving such a request is obliged to enter into consultation without delay and to attempt to seek a mutually acceptable settlement. If such a settlement is not reached, Article 15 paragraph 3 of the Agreement provides that the States Parties involved are obliged to use appropriate peaceful means to settle the dispute.[20]

The Moon Agreement should not be viewed at this time as a dominant force for preventing harms caused by space debris, however. It has been ratified by only eight States, none of which is, thus far, generally considered to be a space power.[21] If the space nations do accede to the Moon Agreement, it is not clear what standards of conduct will be required to disrupt the existing balance of the environment within the scope of the Agreement.

3. Definition of space debris

International space law treaties contain neither a definition nor a description of space debris. While space debris may be divided into four classes for technical purposes,[22] opinion is divided as to whether the legal scope of space debris includes all technical classes[23].

The growing risk of damage in outer space caused by space debris, particularly in low-Earth orbits, the possible confusion over the literal meaning of "debris", and the need to define the scope of debris all suggest the need for a legal term of art. An explicit definition of space debris may not be necessary, however, if that term is found to be subsumed under an existing space law treaty definition. The logical and reasonable concepts for this purpose are either "contamination", found in Article IX of the Outer Space Treaty, or "space object", to which reference is made in several international space law treaty provisions.

It is unclear to what phenomena "contamination" refers.[24] Moreover, international law provides no definition for "space object". The Liability Convention provides that "space object" includes a space object, the launch vehicle and the component parts of both; the Registration Convention also contains this description.[25] As yet, there is no agreement within the legal community as to which classes of space debris, if any, are included implicitly in the terms "space object" and "component parts".[26]

4. Jurisdiction and control over space debris

If effective remedial action is to be included in a regulatory regime for space debris, consideration should be given to the issues of who is authorized to remove space debris from outer space and when such removal is permitted. Article VIII of the Outer Space Treaty provides that the State of registry of a launched object has the right to make and enforce the law in relation to that object, and that ownership of a space object is not affected by its presence in outer space.[27] This provision raises several issues which ought to be resolved.

It is not clear which of the technical classes of space debris fall within the scope of Article VIII of the Outer Space Treaty, and to what extent jurisdiction and control over space objects, and therefore space debris, is permanent. Further, if it is assumed that inactive payloads are included within the scope of Article VIII, there is no agreement on what would be an appropriate method for distinguishing active payloads from inactive ones.

Moreover, given that ownership of a space object also is permanent, regardless of its use and condition, and given that the rights of ownership include possession, use and disposal, consent of the State of registry would seem to be necessary prior to any attempt by any legal entity to interfere in any way with that space object. States, therefore, may wish to consider whether the doctrine of permanency should apply to space debris, and whether and to what extent consent from the State of registry should be required prior to the removal of an item of space debris.

5. International responsibility for space debris

Article VI of the Outer Space Treaty provides that States are internationally responsible for the activities of their nationals in outer space, whether these nationals are individuals, corporations or governmental.[28] This responsibility would seem to include the duty of States to authorize national space activities (licensing power) and to supervise these activities continually (inspection power).

However, general principles of international law appear to mitigate against using Article VI of the Outer Space Treaty as a tool for addressing the risk posed by space debris. In order to attribute international responsibility to a State, that State must be bound by a legal obligation to conduct a given class of activities in a certain manner.[29] Any regulatory regime establishing such obligations should be as specific as possible.[30] Therefore, an international legal regime, binding States with specific legal obligations, would seem to be necessary before

effective international action on the issue of space debris can materialize.

As well, it is not clear to what extent a uniform, international regulatory regime will be affected by the delegation to individual States of authorization and supervisory functions.

6. Detection and Identification of Space Debris

To remove space debris from outer space and to hold States accountable for damage caused by space debris entails a method of identifying a State which can be linked to the debris. In space law, identification of space objects is addressed in the Registration Convention.[31]

Identification of space objects involves two phases: detection and identification. The Registration Convention contains no provisions for detection. It sets out only the most minimal of requirements for establishing a system which would positively identify space objects[32] and makes no provision for compulsory markings, although such markings must be registered if they are used[33].

7. Liability for damage caused by space debris

The Liability Convention sets out the legal regime for providing compensation for damage caused by space objects. Basically, if a space object causes damage on Earth or to an aircraft in flight, the injured party need not establish fault in order to be compensated for the damage.[34] Where damage is caused in outer space, however, liability is based on fault.[35] In this case, the injured party will be required to prove, among other things, that the party responsible for the damage did not take appropriate steps to avoid that damage.

Negotiations for the Liability Convention were not focused on damage arising in outer space, and did not consider the question of the risks posed by space debris.[36] Consequently, States have not been able to express in an international forum their views on several important liability issues arising from the Liability Convention on the question of damage caused by space debris. These include: (1) the meaning of "damage" and whether this meaning includes damage to the outer space environment *per se*; (2) whether the damage caused must be reasonably foreseeable, that is, whether the damage caused by space debris is of a kind that specialists in the field would expect to occur, and (3) whether it is reasonable to rely on a liability regime based on negligence for damage caused in outer space by space debris.

8. Nuclear activities and environmental modification

The Partial Nuclear Test Ban Treaty, Article IV of the Outer Space Treaty and Article 3 of the Moon Agreement, provide for protection to the outer space environment, the Moon and other celestial bodies to the extent that they prohibit nuclear activities in outer space.

Article I of the Partial Nuclear Test Ban Treaty includes a prohibition against all nuclear explosions in outer space.[37] Article IV paragraph 1 of the Outer Space Treaty prohibits the placement of nuclear weapons in orbit around Earth, in outer space or on celestial bodies.[38] Article 3 paragraph 3 of the Moon Agreement clarifies that the Outer Space Treaty prohibition includes the Moon as a celestial body, and expands the scope of that prohibition to include orbits around and other trajectories to or around the Moon.[39]

Taken together, these provisions would appear to prevent the creation of radioactive space debris resulting from deliberate nuclear explosions, whether for military or peaceful purposes. They do not, however, address the potential risks of radioactive space debris, which could arise if active, retired or stored satellites with nuclear power sources on board were involved in collisions or were otherwise fragmented. In addition, these provisions bind only States Parties to the agreements and do not exclude the possible use of space debris as a means of maintaining national security.

The Environmental Modification Convention prohibits military or other hostile uses of techniques which, through deliberate manipulation, could change the dynamics, composition or structure of outer space.[40] It is unclear whether the application of this treaty is restricted to States Parties. If so, the regulatory effectiveness of the Convention could be severely limited. As well, environmental modification techniques may be used for peaceful purposes, as permitted by international law.[41]

9. Harmful Interference with Satellite Telecommunications

The ITU Convention and its accompanying Radio Regulations make no provision for protection of the outer space environment *per se*. While the ITU Convention does provide for avoidance of harmful interference with the radio frequencies of transponders on board space objects, this interference must be caused by the operating radio station of a space object.[42]

Given that it is an operating radio station on board the space object, and not space debris created by that station, which must cause the harmful interference, the ITU Convention does not seem to apply to situations in which interference is caused by space debris.

B. INTERNATIONAL ENVIRONMENTAL LAW

Space debris is a potential environmental harm arising from space activities and therefore should be considered at law as such. Given that principles of international law apply to space law,[43] the principles of customary international environmental law ought to be considered when developing a regime for the management of space debris.

With its official global inauguration at the Stockholm Conference in 1972, international environmental law is a relatively new area of law with very little unification and, hence, few general principles. Today, the vast majority of law in this field consists of principles of customary international law[44], which have developed independently from one another.

Perhaps the best-known principle of customary international environmental law is that States are internationally responsible for protection of the environment beyond their borders. This protection extends not only to the territory of other States, but also to the territories beyond the jurisdiction of any State. These latter territories are known as the global commons, one of which is outer space. This principle developed from the Roman admonition *sic utere tuo ut alienum non laedas* (use your property so as not to injure your neighbour), and was restated in 1949 in the *Corfu Channel* case.

In *Corfu Channel*,[45] two British warships were damaged by mines in the territorial sea of Albania. Albania was held responsible at international law for the damage because every State, the International Court of Justice held, has an "obligation not to allow knowingly its territory to be used contrary to the rights of others".[46] In this case, the State responsible for the act causing the damage (originating State) was the same State in which the damage actually occurred (injured State). Accordingly, on the facts, where the originating State and the injured State are not one and the same, it is arguable that the *Corfu Channel* duty does not apply. The *Trail Smelter* arbitration addresses this argument.

In *Trail Smelter*,[47] sulphur dioxide emitted from a smelter in Canada (originating State) damaged crops and other property of a national of the United States (injured State). In its judgment, the arbitrator held that a State has a duty to avoid acts causing damage to the territory of other States, if the damage is serious and can be established by clear and convincing evidence.[48] In so ruling, the duty of originating States to avoid acts causing damage in their own territory was broadened to include the avoidance of acts causing damage in the territory of other States.

On the question of recovery for damage, the developing custom of international environmental law was based on the principle of compensation by the originating State for the damage caused to the injured State.[49] States were under no legal obligation to attempt to prevent damage until the *Lac Lanoux* arbitration. In that case,[50] Spain was the downstream user of a border river it shared with France. Spain's use of the river was being threatened by an industrial development project of France. The tribunal held that where the use of a shared resource by one State threatens the use of that resource by the other State, the threatening State has a duty to take into account the interests of the State which possibly could suffer damage. This general duty includes a duty to co-operate, a duty to notify and inform and a duty to consult and negotiate.[51]

In June 1972, the principle of *sic utere* became firmly entrenched as a principle of customary international law at the UN Conference on the Human Environment.[52] Principle 21 of the Conference's Stockholm Declaration, while acknowledging the sovereign right of States to exploit their own resources, combines this right with the duty of States to avoid harms to all territories outside their national boundaries,[53] including the global commons.

Together, these principles of customary international law form the general framework for international environmental law, and suggest basic rules for managing environmental harms. There seems to be, as well, some movement toward the development of specific legal regimes for regulating specific activities.

Perhaps the most important international agreements to date for the future development of both general principles of international environmental law and regulatory regimes for environmental management are the Ozone Layer Treaty[54] and its accompanying Montreal Protocol[55]. The former is a framework agreement, containing general principles applicable to the regulation of any substances which States Parties may agree could have adverse effects on the ozone layer. The latter provides specific control measures for chlorofluorocarbons (CFCs), substances which could have adverse effects on the ozone layer.[56]

The Ozone Layer Treaty also is significant for its potential effect in international environmental law on the general rule that a State must be bound by a legal obligation to conduct a given class of activities in a certain manner in order to have international responsibility attributed to it.[57] The exception to this general rule is the doctrine of abuse of rights, which provides that compensation for damage caused by an activity of an originating State may arise as a consequence of activities of that State which are not unlawful.[58]

The doctrine of abuse of rights is not considered to be a general principle of positive law, but rather a useful agent in developing the law.[59] In the realm of international environmental law, this doctrine could prove to be a useful tool for curtailing activities which have adverse affects on the environment because it partially erodes the sovereign (exclusive) right of States over their own resources, if damage occurs outside that State as a result of lawful acts within it. Such an erosion of State sovereignty seems reasonable in an environmental context, given that environmental harms cannot be defined in terms of State boundaries.

In the Ozone Layer Treaty, it would appear that a version of the doctrine of abuse of rights is found in the "General Obligations" section of the agreement. Article 2 paragraph 1 and Article 2 paragraph 2(b), read together, provide that States are obliged under certain conditions to regulate lawful activities that are found to have or could have adverse effects on the ozone layer.[60]

Given the general legal requirement that international regulatory regimes should create specific obligations in order to bind States, and in view of the individualistic nature of both the environmental harms and the international instruments needed to mitigate and possibly eliminate these harms, proposals for general rules intended to cover the field of international environmental law appear to be impractical. Notwithstanding this impracticality, there are two international agreements which suggest a wealth of basic rules which ought to be considered when developing any future international environmental law treaties, including any such instrument for the regula-

tion of space debris: the World Charter for Nature and the legal principles proposed in the Brundtland Report.

The World Charter for Nature,[61] which builds on the 1972 Stockholm Declaration, provides a comprehensive framework for dealing with environmental problems on a global scale. It incorporates all the duties of customary international law discussed above, addresses the vexing problem of national self-interest, and emphasizes in its remedial measures the preventive aspects required for environmental protection.

The Brundtland Report[62] was published in 1987 by the World Commission on Environment and Development. The Commission was established in 1983 by the United Nations Environment Program, in response to the awareness at the time that State activities had the potential to cause substantial harms to the global environment and, hence, to humanity. The Report concludes that sustainable development is necessary for survival of the planet.[63]

The Brundtland Report offers specific environmental management proposals for the global commons, including outer space;[64] accepts Principle 1 of the Stockholm Declaration as an important step toward sustainable development;[65] calls for a Convention on Environmental Protection and Sustainable Development;[66] and proposes 22 legal principles for use in providing for sustainable development[67].

CONCLUSION

Technology is, like Janus, two-faced. It has declared dominion over the natural resources of Earth, has shaped them to humankind's will and, in so doing, has created untold benefits for the inhabitants of this planet. Yet this very same technology carries with it a destructive potential, too often actualized in recent years. Space debris is the space-based manifestation of this dichotomy: the first space environmental problem.

A by-product of the technological magnificence that gave rise to space age, space debris currently is recognized by leaders in the space-user community as posing a genuine risk to the use of outer space, with the potential to render useless the limited natural resource of certain low-Earth orbits, both as things-in-themselves and as arenas for humankind's space-based activities. This symposium, in addressing the question of the preservation of near-Earth space for future generations, is a powerful confirmation of this concern.

International space law and international environmental law are relatively new legal domains. They are still flexible and fresh enough to contribute significantly to the creation of forward-looking, innovative, practical law and policy both of which acknowledge humanity's urge to explore and create, yet at the same time accept the consequences of the ecological reality that humankind and its activities are a part of nature and not above or beyond it.

It is unclear at this time, however, whether the international community of nations has the collective will to avoid in outer space the problems that technology has thus far spawned. If such a will does exist, or if it can be encouraged, then perhaps we will see the development of principles of international environmental space law and the creation of a space debris management regime that will serve as an example to future generations in outer space and on Earth.

ENDNOTES

* M.E.S., LL.M. This paper was prepared by the author in his personal capacity. Accordingly, the opinions and conclusions expressed in it are those of the author and should not be taken to represent the views of the Government of Canada or any of its agencies.

1. For the layman with an interest in the technical aspects of space debris, see N.L. Johnson and D.S. McKnight, *Artificial Space Debris* (Orbit Books 1987).

2. The two major policy statements to date on space debris are European Space Agency, Space Debris Working Group, *Space Debris*, SP-1109 (ESA November 1988) and United States, Interagency Group (Space), *Report on Orbital Debris* (Washington, DC February 1989). But see also, United States, General Accounting Office, Report to the Chairman, Committee on Science, Space, and Technology, House of Representatives, *Space Program: Space Debris a Potential Threat to Space Station and Space Shuttle*, GAO/IMTEC-90-18 (Washington, DC April 1990) and United States Congress, Office of Technology Assessment, *Orbiting Debris: A Space Environmental Problem – Background Paper*, OTA-BP-ISC-72 (Washington, DC September 1990).

3. For the basis for this conclusion, see Howard A. Baker, *Space Debris: Legal and Policy Implications* (Nijhoff 1989) 61–111. For an analysis of current space debris policy initiatives, see Howard A. Baker, "The ESA and US Reports on Space Debris: Platform for Future Policy Initiatives" (1990), 6 *Space Policy* 332.

4. **Treaty on Principles Governing the Activities of States in the Exploration and Use of Outer Space, Including the Moon and Other Celestial Bodies**, UNGA Res. 2222 (XXI) 19 December 1966; 610 UNTS 205, 1967 CanTS 19, 18 UST 2410, TIAS 7762 (opened for signature 29 March 1967, entered into force 10 October 1967).

5. **Agreement on the Rescue of Astronauts, the Return of Astronauts and the Return of Objects Launched into Outer Space**, UNGA Res. 2345 (XXII) 19 December 1967; 672 UNTS 119, 1975 CanTS 6, 19 UST 7570, TIAS 6599 (opened for signature 22 April 1968, entered into force 3 December 1968).

6. **Convention on International Liability for Damage Caused by Space Objects**, UNGA Res. 2777 (XXVI) 29 November 1971; 1975 CanTS 7, 24 UST 2389, TIAS 7762 (opened for signature 29 March 1972, entered into force 9 October 1973).

7. **Convention on Registration of Objects Launched into Outer Space**, UNGA Res. 3235 (XXIX) 29 November 1971; 1976 CanTS 36, 28 UST 695, TIAS 7762 (opened for signature 14 January 1975, entered into force 15 September 1976).

8. **Agreement Governing Activities of States on the Moon and Other Celestial Bodies**, UN GAOR, A/RES/34/68 (5 December, 1979); opened for signature 18 December 1979, entered into force 11 July 1984.

9. **Treaty Banning Nuclear Weapons Tests in the Atmosphere, in Outer Space and Under Water**, 14 UST 1313, TIAS 5433 (5 August 1963).

10. **Convention on the Prohibition of Military or Any Other Hostile Use of Environmental Modification Techniques**, UNGA Res. 31/72 (10 December 1976); 610 UNTS 151, 31 UST 333, TIAS 9614 (opened for signature 18 May 1977, entered into force 5 October 1978).

11. **International Telecommunication Convention – Nairobi, 1982** (Geneva: ITU, 1982).

12. Article IX of the Outer Space Treaty, *supra*, note 4, states:

    In the exploration and use of outer space, including the Moon and other celestial bodies, States Parties to the Treaty shall be guided by the principle of co-operation and mutual assistance and shall conduct all their activities in outer space, including the Moon and other celestial bodies, with due regard to the corresponding interests of all other States Parties to the Treaty. States Parties to the Treaty shall pursue studies of outer space, including the Moon and other celestial bodies, and conduct exploration of them so as to avoid their harmful contamination and also adverse changes in the environment of the Earth resulting from the introduction of extraterrestrial matter and, where necessary, shall adopt appropriate measures for this purpose. If a State Party to the Treaty has reason to believe that an activity or experiment planned by it or its nationals in outer space, including the Moon and other celestial bodies, would cause potentially harmful interference with activities of other States Parties in the peaceful exploration and use of outer space, including the Moon and other celestial bodies, it shall undertake appropriate international consultations before proceeding with any such activity or experiment. A State Party to the Treaty which has reason to believe that an activity or experiment planned by another State Party in outer space, including the Moon and other celestial bodies, would cause potentially harmful interference with activities in the peaceful use and exploration of outer space, including the Moon and other celestial bodies, may request consultation concerning the activity or experiment.

13. The author uses four categories for classifying space debris: inactive payloads (former payloads which can no longer be controlled by their operators), operational debris (objects associated with space activities, remaining in outer space), fragmentation debris (products of explosions and collisions) and microparticulate matter (a catch-all category including micron-sized objects, such as solid-propellant rocket motor effluent, paint flakes and thermal coatings, and spacecraft-induced phenomena such as outgassing of heavy molecules and space glow). See Baker, *Space Debris*, *supra*, note 3, 3–9.

14. UN GAOR, Committee on the Peaceful Uses of Outer Space (COPUOS), *Fifth Session of the Legal Sub-Committee*, A/AC.105/C.2/SR.68 (USSR, 13 July 1966) 7.

15. *Id.*

16. Article 1 of the Moon Agreement, *supra*, note 8, states:

    1. The provisions of this Agreement relating to the moon shall also apply to other celestial bodies within the solar system, other than the earth, except in so far as specific legal norms enter into force with respect to any of these celestial bodies.

2. For the purposes of this Agreement reference to the moon shall include orbits around or other trajectories to or around it.

17. R.T. Swenson, "Pollution of the Extraterrestrial Environment" (1985), 25 *Air Force LR* 70 at 81–82.

18. Article 7 of the Moon Agreement, *supra*, note 8, states in part:

   1. In exploring and using the moon, States Parties shall take measures to prevent the disruption of the existing balance of its environment, whether by introducing adverse changes in that environment, by its harmful contamination through the introduction of extra-environmental matter or otherwise. States Parties shall also take measures to avoid harmfully affecting the environment of Earth through the introduction of extraterrestrial matter or otherwise.

   2. States Parties shall inform the Secretary-General of the United Nations of the measures being adopted by them in accordance with paragraph 1 of this article and shall also, to the maximum extent feasible, notify him in advance of all placements by them of radioactive materials on the moon and the purposes of such placements.

19. Article 15 paragraph 2 of the Moon Agreement, *supra*, note 8, states:
   A State Party which has reason to believe that another State Party is not fulfilling its obligations incumbent upon it pursuant to this agreement or that another State Party is interfering with the rights which the former State has under this agreement may request consultations with that State Party. A State Party receiving such a request shall enter into such consultations without delay. Any other State which requests to do so shall be entitled to take part in the consultations. Each State Party participating in such consultations shall seek a mutually acceptable resolution of any controversy and shall bear in mind the rights and interests of States Parties. The Secretary-General of the United Nations shall be informed of the results of the consultations and shall transmit the information received to all States Parties concerned.

20. Article 15 paragraph 3 of the Moon Agreement, *supra*, note 8, states:
   If the consultations do not lead to a mutually acceptable settlement which has due regard for the rights and interests of all States Parties, the parties concerned shall take all measures to settle the dispute by other peaceful means of their choice appropriate to the circumstances and the nature of the dispute. If difficulties arise in connection with the opening of consultations or if consultations do not lead to a mutually acceptable settlement, any State Party may seek the assistance of the Secretary-General [of the United Nations], without seeking the consent of any other State Party concerned, in order to resolve the controversy. A State Party which does not maintain diplomatic relations with another State Party concerned shall participate in consultations, at its choice, either itself or through another State Party or the Secretary-General as intermediary.

21. As of 20 February 1992, the eight States Parties to the Moon Agreement were Australia, Austria, Chile, Mexico, The Netherlands, Pakistan, Philippines and Uruguay.

22. See, *supra*, note 13.

23. See Baker, *Space Debris*, *supra*, note 3 at 61–62.

24. *Supra*, text accompanying note 13. For an analysis of the concept of "contamination", see Baker, *Space Debris*, *supra*, note 3, text accompanying notes 358–364 and 410–416.

25. Both Article I(d) of the Liability Convention, *supra*, note 6, and Article I(b) of the Registration Convention, *supra*, note 7, state:
   The term "space object" includes component parts of a space object as well as its launch vehicle and parts thereof.

26. K. Spradling, "Space Debris: The Legal Regime, Policy Considerations and Current Initiatives". Prepared for the 28th Aerospace Sciences Meeting, American Institute of Aeronautics and Astronautics, 8-11 January 1990, Reno, Nevada, at 4.

27. Article VIII of the Outer Space Treaty, *supra*, note 4, states:
   A State Party to the Treaty on whose registry an object launched into outer space is carried shall retain jurisdiction and control over such object and over any personnel thereof, while in outer space or on a celestial body. Ownership of objects launched into outer space including objects landed or constructed on a celestial body, and of their component parts is not affected by their presence in outer space or on a celestial body or by their return to earth. Such objects or component parts found beyond the limits of the State Party to the Treaty on whose registry they are carried shall be returned to that State Party, which shall, upon request, furnish identifying data prior to their return.

28. Article VI of the Outer Space Treaty, *supra*, note 4, states:
   States Parties to the Treaty shall bear international responsibility for national activities in outer space, including the Moon and other celestial bodies, whether such activities are carried on by governmental agencies or by non-governmental entities, and for ensuring that national activities are carried out in conformity with the provisions set forth in the present Treaty.

29. I. Brownlie, *Principles of Public International Law*, 3d ed. (Clarendon Press 1979) 434–435.

30. Brownlie, *ibid.*, at 441.

31. For an analysis of the Registration Convention, see A.J. Young, "A Decennial Review of the Registration Convention" (1986), 11 *Annals Air & Space L* 287.

32. Only Article IV of the Registration Convention, *supra*, note 7, indicates the nature of the information to be registered with the United Nations. It states:

   1. Each State of registry shall furnish to the Secretary-General of the United Nations, as soon as practicable, the following information concerning each space object carried on its registry:
      (a) Name of launching State or States;
      (b) An appropriate designator of the space object or its registration number;
      (c) Date and territory or location of launch;
      (d) Basic orbital parameters, including:
         (i) Nodal period;
         (ii) Inclination;
         (iii) Apogee;
         (iv) Perigee;
      (e) General function of the space object.

   2. Each State of registry may, from time to time, provide the Secretary-General of the United Nations with additional information concerning a space object carried on its registry.

   3. Each State of registry shall notify the Secretary-General of the United Nations, to the greatest extent feasible and as soon as prac-

ticable, of space objects concerning which it has previously transmitted information, and which have been but are no longer in earth orbit.

33. Article V of the Registration Convention, *supra*, note 7, states:

Whenever a space object launched into earth orbit or beyond is marked with the designator or registration number referred to in article IV, paragraph 1(b), or both, the State of registry shall notify the Secretary-General of this fact when submitting the information regarding the space object in accordance with article IV. In such case, the Secretary-General of the United Nations shall record this notification in the Register.

34. Article II of the Liability Convention, *supra*, note 6, states:

A launching State shall be absolutely liable to pay compensation for damage caused by its space object on the surface of the earth or to aircraft in flight.

35. Article III of the Liability Convention, *supra*, note 6, states:

In the event of damage being caused elsewhere than on the surface of the earth to a space object of one launching State or to persons or property on board such a space object of another launching State, the latter shall be liable only if the damage is due to its fault or the fault of persons for whom it is responsible.

36. See B. Cheng, "Convention on International Liability for Damage Caused by Space Objects" in N. Jasentuliyana and R.S.K. Lee (eds), *Manual on Space Law: Volume I* (Oceana 1979) 83 at 83–84, and H. Reis, "Some Reflections on the Liability Convention for Outer Space" (1978), 6 *J Space L* 126 at 127.

37. Article I paragraph 1(a) of the Partial Nuclear Test Ban Treaty, *supra*, note 9, states:

Each of the Parties to this Treaty undertakes to prohibit, to prevent and not to carry out any nuclear weapon test explosion, or any other nuclear explosion, at any place under its jurisdiction or control in the atmosphere; beyond its limits, including outer space; or under water, including territorial waters or high seas.

38. Article IV paragraph 1 of the Outer Space Treaty, *supra*, note 4, states:

States Parties to the Treaty undertake not to place in orbit around the earth any objects carrying nuclear weapons or any other kinds of weapons of mass destruction, install such weapons on celestial bodies, or station such weapons in outer space in any other manner.

39. Article 3 paragraph 3 of the Moon Agreement, *supra*, note 8, states:

States Parties shall not place in orbit around or other trajectory to the moon objects carrying nuclear weapons or any other kinds of weapons of mass destruction or place or use such weapons on or in the moon.

40. Article I paragraph 1 of the Environmental Modification Convention, *supra*, note 10, states:

Each State Party to this Convention undertakes not to engage in military or other hostile use of environmental modification techniques having widespread, long-lasting or severe effects as the means of destruction, damage or injury to any other State Party.

Article II states:

As used in article I, the term "environmental modification techniques" refers to any technique for changing – through the deliberate manipulation of natural processes – the dynamics, composition or structure of the earth, including its biota, lithosphere, hydrosphere and atmosphere, or of outer space.

41. Article III paragraph 1 of the Environmental Modification Convention, *supra*, note 10, states:

The provisions of this Convention shall not hinder the use of environmental modification techniques for peaceful purposes and shall be without prejudice to generally recognized principles and applicable general rules of international law concerning such use.

42. Article 35 paragraph 1 of the ITU Convention, *supra*, note 11, states:

All stations, whatever their purpose, must be established and operated in such a manner as not to cause harmful interference to the radio services or communications of other Members or of recognized private operating agencies, or of other duly authorized operating agencies which carry on radio service, and which operate in accordance with the provisions of the Radio Regulations.

Harmful interference has been defined as "[a]ny emission, radiation, or induction which endangers the functioning of a radio navigation service or of other safety services, or seriously degrades, obstructs or repeatedly interrupts a radiotelecommunication service operating in accordance with the Radio Regulations", in R.S. Jakhu, *The Legal Regime of the Geostationary Orbit* (Doctoral dissertation, McGill University, Montreal, Canada, 1983) note 141.

43. Article III of the Outer Space Treaty, *supra*, note 4, states:

States Parties to the Treaty shall carry on activities in the exploration and use of outer space, including the moon and other celestial bodies, in accordance with international law, including the Charter of the United Nations, in the interest of maintaining international peace and security and promoting international co-operation and understanding.

44. Generally, customary international law arises when there is a general recognition among the States in question that a certain practice is obligatory. See Brownlie, *supra*, note 29 at 4–12.

45. *United Kingdom v. Albania* (Corfu Channel – Merits), [1949] ICJ Rep. 4.

46. Corfu Channel, *ibid.*, at 22.

47. *United States and Canada* (Trail Smelter Arbitration) (1931–1941), 3 RIAA 1905.

48. Trail Smelter, *ibid.*, at 1965.

49. In *Chorzow Factory* (Indemnity) (1928), PCIJ, Ser. A, no. 17, the Permanent Court of International Justice stated at 47:

The essential principle contained in the actual notion of an illegal act – a principle which seems to be established by international practice and in particular by the decisions of arbitral tribunals – is that reparation must, as far as possible, wipe out all the consequences of the illegal act and re-establish the situation which would, in all probability, have existed if that act had not been committed. Restitution in kind, or, if this is not possible, payment of a sum corresponding to the value which a restitution in kind would bear; the award, if need be, of damages for loss sustained which would not be covered by restitution in kind or payment in place of it – such are the principles which should serve to determine the amount of compensation for an act contrary to international law.

50. *France v. Spain* (Lac Lanoux Arbitration) (1957), 12 RIAA 281, transl. in (1959), 53 *Am J Int'l L* 156.

51. It should be noted that, in both *Trail Smelter* and *Lac Lanoux*, an agreement was already in place for the settlement of disputes arising during arbitration.

52. For an overview of the Conference, see L.B. Sohn, "The Stockholm Declaration on the Human Environment" (1973), 74 *Harvard Int'l LJ* 423.

53. United Nations, Conference on the Human Environment, *Report of the UN Conference on the Human Environment*, "Declaration of the United Nations Conference on the Human Environment" (Stockholm Declaration), A/CONF.48/14 (1970), 2-65 and Corr 1 (1972). The Stockholm Declaration is reprinted in (1972), 11 *Int'l Leg. Materials* 1416. Principle 21 of the Stockholm Declaration, *ibid.*, states:

    States have, in accordance with the United Nations and the principle of international law, the sovereign right to exploit their own resources pursuant to their own environmental policies, and the responsibility to ensure that activities within their jurisdiction or control do not cause damage to the environment of other States or of areas beyond the limits of national jurisdiction.

54. **Vienna Convention for Protection of the Ozone Layer**, UN GAOR, UNEP/IG.535 (opened for signature 22 March 1985, entered into force, 22 September 1988), reprinted in (1987), 26 *Int'l Leg. Materials* 1516.

55. **Protocol on Substances that Deplete the Ozone Layer** (opened for signature 16 September 1987, entered into force 1 January 1989), reprinted in (1987), 26 *Int'l Leg. Materials* 1541.

56. The Montreal Protocol was amended in June 1990 to accelerate the phase-out of CFCs, to add other ozone depleters to the ban list and to provide financial support to participating Third World countries. See **Adjustments and Amendments to the Montreal Protocol on Substances that Deplete the Ozone Layer**, UN GAOR, UNEP/OzL. Pro.2/3 (opened for signature 29 June 1990, not yet in force; 20 ratifications required, 2 States have ratified (Canada, New Zealand)), reprinted in (1991), 30 *Int'l Leg. Materials* 537. See also, L.B. Talbot, "Recent Developments in the Montreal Protocol on Substances that Deplete the Ozone Layer: The June 1990 Meeting and Beyond" (1992), 26 *Int'l Lawy.* 145.

57. See, *infra*, section A.5 – International Responsibility for Space Debris.

58. Brownlie, *supra*, note 29 at 443.

59. Brownlie, *supra*, note 29 at 445.

60. Article 2 paragraph 1 and Article 2 paragraph 2(b) of the Ozone Layer Treaty, *supra*, note 54, state:

    1. The Parties will take appropriate measures in accordance with the provisions of this Convention and those protocols in force to which they are party to protect human health and the environment against adverse effects resulting or likely to result from human activities which modify or are likely to modify the ozone layer.

    2. To this end the Parties shall, in accordance with the means at their disposal and their capabilities:

        (b) Adopt appropriate legislative or administrative measures and co-operate in harmonizing appropriate policies to control, limit, reduce or prevent human activities under their jurisdiction or control should it be found that these activities have or are likely to have adverse effects resulting from modification or likely modification of the ozone layer ... .

61. UNGA Res. 37/7, 28 October 1982, reprinted in (1983), 22 *Int'l Leg. Materials* 455.

62. UN GAOR, *Report on the World Commission on Environment and Development*, A/42/427 (4 August 1987) (Brundtland Report), reprinted as World Commission on Environment and Development, *Our Common Future* (Oxford University Press 1987).

63. Sustainable development is viewed as development which meets human needs, while at the same time protects and conserves the natural environment. See *Our Common Future*, *supra*, note 62 at 43-46.

64. *Our Common Future*, *ibid.*, at 274–279. At 277, space debris is stated to be a growing threat to humankind.

65. *Our Common Future*, *ibid.*, at 330. Principle 1 of the Stockholm Declaration, *supra*, note 53, states in part:

    Man has the fundamental right to freedom, equality and adequate conditions of life, in an environment of a quality that permits a life of dignity and well-being, and he bears a solemn responsibility to protect and improve the environment for present and future generations.

66. *Our Common Future*, *supra*, note 62 at 64.

67. These principles, set out in Annex I to the Brundtland Report, *supra*, note 62, also have been published, with full legal commentary, in World Commission on Environment and Development, *Legal Principles for Environmental Protection and Sustainable Development* (Nijhoff 1987).

22: Who Should Regulate the Space Environment: the Laissez-Faire, National, and Multinational Options

Diane P. Wood[*]

Harold J. and Marion F. Green Professor of International Legal Studies, The University of Chicago Law School

As near earth space becomes increasingly crowded with man-made debris, the risks to further peaceful uses of that region have begun to capture world-wide attention. It seems plain that some kind of regulation, by some sort of entity, is essential to several goals: to name a few, the maintenance of a safe environment for human beings in space; the avoidance of environmental damage to the surface of the Earth; and the continuation of economically beneficial uses of near earth space. Regulation, however, often has unintended consequences, and upon further examination sometimes proves to be more costly than the *laissez faire* alternative. Even when it is clear that an unregulated environment is inappropriate, important decisions must be made about the level at which regulation ought to occur: local, regional, national, or international.

This paper examines the related questions of (1) the desirability of regulating some or all aspects of the space debris problem, and (2) the optimal source of such regulation. It does so in the light of studies of the economics of regulation that have been conducted principally within the United States, and in the light of international experience in regulating other issues similar to the problem of space debris. In the final analysis, a multilateral approach is preferable to either national regulation or a global free-for-all. Nevertheless, prior efforts at multilateral regulation have not always produced successfully functioning regimes. A close look at the reasons why this is so may help drafters of a space debris treaty to avoid repeating the mistakes of the past.

I. *LAISSEZ FAIRE*

At the present time, there is no comprehensive regulatory system that addresses the problem of space debris, although one can find some general regulatory principles in existing space conventions.[1] Thus, the regulatory environment now approximates the condition of *laissez faire:* each nation or entity that launches an object into space decides for itself how many precautions it will take against the creation of additional orbital debris; and, apart from potential liability due to a collision with an identifiable object, each launching entity can safely disregard the risks that its own debris may generate.

As the aggregate amount of debris increases, however, calls for effective regulation of the space environment have grown more urgent. This therefore seems like an ideal time to examine the question whether any of the usual reasons advanced to regulate an activity apply here. If they do, the next step is to ask which ones have the greatest force, since the eventual regulatory scheme adopted will be different depending upon what evils it is intended to address.

In a leading book written early in the wave of deregulation that swept through the United States during the early 1980s, Stephen Breyer identified a number of possible justifications for regulation.[2] They included the following points: (1) compensation for undesirable spillovers or externalities, which are the costs of an activity that are not normally borne by the producer in the absence of regulation (for example, polluting the air or water as a result of industrial processes); (2) control of monopoly power, usually of a public utility; (3) correction of a lack of adequate information in a market that prevents efficient transactions (for example, requiring the disclosure of ingredients in food, or requiring the licensure of highly skilled professionals like physicians); (4) equalization of the bargaining power between buyers and sellers, when, for example, a group of small, powerless buyers face a monopolist as seller, or vice versa; (5) the avoidance of a "moral hazard," in which the person consuming the product or service does not bear its full cost, and hence demands too much (as is arguably the case in first dollar medical insurance plans, which induce people to visit the doctor too often); and (6) governmental paternalism (in Breyer's terms), which might also be termed noneconomic public interest.[3]

Of those justifications, the one that deserves the closest scrutiny in the present context is the first, dealing with spillovers or externalities. To take an example, consider the recent trip of the space shuttle Explorer, during which the astronauts captured a telecommunications satellite that had been dysfunctional and launched it into the proper orbit. In calculating the costs of the mission, one would consider the costs of building the shuttle, the costs of fuel, supplies, and other variable costs of the mission, and the wear and tear on the shuttle. However, the cost calculation is not likely to include any component for future damage that debris from the shuttle might inflict on other functioning objects in space, just as the environmental cost of automobile emissions was not reflected in car prices until emission control devices were required by law.

Several factors make some type of government intervention necessary in this case. First, it is unlikely that most damage eventually caused by debris could be traced back to the source, such that a proceeding for damages would eventually require the one who caused the harm to pay for it.[4] This possibility of a later damage suit is what normally induces firms to consider the social costs of their activities, and to internalize those costs (often through procurement of insurance). Typically, however, environmental problems do not lend themselves to this kind of private solution. The environmental problems that give rise to externalities listed by William Baumol and Wallace Oates in their text on the theory of environmental policy illustrate the point well.[5] They identify the following seven:

a.   Disposal of toxic wastes,
b.   Sulfur dioxide, particulates, and other contaminants of the atmosphere,
c.   Various degradable and nondegradable wastes that pollute the world's waterways,
d.   Pesticides, which, through various routes, become imbedded in food products,
e.   Deterioration of neighborhoods into slums,
f.   Congestion along urban highways,
g.   High noise levels in metropolitan areas.

In these cases, the authors point out, private negotiations will not be successful, because of the numbers of people involved on one or both sides of the equation.[6] One could take the point further, looking at the problem of acid rain, the hole in the ozone layer, or global warming, and note that someone who wanted to negotiate might not even know with whom he or she should be talking.

Another problem with reliance on private decisionmaking (by which I mean decisions made by the actor engaging in the space activity, whether a government agency or a private entity) relates to the absence of clearly defined property rights in near earth space. In his Nobel-prize winning essay on *The Problem of Social Cost*, Ronald Coase addressed the problem of "those actions of business firms which have harmful effects on others" — exactly the problem at hand.[7] With respect to pollution in space, Glenn Reynolds and Robert Merges noted that the Coase Theorem predicts that one can obtain an efficient allocation of resources "either by assigning property rights to private firms, or by shaping a liability rule making transgressors liable for damages inflicted on existing resource users. Either way would eliminate the necessity for a large-scale centrally administered regulatory agency."[8] However, the authors then said in passing that "it is less obvious how such rights should be assigned in the space context."[9]

In fact, experience has shown that it is difficult at best to assign clear property rights for any area beyond the common

jurisdiction of nation-states, whether it is space, the deep sea-bed,[10] Antarctica,[11] or the Moon.[12] Yet without such rights, no one nation alone, and no one firm alone, would have an adequate incentive to incur the expense necessary to prevent (or at least reduce to the maximum feasible degree) the amount of debris generated by its space objects. This point was perhaps less clear when space seemed infinite and the risks of damage from debris remote. Now, however, the debris problem must take its place beside other better known environmental issues that have given rise to the need for some kind of regulatory measures to protect "the commons."

Although the other justifications for regulation listed above provide a less compelling rationale for a space debris regime, each of them has some bearing on the problem and deserves brief consideration.

Pollution-control regimes are rarely justified by a fear of monopoly power, since it is slightly absurd to think of a monopoly in waste products. However, if the regulation of space debris were part of a broader regulatory system for space resources that relied on the idea that space is part of the common human heritage, the fear of smaller nations that more powerful countries would monopolize space might play some role. On the assumption that the time is not ripe for any such comprehensive scheme, however, it is safe to disregard this rationale.

Both the correction of a lack of information and the equalization of bargaining power occasionally contribute to environmental regulations. For example, firms using potentially dangerous substances in their products have been required in the State of California to disclose the presence of those products to consumers, so that each consumer may make an informed choice in his or her purchases.[13] It is hard to imagine how the greater availability of information about the potential for debris would alter existing market arrangements, either by inducing the owners of satellites to use different suppliers or different designs, or by changing the demand for space-based services in general. Bargaining power may play some role in rules requiring the clean-up of toxic waste sites, since individuals affected by the presence of the site may have difficulty organizing effectively so that negotiations with the polluting firm can work. Again, this is simply not an important problem for space debris, and one would therefore not need to create regulations that respond to it.

The last two points from the earlier list are the moral hazard problem and the paternalism point. Both of these are related to the externalities point. If satellite use is actually costlier over time than present pricing reflects, people will demand too much of it. The only question here is whether the moral hazard can be avoided in any way other than by means of governmental intervention; for the reasons noted above, the answer is probably no. Paternalism, or public interest regulation, is different in that it does not need to be efficiency-based. Society may subsidize minority businesses in the public interest for reasons that are essentially moral; we set aside national parks so that future generations may enjoy their beauty. While this may play some role in the efforts to reduce space debris, it would only be dispositive if the regulatory regime actually went beyond optimal pollution control and was used (for example, by some kind of tax) for other public purposes, such as wealth redistribution. This, in my view, would not be an appropriate use of space debris regulation, and it would be important to ensure that the rules adopted did not lend themselves to such a diversion.

To conclude this section, although it is clear that the problem of space debris cannot and will not be solved by a pure *laissez faire* system, it helps to identify the various justifications for regulation to see which ones have the greatest applicability to this problem. By doing so, one can begin to see what kinds of regulations will address the issue most effectively, and what kinds of regulations are either unnecessary or undesirable. With this in mind, we can turn now to the second major question: who should do the regulating?

II. NATIONAL REGULATION

The two principal options for the regulation of pollution in space are to leave matters up to each nation (properly defined), or to create some kind of supranational authority. In this section, I examine the pros and cons of regulation strictly at the national level.[14] By this, I mean national regulation in its purest form, rather than national regulation undertaken to implement some international obligation, as exists in many areas, ranging from health regulations issued to comply with World Health Organization standards, to trade rules that comply with the General Agreement on Tariffs and Trade. The latter kind of regulation will be considered below, as a form of international law.

What can possibly be said for national regulation when the problem so plainly occurs in a territory beyond any nation's jurisdiction? Environmental policy normally should be made by a governmental unit that meets two criteria: first, it should have the ability to enforce rules against the polluters, and second, it should suffer most of the detriments of the pollution.[15] Without effective enforcement power, the rules will be of no effect; without a sufficient congruence between the territory that suffers and the territory that regulates, the proper incentive to regulate will not be present. Most people have focused on the second of these criteria in concluding that the regulation of space debris must be accomplished internationally, and, as far as it goes, this seems indisputable. However, the problems posed by the first criterion deserve closer examination, since they are ultimately responsible for the difficulties that have been experienced in other areas of international environmental cooperation.

We can begin with a simple proposition: those who bear the burdens of environmental regulation of activities should also enjoy whatever benefits those activities create. Normally, the way to ensure that one enjoys the benefits of economic activity

is to make sure that property rights in that activity are secure. If a farmer grows wheat, he has the legal right to exclude others from coming onto his field and taking the harvest away; if a company drills an oil well, it may sell the oil to whomever it wishes; if someone invents a sufficiently original product or process, she may obtain a patent for it.

All these property rights – from the simple right to exclude, to the right to dispose of a depletable nature resource, to the intangible rights associated with a patent – are a function of national law. Although nearly every country's laws would recognize the first kind of property right, the consensus breaks down as we move down the line. In some places, such as Mexico and Saudi Arabia, the right to drill for oil and the rules that govern its sale allow for much more restricted private property rights. This is true both because of the depletable nature of the resource, and because of the perceived public interest in the wealth that the resource represents on world markets. Even greater variety exists with respect to patent laws and other intellectual property laws around the world. Some countries refuse to recognize private property rights in pharmaceutical products, for example; others have patent schemes that include compulsory licensing of patents that are not exploited by the owner after a certain period of time.

The variety in national conceptions of property rights has several consequences for the problem of property rights in space. In the first instance, it suggests that different nations have come to different political conclusions about the trade-off between private and public control of resources, and that these nations may see no reason to change their regime for space activities. It also indicates that the effort to come to a consensus position on the right to exploit near earth space will face the same kind of difficulties that the deep sea-bed posed during the Law of the Sea negotiations.

The strength accorded to property rights is not a question that is susceptible to scientific resolution; it is instead the most political of issues. If, either because nations continue to prefer a diversity of property rights in their space-faring activities, or because it is simply too hard to move beyond diversity, the present system that governs the enjoyment of benefits continues, this will have consequences for the burden side of the equation: the side that imposes controls to limit orbital debris. Nations with a strong benefit rule (*i.e.* strong and clear property rights) may be willing to impose more stringent environmental regulations, since they and their firms will enjoy relatively more of the overall wealth generated by space activities. Other nations with weaker benefit rules may find it impossible to tax their own activities with environmental rules without reducing overall profit to an unattractively low level.

One might respond with two points: first, why should the nations with the strong property rules confer a benefit on the world community as a whole through strict regulation of space debris, if other nations are not similarly contributing; and second, would it not be better to try to overcome the difficulties in finding a consensus position internationally, so that both the cost and the benefit sides of space activity accurately reflect its global nature?

It would be foolish to deny the force of the first point, since as a rule nations generally like to serve their self-interest just as much as the individual actors discussed in Part I do. Nonetheless, instances of altruistic behavior occur with surprising frequency. Examples may include the renunciation of certain methods of waging war (biological weapons, chemical weapons), even with the knowledge that outlaw nations may not observe these limits;[16] enforcement of the Geneva High Seas Convention rules concerning pollution prevention in areas where monitoring is as a practical matter difficult; and measures taken to secure or improve the condition of human rights in other countries. Add to the general altruistic motives the attraction of acquiring the moral force of world leadership on an issue like space debris, and the economic benefits of an advantage in debris-reducing technologies, and it is not out of the question to imagine that individual nations might rise to the occasion. This would, however, only solve the debris problem to the extent that the regulating nations were responsible for it; sooner or later, expansion to the international level would be inevitable.[17]

The preceding point leads to the second question: is anything at all lost by attempting right now to move to a global consensus on the issue of the property rights to be enjoyed by those who use the resources of space, and the concomitant obligations they should incur to prevent the accumulation of space debris? The answer to this question may lie in the extensive discussion over the American decision not to sign the 1982 Law of the Sea Convention, and in more recent signals that the U.S. may refuse to sign the biodiversity treaty at the Earth Summit in Rio de Janeiro at the beginning of June 1992, which is designed to slow the extinction of plant and animal species. If those policy positions were due to nothing more or less than unvarnished greed and selfishness, then little can be said in their defense. However, as is often the case, the real situation is more complex, and the concerns raised by U.S. policy-makers require serious attention.

In the case of the Law of the Sea Convention, the Reagan Administration officials who instigated the policy review of the draft treaty in 1981 raised a number of points. They argued, for example, that access to the resources of the deep sea-bed was essential to U.S. national security; that the LOS regime would provide inadequate incentives for the mining industry to conduct research and development; that the Authority envisioned in the treaty would impose unjustified taxation and/or production controls; and that technology developed by U.S. firms would receive inadequate protection.[18] Similar points are being raised about the biodiversity treaty, as reported by the press:

> Specifically, U.S. officials said that the language in the biodiversity treaty doesn't go far enough in requiring developing nations to make inventories of their plant and animal species and undertake conservation, doesn't pro-

vide adequate patent and copyright protection for U.S. biotechnology and other products, unfairly treats biotechnology as inherently unsafe, and doesn't ensure an equitable financing mechanism for financial and technical aid flowing from rich to poor nations.[19]

The points that require serious attention relate to the level of protection that investors in the relevant areas (deep sea-bed, species protection, outer space) legitimately may demand.

Within any well established national system, investors know what risks they are facing when they decide to enter the market.[20] They can therefore estimate the likely profitability of the enterprise, and make an informed decision on the costs, including the costs of local regulations. In any system that replaced a national system, it would take some time before similar confidence would be justified, even if the system had made every effort to assure adequate protection for capital investments and technology. If the system did not contain such assurances, but instead involved compulsory transfers of income or technology, it would appear inferior to an investor whose national system was more solicitous of these rights. In the end, less investment would occur – an overall social loss, if development of the area in question was thought to have potential for human welfare.

Returning more specifically to the benefits and costs of exclusive national regulation of space debris, the principal benefit of rejecting an international regime would be the ability of each nation to balance the degree of property protection for its own (or its nationals') activities in space against the costs of this kind of pollution control regulation. On the cost side, the problems of national regulation include the lack of an incentive to restrict debris-producing activities, the difficulty of making individual nations accountable through ordinary methods of assigning responsibility for damages, and the fact that the damage quite literally affects the entire orbital region, beyond the jurisdiction of any one nation. Taken together, these points argue strongly for an international regime. In order to be effective and politically acceptable, such a regime will have to be as responsive as it can be to the loss of the advantages of national regulation.

III. MULTILATERAL REGULATION

In considering the prospects for multilateral regulation, it is useful to approach the problem by looking at relatively successful examples of such regulation, and to see whether it would be possible to build on those successes in the special case of space environmental law. The two examples I have chosen represent two different multilateral alternatives: the first is the Antarctica Treaty, which involves as voting members only the twenty-six countries with a significant investment in Antarctica, but which makes provision for broader participation in decisionmaking by additional interested countries; the second is the General Agreement on Tariffs and

Trade, which includes among its Contracting Parties more than one hundred countries.

A. The Antarctica Treaty

The original Antarctica Treaty of December 1, 1959, emphasized international security and scientific cooperation, rather than environmental protection, reflecting the needs of the time.[21] Article I states that "Antarctica shall be used for peaceful purposes only," and it prohibits any measures of a military nature. Article V prohibits nuclear explosions and the disposal of radioactive waste material in Antarctica. Articles II and III both address scientific cooperation, by respectively assuring the continuation of freedom of scientific investigation on the continent, and establishing procedures for international cooperation, pursuant to which the Contracting Parties agreed as follows:

(a) information regarding plans for scientific programs in Antarctica shall be exchanged to permit maximum economy and efficiency of operations;

(b) scientific personnel shall be exchanged in Antarctica between expeditions and stations;

(c) scientific observations and results from Antarctica shall be exchanged and made freely available.

The Treaty was also sensitive to a variety of political requirements. First, in Article IV, it preserved all pre-existing claims to the territory of Antarctica, by carefully taking no position on any of them. Second, through Article VII it made Antarctica an open continent, by assuring access to any or all areas by observers of each of the Contracting Parties at all times. It did not establish any kind of supranational governing authority, but instead, in Article VIII, it made observers subject only to the jurisdiction of the Contracting Party of which they were nationals, and it required consensual resolution of any disputes that might arise among the Parties. Finally, it created a follow-up mechanism in Article IX, pursuant to which the Contracting Parties would consult together on matters of common interest, and recommend further measures in furtherance of the principles and objectives of the Treaty, including:

(a) use of Antarctica for peaceful purposes only;

(b) facilitation of scientific research in Antarctica;

(c) facilitation of international scientific cooperation in Antarctica;

(d) facilitation of the exercise of rights of inspection provided for in Article VII of the Treaty;

(e) questions relating to the exercise of jurisdiction in Antarctica;

(f) preservation and conservation of living resources in Antarctica.

The last of those purposes gave rise to the recently signed Protocol on Environmental Protection to the Antarctica Treaty, which President George Bush forwarded to the United States Senate on February 14, 1992.

The Antarctica Treaty has been a success for more than thirty years.[22] As we consider similar possibilities of international cooperation for near earth space, it therefore deserves close attention, to see if the characteristics that have led to its success can be replicated in other areas. Three reasons seem central: (1) the Treaty has clearly defined purposes, and hence is not subject to widely varying interpretations by different Parties; (2) all of the Consultative Parties (as noted, twenty-six at present) have a substantial stake in the region, as demonstrated by the requirement that a country must conduct substantial scientific research activities in Antarctica to be entitled to full voting status; and (3) the enforcement and decisionmaking machinery established by the Treaty has thus far worked effectively, weathering close calls such as the last-minute opposition of the United States to the Environmental Protection Protocol (which was eventually overcome).

International cooperation on specific subjects, whether the use of Antarctica, limitations on the production of chemicals that deplete the ozone layer, the control of narcotic drugs, hijacking and international terrorism, or marine mammal protection, is more likely to be successful than comprehensive new systems of global governance, like the Law of the Sea Convention. This is particularly true if, as will surely be the case with the space debris problem, a consensus can develop about the desirable or undesirable nature of the activity. No one is likely to argue seriously that space debris is a positive good; the most that might be asserted is that it is not bad enough *yet* to justify extremely expensive regulatory schemes. In the case of the Environmental Protocol to the Antarctica Treaty, all Consultative Parties were eventually convinced that the risks to the Antarctic environment of mining activities were too great to justify those activities for the foreseeable future.

The existence of a stake in the region is also very important. It creates a sense of reciprocity among the countries creating the regulatory regime, so that all suffer or all benefit in roughly the same way. No country is excluded from joining the group, and acceding to the Treaty, except insofar as the substantive requirements may or may not be met. For a time, the United Nations debated the question whether it should take over the Antarctic Treaty and substitute a broader multilateral regime.[23] The developing countries suggested, among other things, that the United Nations should reject once and for all the territorial claims that the Antarctic Treaty had finessed, that it should adopt a more democratic regime that included all countries in the governing structure, that there should be a centralized professional agency charged with protecting the environment, and that Antarctica (like the deep sea-bed and the Moon) should be declared part of the "common heritage of mankind."[24]

The Consultative Parties rejected all these arguments, in some instances on the ground that the existing arrangements satisfied all legitimate concerns (*e.g.*, the territorial claim point, the democracy point, and the environmental protection point), and in other instances on the ground that the suggestion was inappropriate for Antarctica (the "common heritage" point). More generally, efforts to impose a one-nation, one-vote style of "democracy" on the international community under circumstances where the burdens of an agreement do not fall equally on all nations have tended to fail. Examples include the efforts to create a Common Fund for commodity stabilization agreements; the stalled Code of Conduct for the Transfer of Technology that has been under consideration for more than two decades in the United Nations Conference on Trade and Development (UNCTAD); and, of course, the provisions of the Law of the Sea Convention for an Enterprise and an Authority to regulate the deep sea-bed. More targeted efforts (usually bilateral, but occasionally regional) to address issues of trade with developing countries and protection of intellectual property have been more successful.[25]

The last point relates to the enforcement and decisionmaking machinery of the Antarctica Treaty. Given the relatively small number of Consultative Parties, and the narrowness of the Treaty's goals, a rule of unanimity has worked thus far. Unanimity has the virtue of assuring the member states that their interests will not be ignored, whether or not they are major powers in the area; it has the obvious vice of paralysis, if one country holds out for a position that no others support. Just as the European Community has found over its thirty-plus years of existence that effective governance eventually requires limiting a requirement of unanimity to the most important questions, the Antarctic Parties are moving in the same direction. Under pressure from the United States, the Environmental Protocol provides that three-quarters of the Consultative Parties may bring the mining question up again after the fifty-year moratorium expires. It is clearly speculative at this point to predict who, if anyone, will want to revisit the issue in the year 2042. The point for present purposes is that over time, the pressure builds to create international structures under which a nation consents to a *procedure* under which it may find itself bound to an international rule without its agreement to the particular rule. The Antarctic regime has weathered the first step of the transition away from unanimity.

B. The GATT

The other example I have chosen is a broad multilateral agreement, which includes nations of nearly every level of development and every political system. The General Agreement on Tariffs and Trade was one of a series of post-World War II economic agreements (others included the agreements establishing the International Monetary Fund and the World Bank) which were designed to place the world back on the road to prosperity. Although the GATT system has had its share of strains over the forty-five years of its operations, it is a remarkable testament to its resiliency that it has operated with any success at all for such an extended period.

Before indicating the reasons that underlie the GATT's endurance, it may be useful to summarize very briefly what rules the General Agreement creates and how they are

enforced. Essentially, the GATT can be reduced to a few general principles: (1) all signatories promise to give "most favored nation" treatment to other signatories, with no strings attached; (2) when tariff reductions on specified items are agreed among the Parties, the new lower tariffs are "bound," and cannot be changed without giving rise to certain enforcement rights for other affected Parties; (3) in general (although exceptions exist), the Parties must not discriminate between their own nationals and other GATT Member State nationals in any of their domestic legislation that affects international trade in goods; and (4) the General Agreement as a whole exhibits a preference for trade restrictions in the form of tariffs, as opposed to quotas or other types of so-called "non-tariff barriers."[26] Two principal enforcement mechanisms exist. First, the General Agreement establishes a system of consultation and dispute resolution for all circumstances in which a country believes that its rights have been nullified or impaired. Second, usually after a panel adjudication of such nullification and impairment, the aggrieved country is entitled to retaliate against the offender by cancelling trade concessions or taking similar economic measures. Thus, for an international agreement, the GATT has in principle an enforcement structure with teeth.[27]

Unlike the Antarctica Treaty, which as an instrument submitted to the U.S. Senate for advice and consent, and duly ratified by the President, needs no further implementing legislation, the GATT represents an international agreement that is given effect principally through national legislation. It thus offers an illustration of a hybrid system that takes advantage of national rules with international machinery that exists only for support and oversight. Each country is responsible for monitoring its own legislation to ensure that it does not discriminate against foreigners, or among different foreign member countries; each country must adjust its tariff schedules so that they comply with the GATT bindings.

Why, then, has GATT succeeded? For the same reasons the Antarctica Treaty did: (1) its subject matter is, or at least was, relatively clear: the rules for international trade in goods; (2) all Contracting Parties have a stake in the system, both to keep tariffs as low as possible for their own exports, and to enjoy the greater worldwide prosperity that open international trade creates; and (3) the decisionmaking and enforcement apparatus, while clearly not perfect, worked fairly well to enforce both the letter and the reality of the rules requiring nondiscrimination and adherence to tariff commitments. Furthermore, to the extent that the GATT system is experiencing strains at the present time, those strains relate to these three points. As tariffs have been lowered over many years, non-tariff barriers to international trade, such as explicit or hidden quotas on imports, alleged product standards that disproportionately harm imports, and burdensome customs formalities, have taken their place as the most important clogs to the system. In addition, international trade itself has changed since the late 1940s, from trade principally in goods, to a nearly equal mixture of trade in goods and trade in services. As the world has shrunk, and firms facing high tariffs or other trade barriers find it easy to respond by establishing a subsidiary in the protected country, it has become clear that investment rules cannot be kept hermetically sealed off from trade rules. All three of these problems have greatly complicated the GATT's agenda – some would say far beyond its ability to cope.

Related to the second general problem is the commonality of interest among GATT Parties, or its lack. As the substantive agenda of GATT has expanded into areas that were traditionally thought to be matters of domestic policy (for example, what kind of development assistance will certain industries receive? on what terms will foreign investment enter? what criteria will govern the issuance of patents?), the interests of the developed countries and the developing countries have diverged. This, too, is causing severe strains in the system. Finally, as countries have become dissatisfied with the GATT's institutional machinery, they have increasingly turned to self-help, or to bilateral or regional arrangements. The United States has done all three, through its domestic authority under section 301 of the Tariff Act, through its network of "voluntary" restraint agreements that limit "excessive" imports, and through its exploration of a possible North American Free Trade Agreement. The GATT structure is in danger of becoming irrelevant, because it has not responded (and perhaps could not have responded adequately) to the demands for dispute resolution that have been placed upon it.

The lessons from the GATT experience to the problem of space debris are straightforward. It would be inadvisable, in light of the present disparity of space experience among the countries of the world, to attempt to negotiate a detailed U.N.-wide multilateral. At the most, an agreement involving all U.N. members might create a framework for cooperation, information-sharing, and research, which does not impose strict regulatory obligations upon signatories. GATT also suggests that it is risky at best in the international arena to bite off too much at one time. A treaty addressing the problems of orbital debris should obviously be coordinated with the other instruments of space law, but it need not solve all problems for all time. Finally, on a more positive note, the interrelationship between national law and the international machinery of GATT stands as a reminder that international law and national law can and should work together.

IV. A TREATY THAT WILL SUCCEED

What, then, will be the characteristics of a treaty on the subject of space debris that can avoid the fate of Part XI of the Law of the Sea Convention and the Moon Treaty? Bearing in mind that it will be impossible to please the entire world community, I offer the following thoughts.

First, one would want to restrict active participation in the treaty to the space-faring nations. Following the Antarctica example, one would include as voting members any and all nations presently active in space, and make provision for new membership as and when new space-faring nations emerge. In addition, it would be important to provide for other forms of participation for non-space-faring nations, such as observer status or another type of non-voting membership, so that questions affecting global welfare will not be discussed without input from all.

Second, an effective agreement would limit itself to the problem of space debris, however it may arise, developing definitions that are as clear as possible in light of technical and scientific knowledge about the issue. It may be desirable to limit it to debris whose origin is unknown and unknowable, and to treat collisions in space or damage on Earth from identifiable objects (like the Soviet satellite Cosmos 954 or the American Skylab) differently, either under different articles of a debris treaty or outside its framework altogether.

In this connection, the interaction between treaty law and national law will be important. At a minimum, the treaty would need to accept as a given whatever property interests the launching State claimed in its spacecraft (plus cargo, etc.), and whatever property interests the launching State conferred on any of its nationals authorized to engage in space activities. For sound economic reasons, as well as practical political realities, the alternative of creating counterparts to the Law of the Sea Authority and Enterprise, which would impress public responsibilities and financial taxes directly on space objects, will not work. Whether a phrase like "common heritage of mankind" is included in the treaty or not may be beside the point, except to the extent that it has come to mean something like the Law of the Sea deep sea-bed system. Space is an area that, with the possible exception of the geostationary orbit, is simply not amenable to the traditional kinds of jurisdictional line-drawing. To that extent, it physically must be shared by all nations. Just as traditional property law requires users not to waste common resources, and not to deplete them excessively, there is a moral and economic obligation on space users to keep the space environment safe and available for all.

Finally, experience shows that the kind of institutional machinery established can make a significant difference to the long-term success of an agreement. Although it may be difficult to devise a system that can be monitored as easily as the Antarctica Treaty's rules, or even the GATT's, levels of compliance will depend on how much others know about their practices. Given the nature of the problem, monitoring by an international agency (such as function performed by the International Atomic Energy Agency for nuclear devices) is probably preferable to individualized monitoring by all member States. International agreements often use simple reporting and publicity requirements effectively, even when the institutional machinery would not be well suited to "punish" offenders. If one wanted to go further, it might be possible to establish some kind of insurance fund against damage caused by space debris, contributions to which would be made proportionally by all space users. The ambition of the agreement and the complexity of the institutional machinery are directly related to one another. Both are eventually dependent on the seriousness with which the negotiating States approach the problem, and how effectively technology permits them to solve it.

ENDNOTES

*   I am grateful for the support of the Arnold and Frieda Shure Research Fund and the Russell Baker Scholars Fund of the Law School.

1.   See, for example, Treaty on Principles Governing the Activities of States in the Exploration and Use of Outer Space, including the Moon and other Celestial Bodies, January 27, 1967, 18 UST 2410, TIAS 6347, 610 UNTS 205 (entered into force Oct. 10, 1967) (referred to as "the Outer Space Treaty"), Art. VII; Convention on International Liability or Damage Caused by Space Objects, March 29, 1972, 24 UST 2389, TIAS 7762 (referred to as "the Liability Convention"), Arts. II, III; Convention on Registration of Objects Launched into Outer Space ("the Registration Convention"), January 14, 1975, 28 UST 695, TIAS 8480, Art. IV. See generally Howard A. Baker, Space Debris: Legal and Policy Implications (1989); Stephen Gorove, Developments in Space Law: Issues and Policies, ch. 14, "Space Debris and International Space Law," pp. 163-173 (1991); Glenn H. Reynolds and Robert P. Merges, Outer Space: Problems of Law and Policy, ch. 5, "Other Treaties, Agreements, and Issues," pp. 167-98 (1989).

2.   Stephen Breyer, Regulation and Its Reform (1982).

3.   Breyer (1982) at 15-34.

4.   In the rare case where causation is clear, such as the Cosmos 954 incident, this point would not hold. An eventual international regime governing space debris would appropriately distinguish between these two kinds of situations.

5.   William J. Baumol and Wallace E. Oates, The Theory of Environmental Policy 11–12 (2d ed. 1988).

6.   Id. at 12.

7.   Ronald H. Coase, The Firm, the Market, and the Law, ch. 5, at 95 (1988), reprinting the article of this name that originally appeared in 3 J. Law & Econ. 1 (1960).

8.   Reynolds and Merges, note 1 above, at 158.

9.   Id.

10. See the 1982 United Nations Convention on the Law of the Sea, Part XI (concerning activities on the deep sea-bed), reprinted at 22 ILM 1293-1308. For two comprehensive treatments of the political difficulties caused by the deep sea-bed problem, see James B. Morell, The Law of the Sea: An Historical Analysis of the 1982 Treaty and Its Rejection by the United States (1992), and Markus G. Schmidt, Common Heritage or Common Burden? The United States Position on the Development of a Regime for Deep Sea-Bed Mining in the Law of the Sea Convention (1989).

11. The Antarctic Treaty of December 1, 1959, 12 UST 794, TIAS 4780, 402 UNTS 71, does not specifically address the question of mineral exploitation. See Francesco Francioni, Legal Aspects of Mineral Exploitation in Antarctica, 19 Cornell Int'l L J 163 (1986). To fill that gap, the Antarctic Treaty Consultative Parties developed a Convention on the Regulation of Antarctic Mineral Resource Activities, June 2, 1988, reprinted at 27 ILM 859 (1988), which prohibited mineral resource activities in the area outside the framework of the Convention and established machinery for approving such activities. In 1991, the Consultative Parties took this one step further, by agreeing to a Protocol on Environmental Protection to the Antarctic Treaty, which was signed on October 4, 1991, in Madrid. The Protocol designates Antarctica as a natural reserve and provides for a ban on exploiting its mineral resources, which will last for fifty years. The Protocol would have been signed in June 1991, but the United States expressed last-minute reservations that delayed matters by four more months.

12. The Moon Treaty, UN Gen Assembly Res 34/68, approved December 5, 1979, was never approved by the United States Senate, largely because of the language in Article 11 of the Treaty stating that the Moon and its natural resources are part of the "common heritage of mankind." See Reynolds and Merges, note 1 above, at 115–123.

13. This is one of the principal effects of Proposition 65, which was passed by California voters in 1986. See "Inspired Prop. 65 Authors Take Aim at Air Pollution," Los Angeles Times, Thurs., Nov. 6, 1986, pt. 1, p. 3, col. 4, for a description of the scope of the law.

14. For my purposes, it does not matter which nation is doing the regulation, although I recognize that the question of assigning responsibility to a particular country is a difficult one. Under the Registration Convention, note 1 above, the state of registration might bear all responsibility. Alternatively, if the debris problem is principally caused by cargo clearly owned by another state or its national, one might assign responsibility to that state. It should be clear that to the extent we rely on national regulation, something akin to a conflict of laws regime that clearly allocates jurisdiction would be highly desirable.

15. Baumol and Oates, note 5 above, make a similar point when they say that the optimal size of jurisdiction for the setting of environmental standards depends on the nature of the pollutant, and that "the jurisdiction must be of sufficient size to internalize the great bulk of the pollution." Id. at 295.

16. See Geneva Protocol of 1925 on biological and chemical weapons; Biological Weapons Convention of 1972.

17. Compare the development of the international arrangements protecting trade in endangered species, marine mammal protection, and chloroflurocarbon usage. In each of these cases, international agreements have been developed, in recognition of the global scope of the problem.

18. See Schmidt, note 10 above, at 217–218, 230.

19. "U.S. May Refuse To Sign Treaty on Saving Species," The Wall Street Journal, Tuesday, May 26, 1992, at C12, col. 5.

20. The corollary to this point is that when investors have no idea what legal regime they are facing, as is presently the case in many of the former republics of the Soviet Union, and in some of the Central and East European countries, they will be reluctant to participate in that market. This, of course, is what is happening in those areas. Similarly, when the restrictions on foreign investment and the quality of property protection accorded to foreigners become too burden-

some, the flow of foreign investment will be restricted, as happened during the 1970s in many developing countries with restrictive laws (*e.g.*, Mexico, the Andean Pact countries, India).

21. For citations to the treaty and subsequent changes, see note 11 above.

22. This has consistently been the view of the United States, see David A. Colson, "The United States position on Antarctica," 19 Cornell Int'l L J 291, 295 (1986), as well as the other Consultative Parties, see Moritaka Hayashi, "The Antarctica Question in the United Nations," 19 Cornell Int'l L J 275, 282–283 (1986).

23. See Hayashi, note 22 above, at 275, 277–279.

24. See generally Hayashi, note 22 above.

25. With respect to trade, the Lomé Conventions between the European Community and the African, Caribbean, and Asian countries are a counter-example. Most of the developed countries have extensive networks of bilateral investment treaties with developing countries that establish the ground rules for foreign direct investment. These treaties were often negotiated at the very same time the more ambitious multilateral efforts were failing.

26. Of the voluminous literature on the GATT, the best overall studies of GATT's basic structure remain John Jackson, World Trade and the Law of GATT (1969), and Kenneth Dam, The GATT: Law and International Economic Organization (1970).

27. Candor compels the qualification here that the dispute settlement mechanism has not operated as well as many believe it ought to have done, and an important part of the on-going negotiations by the GATT Contracting Parties is an effort to improve them.

VI. A Multilateral Treaty

23: Orbital Debris: Prospects for International Cooperation

Jeffrey Maclure

Foreign Affairs Officer, Office of Science, Technology and Health, U.S. Department of State, Washington, D.C.

William C. Bartley

Senior Adviser for International Affairs, Office of Science, Technology and Health, U.S. Department of State, Washington, D.C.

International awareness of orbital debris is at least a decade old. The United Nations "UNISPACE '82" report characterized the increase in space science, technology and applications as a mixed blessing, as this activity had generated a vast amount of "space debris." (1)

Prophetically, at that time the U.N. report stated that while the "probability of accidental collision with a live space object is yet statistically small, it does exist and the continuation of present practices ensures that this probability will increase to unacceptable levels." (2)

In the intervening ten years, spacefaring nations have continued to generate debris in the orbital environment. It now appears that if left unattended, additional spacecraft launches, crowded collision-prone orbits, on-orbit construction, and in-space explosions and fragmentation could create a serious orbital debris problem that may increasingly and detrimentally affect activity in outer space.

IDENTIFICATION OF THE PROBLEM

The United States Government – led by NASA and the Departments of Defense and State – began to focus its attention to the problem of orbital debris in the mid-1980's. Initial interagency concepts were incorporated during the 1987 review of national space policy commissioned by the White House. (3) In 1988, President Reagan authorized a new space policy that included the following statement on debris:

All space sectors will seek to minimize the creation of space debris. Design and operations of space tests, experiments and systems will strive to minimize or reduce accumulation of space debris consistent with mission requirements and cost effectiveness. (4)

The 1988 directive also mandated that a Government interagency working group provide recommendations on how best to implement the new orbital debris policy.

The resulting interagency *Report on Orbital Debris* was released in 1989. Not surprizingly, international linkages figured prominently. The report found that the "causes and consequences of orbital debris are global in scope..." and that "international cooperation is essential to a satisfactory solution..." (5) One of its recommendations asserted that

the U.S. should inform other space-faring nations about the conclusions of this report and seek to evaluate the level of understanding and concern of other nations and relevant international organizations about orbital debris issues. Where appropriate, the U.S. should enter into discussions with other nations to coordinate minimization policies and practices. (6)

MINIMIZATION OF DEBRIS

Orbital debris may be minimized both through prevention of further debris generation, and mitigation of existing debris conditions.

The 1989 Report on Orbital Debris has led to the existing U.S. effort – in concert with other spacefaring nations – to monitor and manage data and information on debris, minimize its generation, and survive contact with debris in the space environment.

Countries around the world are participating in an international re-examination of the design and operation of launch vehicles, spacecraft buses, and onboard spacecraft instrumentation. Rockets are being designed to vent residual fuel from spent upper stages that otherwise might explode and fragment the surrounding body. Energy storage devices such as batteries and high pressure containers are being improved to minimize the possibility of explosion. The potential for boosting satellites to escape orbits, as well as their deorbit and re-entry, are being assessed. The designation of dedicated graveyard orbits also is being considered, as is on-orbit collection and salvage of spacecraft and other objects.

(The opinions expressed in this paper are the personal views of the authors, and are not statements of official policy of the United States Government or the U.S. Department of State.)

INTERNATIONAL COOPERATION

Orbital debris minimization will be complex, expensive, and achieved only through coordinated technological innovation and application. Spacefaring nations will need to cooperate to achieve a better understanding of the problem, to avoid the further generation of debris, and to provide each other with mutually beneficial technical information and assistance. The development of appropriate and equitable standards of international conduct in debris minimization may follow as a result.

The contact, consultation and coordination recommended in the 1989 Report on Orbital Debris already is underway. With U.S. Government interagency coordination, NASA is conducting informal technical discussions with Japan, Russia, China, and the European Space Agency, a regional organization that produced its own orbital debris report in 1988. (7) Contacts with France, Germany, Canada, Australia, and India have been made as well.

Notwithstanding ongoing international information exchange, there is speculation on whether the existing mode of informal technical dialogue should expand into a more structured, institutionalized international regime for orbital debris management. It appears that the time is not yet ripe.

As outlined in the 1989 Report, each spacefaring nation wants to avoid imposing criteria for debris minimization on itself that, if not followed by others, would disproportionately constrain its individual national civil, military and commercial space programs. (8) Political considerations and concerns for sensitive technology transfer also weigh in. Although good informal progress toward international debris prevention standards has been made in a relatively short time, a "consistency of policies, standards and practices" (9) has yet to be achieved.

As well, while the problem of orbital debris is recognized as serious, it is believed that the level of scientific and technical understanding of debris generation and minimization is not at the point where formalized international policies and practices can accurately be formulated.

A good example of how current knowledge and practice can be significantly modified as a result of greater understanding are the new considerations of debris minimization at geosynchronous altitudes. It is accepted that explosions in this orbit can be minimized, yet in-orbit high velocity collisions will still result in fragmentation. Until now, boosting debris to an adjacent graveyard orbit was thought to reduce the threat of collision by providing spacecraft with additional breathing room.

Yet debris buried in an adjacent graveyard orbit can have the same potential for high velocity collision and fragmentation, and resulting detritus can pass through the operational orbit and impact on spacecraft. However, a recent International Telecommunications Union delimitation on orbital inclinations now will permit the possible use of both a low collision velocity "stable plane orbit" at geosynchronous altitude, and an accompanying low velocity graveyard orbit. Utilization of these orbits would require modification of ground segment technologies and procedures, but could minimize additional debris generation. (10)

In the forum of the United Nations Committee on the Peaceful Uses of Outer Space, the United States has taken the position that greater knowledge of debris is required before adding the issue to the Committee's agenda, or before beginning to discuss the establishment of international regulatory measures. (11) Debris continues to be an issue of interest, however, and the United States and other spacefaring nations inform the Committee of progress in their research and cooperation.

Deferring debate within the United Nations while expert, yet informal, international research and coordination continues could help avoid politicization of the debris issue, and possible premature constraints on the development of technology and cooperative programs.

Given the dearth of knowledge, incomplete technological development, untested preventive mission operations, and technical and political sensitivities, it appears that the current course of informal international consultation and information exchange will continue for at least the near term. The time will come, though, when additional structure may be appropriate to help direct international communication and cooperation, establish standards, and regulate activity.

POTENTIAL FOR REGULATION

While relevant, current international space law, treaties and conventions do not specifically refer to debris and are inadequate for its regulation. As such, the negotiation of a new multilateral treaty or convention has long been cited as a promising mechanism for international cooperation in debris minimization.

Prospects for successful negotiation and conclusion of such a convention in the near term are slim. Involvement of non-spacefaring nations could inject politicized, non-technical issues into negotiations. International legal and economic issues relevant to a new debris regime – such as definition of debris, regulatory jurisdiction and control, and expanded liability – will also require attention, development and resolution. (12) And again, the baseline of a much better understanding of the debris environment and minimization technologies will be needed as well to guide the drafting of an effective regulatory convention.

The potential exists for the development of non-binding principles on orbital debris similar to those developed in the United Nations addressing satellite remote sensing of the earth. (13) However, it should be remembered that debate on the remote sensing principles took many years, and the same legal, technical and political no-man's land would have to be traversed before reasonable principles on debris could be agreed upon.

With these realities in mind, the first formal international cooperative activity in debris minimization is likely to take place under the terms and conditions of individual bilateral arrangements. Such arrangements could encourage mutually advantageous research, development and exchange of information, and would be essential to any cooperation in future mitigation such as on-orbit sweep, transfer, and retrieval missions.

Bilaterals have the advantage of addressing the specific interests of only two parties, and can be tailored to the level of technical and programmatic coordination desired. If two major spacefaring, debris-generating nations or regional space agencies could reach agreement on a framework arrangement, it could serve as a model that others could adapt to their own circumstances. The development of several bilateral arrangements could be the beginning of de facto standards of conduct.

Finally, scholars have proposed that, to save time and avoid politicization, an informal multilateral debris working group might be established whose activities would lead to greater international information dissemination, program coordination, and the development of appropriate standards. Representation in such an international group could either be governmental or on a more advisory, private sector level. (14)

As individual national governments have ultimate authority, responsibility, and regulatory oversight of space systems, an international debris coordination group probably would have the best assurance of success through participation by government representatives.

A possible organizational model for a new international debris group exists in the Consultative Committee for Space Data Systems (CCSDS), an informal forum established in 1982 to discuss common problems relative to space information and data systems. The CCSDS is guided by a charter, and current members are primary space agencies from the United States, Canada, Brazil, the United Kingdom, France, the European Space Agency, Germany, India and Japan. Other space-related agencies in China, Canada, Brazil, Japan, the United States and Sweden have CCSDS Observer status. (15)

Based on this existing template, a Consultative Committee for Orbital Debris could be established by spacefaring nations to serve as a focused, technical, voluntary, and informal multilateral forum for the cooperative development of international standards for debris minimization.

Like the CCSDS, a Consultative Committee for Orbital Debris would help begin to institutionalize the international exchange of information on technologies and operations. Primary products of the Committee would be technical, non-binding recommendations, reached through consensus and providing guidance for individual development of interoperable debris minimization standards.

Membership on the Debris Committee could be open to individual spacefaring nations, and also to regional and international organizations involved in space system operations such as EUMETSAT, INTELSAT, and INMARSAT. Provision might be made for other government, organization, and interest group observers as well. The interests of private, commercial launch vehicle and satellite firms would be represented by

national delegations. A secretariat and various working groups would coordinate and carry out Committee activities.

The establishment of such a Committee could have merit as an intermediate step between informal contacts and more formal international arrangements for the regulation of debris.

CONCLUSIONS

Given current understanding of orbital debris and the state of existing space law, informal efforts in international debris minimization continues to be the most productive mode of international cooperation. Governments should anticipate, however, the development of internationally-coordinated debris minimization standards.

Further international cooperation will be dependent on continued research and development, the exchange of information on debris minimization, and expanded understanding of the debris environment.

The impetus toward international cooperation in orbital debris is strong, and prospects for the future are excellent if nations build on already productive relationships. Without attention, orbital debris has the potential to hinder progress in space science and applications. But with anticipatory action on earth, the world community can avert a tragedy of the commons (16) in space.

NOTES

1. *Report of the Second United Nations Conference on the Exploration and Peaceful Uses of Outer Space*, Vienna, August 9-21, 1982, A/CONF.101/10, August 31, 1982.

2. Ibid.

3. Michaud, Michael, "U.S. and International Policy Perspectives and Implications Emerging from Space Debris," *Space-A New Era: Fifth National Space Symposium Proceedings Report*, April 4, 1989; Remarks by W. Kendall, J.D., Office of Advanced Technology, U.S. Department of State, before an Office of Technology Assessment Workshop on Orbital Debris, Washington, D.C., September 25, 1989.

4. National Space Policy, 1988.

5. *Report on Orbital Debris*, Interagency Group (Space) Working Group on Orbital Debris, Washington, D.C., February 1989, p. 51.

6. Ibid.

7. *Space Debris: A Report from the ESA Space Debris Working Group*, November 1988.

8. *Report on Orbital Debris*.

9. *Ibid*

10. Kessler, Donald, "Management of Orbital Debris in the Geosynchronous Orbit," paper for the International Radio Consultative Committee of the International Telecommunications Union, 1992.

11. Statements by United States Representatives to the Twenty-Eighth and Twenty-Ninth Sessions of the Scientific and Technical Subcommittee of the United Nations Committee on the Peaceful

Uses of Outer Space, Press Releases USUN 03-(91) and USUN 05-(92), February 19, 1991, and February 25, 1992, respectively.

12. Sterns, Patricia M. and Tennen, Leslie I., "Orbital Sprawl, Space Debris and the Geostationary Ring," *Space Policy*, vol. 6, August 1990; *Orbiting Debris: A Space Environmental Problem*, Office of Technology Assessment, U.S. Congress, USGPO, September 1990; Roberts, Lawrence D., "Addressing the Problem of Orbital Space Debris: Combining International Regulatory and Liability Regimes," *Boston College International and Comparative Law Review*, vol. 15, Winter 1992.

13. "Principles Relating to Remote Sensing of the Earth from Outer Space," United Nations General Assembly, A/41/751, December 1, 1986.

14. He Qizhi, "Environmental Impact of Space Activities and Measures for International Protection," *Journal of Space Law*, vol. 16, no. 2, 1988; McKnight, Darren S., "Track Two Diplomacy: An International Framework for Controlling Orbital Debris," *Space Policy*, vol. 7, February 1991; Williamson, Ray A., "The Growing Hazard of Orbiting Debris," *Issues in Science and Technology*, vol. 8, Fall 1991.

15. "Consultative Committee for Space Data Systems: Achievements and Products," CCSDS Secretariat, November 1989; Charter of the Consultative Committee on Space Data Systems, CCSDS, September 1990.

16. Hardin, Garrett, "The Tragedy of the Commons," *Science*, December 13, 1968.

24: Preservation of Near Earth Space for Future Generations: Current Initiatives on Space Debris in the United Nations[+]

Stephen Gorove

Director of Space Law and Policy Studies, University of Mississippi Law Center; Vice President, International Institute of Space Law, International Astronautical Federation (IAF); Member, International Academy of Astronautics; IAF Representative Before the U.N. Committee on the Peaceful Uses of Outer Space

INTRODUCTION

The unique scientific and technological achievements of the mid-twentieth century which enabled mankind to leave mother Earth and opened the door for the exploration and use of outer space have not only brought about many unexpected benefits and potentials for further advancement, both outward- and inward-looking, but have also carried with them many new responsibilities.

At the time when the first artificial satellites completed their initial orbits around the earth few people, if any, would have surmised that within the relatively short span of a little over three decades, there would be serious concerns about an increasing number of nonfunctioning, uncontrollable or abandoned man-made space objects circling the Earth. These pose an ever-increasing risk of collision with active, manned and unmanned spacecraft, and carry potentially fatal consequences for spacefarers and objects of present and future generations. Yet as we are approaching the twenty-first century, it has become eminently clear to a number of competent scientists and other experts that unless appropriate measures are promptly taken the continued, unchecked proliferation of useless space objects, commonly referred to as space debris, will indeed become a major threat to space activities and an impediment to the exercise of the freedom of exploration and use of outer space, a cardinal principle of international space law.

The overview of the nature, status and scope of the space debris problem, including its sources and effects, as well as the current national techniques and practices, together with the interdisciplinary implications focusing on economic and legal issues, are set forth in other segments of this interdisciplinary symposium. The assigned task and purpose of this presentation is to review the current initiatives in the United Nations and evaluate the chances for the conclusion of a multilateral treaty to deal with the space debris problem.

INITIATIVES LEADING TO THE FIRST U.N. GENERAL ASSEMBLY RESOLUTION ON SPACE DEBRIS

While sporadic manifestations of the recognition of risks created by man-made space debris may be found in the records of the U.N. Committee on the Peaceful Uses of Outer Space more than a decade ago, it was only in recent years that the matter of space debris received more consistent attention by member countries in the hope that it would be placed on the agenda of the Committee and its two Subcommittees, the Scientific and Technical and the Legal Subcommittee.

One of the specific occasions prompting a discussion of space debris issues arose during the twenty-fifth session of the Scientific and Technical Subcommittee where concern was expressed about the possible consequences of a collision between a space object carrying nuclear power source (NPS) on board and particles of space debris. It was pointed out that this may involve the disintegration of NPS into a large number of fragments, some of which would be radioactive; that the distribution of the resulting debris cloud in a shell around the Earth could pose a hazard to other space missions, especially manned missions, and that the collision might result in the early re-entry of the radioactive debris into the Earth's atmosphere.[1]

To assess the risks of a collision between a space object carrying NPS on board and particles of space debris, some delegations expressed the opinion that a State should include specific information, including the generic classification of the NPS, when notifying the U.N. Secretary-General pursuant to article 4 of the Convention on Registration of Objects Launched into Outer Space. Other delegations were of the view that such notification is not germane to the safety of NPS in space and that prompt and full provision of information in the event of uncontrolled NPS re-entry is the central issue.[2]

A more coordinated effort focusing attention on the space debris problem occurred during the full Committee's 1989 meeting, when Sweden together with Australia, Belgium, Canada, the Federal Republic of Germany, the Netherlands and Nigeria, proposed that the issue of space debris be placed on the agenda of the Scientific and Technical Subcommittee.[3] Although no consensus was reached on that specific proposal, the Committee agreed that space debris was an issue of concern to all nations and considered it essential that more attention be paid by Member States to the problems of collisions with space debris and other aspects of space debris. The Committee also called for the continuation of national research on this question. This action turned out to be important because it led to the first major step by the U.N. General Assembly with respect to space debris. In its resolution 44/46 dealing with international cooperation in the peaceful uses of outer space, the General Assembly recommended that "more attention be paid to all aspects related to the protection and preservation of the outer space environment, especially those potentially affecting the Earth's environment" and that it was "essential that Member States pay more attention to the problem of collisions with space debris, and other aspects of space debris," and that "national research on that question" be continued.

The Committee's consensus expressed in its language on space debris and the subsequent incorporation of that language into General Assembly resolution 44/46 did not occur in a vacuum. Apart from prior U.N. discussions, the increasing worldwide concern over the protection of the environment, the multitude of studies, expert opinions, scholarly gatherings, and symposia at national and international levels were no doubt helpful in paving the way toward its adoption.[4] The resolution also provided a legitimate ground for delegations to use it, in subsequent sessions, as a point of reference when addressing space environmental issues, in general, and space debris concerns, in particular.

FOLLOW-UPS IN 1990

Already during the 1990 session of the Scientific and Technical Subcommittee, some delegations took the occasion to elaborate on the mandate of resolution 44/46 in their own terms by emphasizing the fact that outer space was progressively becoming an integral part of the human environment, providing an increased sense of urgency to the proper treatment of the questions relating to the threats posed by space activities to the Earth environment and to the preservation of the space environment itself. In their view, increased knowledge regarding the space environment was necessary to control the amount of space debris. They felt that the tracking and monitoring of the space debris environment could not be done without international cooperation. Thus the emphasis on urgency, the necessity for increased knowledge and the need for international cooperation in tracking and monitoring space

debris were brought up as additional suggestions for possible consideration.[5] Although not shared by some delegations (*e.g.* the U.S. and France), the view already advanced in 1989 was reiterated, that the question of space debris should be included as an item on the Subcommittee's agenda.[6]

While the increasing quantity of space debris continued to be regarded as a threat to the exploration and utilization of outer space and required steps to be taken against the generation of further debris, a suggestion was made by Czechoslovakia during the 1990 session of the Legal Subcommittee that the Scientific and Technical Subcommittee could consider the scientific and technical aspects while the Legal Subcommittee could study the legal aspects of the problem. By joining forces, the two subcommittees would no doubt succeed in developing universally accepted principles aimed at protecting the space environment against pollution by debris.[7] This was an important idea demonstrating the recognition of the interdisciplinary nature of the subject matter of space debris and, at the same time, setting a specific objective of developing universally accepted principles.

A further significant manifestation of closer attention to the space debris issue was reflected during the Committee's 1990 meeting in the chairman's introductory statement which, because of its tone setting initiative and relevance, may be highlighted as follows:

An extraordinary variety of man made objects has been injected into orbits around the Earth. Many are still intact units, but collisions between some others, as well as accidental and deliberate explosions of rocket components, have created a huge number of fragments. Space objects can therefore be divided into two catagories: firstly, satellites which are active or under control, and secondly, space debris, meaning dead satellites and mission-related objects like spent rocket stages, fragments of disintegrated rockets and satellites, engine exhaust particles, paint flakes, etc.

According to a periodic NASA satellite situation report, there are well over 7,000 known and trackable space objects presently circling the Earth, and certainly several times as many too small to be detected with present technology. No more than about 350 are active satellites, while the rest do not fulfill any useful function....

This increasing population of orbiting objects is causing concern for a number of reasons: the safety of manned space flight; accidental re-entry of space hardware; contamination by nuclear material in space and on the ground; damage to or loss of active satellites through collisions; debris proliferation through secondary collisions; crowding in the geostationary ring; interference with astronomical observations on the ground and in space; and interference with experiments in space.

Space debris constitutes an unacceptable risk to man and material in space and on the ground. Little can be done about man-made objects and debris already in orbit.

Man-made cleaning operations are beyond the capabilities of present technology. The only natural cleaning effect, the atmospheric drag enhanced by solar activity, cannot cope with all the debris generated in the course of space operations. All that can be done is to minimize collision risks and the future proliferation of debris by preventive measures at the planning stage. This would involve improved design, and the avoidance of intentional and accidental explosions in space. Space hardware could be programmed to remove to extremely high disposal altitudes at the end of its working life, or be propelled down into the atmosphere, where it would burn up. There is however neither an international agreement, universal application nor even a recommendation of such measures. And, of course, they would involve the expenditures of considerable time, effort and money, with resistance to be expected from States with substantial space programmes, which would have to meet the cost.[8]

Following the tone setting statement of the Committee's chairman, the Swedish delegate took the lead in drawing attention to recent studies in the Federal Republic of Germany, in the United States by a Government interagency group cochaired by NASA and the Department of Defense, by COSPAR and the International Astronautical Federation which indicated that space debris has become a matter of increasing concern and has been identified as a serious threat to man's exploration and utilization of outer space. He noted that according to experts the international space station planned for the late 1990s faced serious safety risks from intermediate-sized pieces of orbital space debris too large to be deflected by protective metal shielding on the station and too small to be tracked by radar from Earth and avoided. He emphasized that while the spacefaring nations had to assume a special responsibility in addressing the problem, it appeared clear that space debris is a global concern, requiring the attention and involvement of the international community. This was evident from the fact that pollution of the space environment and the increasing risks of collisions between space debris and spacecraft reduce the accessibility of space for all countries. Sweden shared the view, expressed both in the United States interagency study and the report by the European Space Agency, that more knowledge was needed, particularly concerning smaller space debris and that a better scientific and technical understanding of the problem was essential, not only among the major spacefaring nations, but in the international community as a whole. Therefore, the Subcommittee should focus on exchange of information aimed at enhancing the understanding of the problem.[9]

While some countries, like Italy, expressed a general desire to see progress on the question of protection and preservation of the space environment, including the challenge created by the phenomenon of space debris environment, others (*e.g.* the Netherlands) called for urgent action involving an improvement in the understanding of the problem, besides action to minimize risks by preventing further accumulation of the debris and introducing changes in the design and operational practice of spacecraft in order to avoid collisions.[10] Several other countries voiced their support for the discussion of the subject of space debris.[11] Still others pointed to the value of completed studies on space debris,[12] with some stressing the need for further studies.[13] In connection with the latter, Japan felt it would be advisable if the information accrued from national research would be shared in an exchange of views at the future session of the Scientific and Technical Subcommittee.[14]

As to the attitudes of the spacefaring nations, the Soviet Union appeared to take no concrete position apart from a general remark that it welcomed the chairman's statement on such questions as space debris along with other issues. At the same time, the United States indicated that it did not share the view that space debris should be on the Scientific and Technical Subcommittee's agenda as a separate item. In the U.S. view, more attention to the problem of space debris and continuation of national research, as called for by General Assembly resolution 44/46, was the proper approach. Building on previous research, the United States was in the final stages of preparing a multi-year, two-phase orbital debris research plan, focusing on research in two rival areas: characterization of the debris environment and methods of minimizing and surviving debris.[15] France shared the U.S. view in opposing a discussion of the space debris issue as a separate agenda item.[16] Ultimately, all members of the Committee agreed that space debris was an issue of concern to all nations and that it could be an appropriate subject for discussion by the Committee in the future.[17] This consensus was reflected in the ensuing General Assembly resolution 45/72 which added the phrase to the text of its prior year resolution that space debris could be an appropriate subject for discussion by the Committee in the future.

DEVELOPMENTS IN 1991

(a) Views in the Subcommittees

The Scientific and Technical Subcommittee in its 1991 session noted the importance of reducing the generation of space debris and of international cooperation in addressing these issues. More delegations expressed the view that the question of space debris should be included on the Subcommittee's agenda.[18] Other delegations (*e.g.* U.S. and France), while recognizing the importance of the subject, regarded such consideration as premature until further national research on the problem of space debris had been completed.[19]

Notwithstanding the divergent views, the Subcommittee agreed that further studies should be conducted on the problem of collision of NPS with space debris and that it should be kept informed of such studies.[20] There was also a suggestion that an international expert group should be established to consider the problem.[21] In addition, the Subcommittee noted that there was a need for further research concerning space debris, for the development of improved technology for monitoring space debris, and for the compilation and dissemination of data on it.[22]

Encouraged by the General Assembly's position that space debris could be an appropriate subject for discussion, in the course of the Legal Subcommittee's meeting, a growing number of delegations expressed themselves in favor of it.[23] One of the serious and urgent concerns was that the unchecked growth of space debris was liable to hamper space exploration and other peaceful uses of outer space and jeopardize space activities of all States and in particular, of developing countries which had not yet been able to undertake such activities.[24] Also, the increasing risks of collision of debris in certain orbits with NPS-carrying satellites created an unacceptable risk to the safety of human beings and material objects in space. As more and more objects carrying NPS were launched, and those no longer in operation were abandoned or left in orbit, the chances of collision would only increase. Radioactive contamination of the environment was no longer merely a hypothetical matter. Not only was the safety of certain orbits thus becoming increasingly threatened, but also the safety of the Earth itself. Accordingly, there was a pressing need to find a solution in order to remove the debris from outer space and, in particular, the geostationary orbit.[25]

As to posssible remedial measures, it was suggested by Argentina that one of the most appropriate ways of resolving the issue would be to develop norms stipulating the scope and content of article IX of the 1967 Outer Space Treaty. On the basis of that norm, effective machinery should be established for prior consultation in cases when the outer space activities of one State threatened to impede the exercise by other States of their right to conduct such activities. In this connection, it was stressed that the question whether the activities of one State might constitute an obstacle to the activities of other States should not be left solely to the State concerned, but should be considered objectively.[26]

The repeated references to the need to set up rules and regulations to govern space debris added a more concrete element to what was hoped to be accomplished as a result of future discussions of the debris problem.[27]

Notwithstanding the swelling tide of delegations supporting the inclusion of space debris as an item on the agenda, the United States and France continued to oppose such a move at the time. For its part, the U.S. recognized that space debris was an extremely important and complex question but held the view that the Legal Subcommittee should deal with it at a later date because it believed that it was still too soon for the item to be placed on the Subcommittee's agenda. The U.S. noted that various national and international organizations were at that time carrying out important research into the matter. The United States was cooperating with other countries on the study of space debris and the results of that research would have to be made available to other countries before the issue of

the potential ramifications of space debris could be addressed. The benefit of such research would accrue to all countries whether presently in space or not. The United States would continue to consult with other governments to determine which areas might give rise to cooperation or discussion on a global level.[28] In a similar vein, the French delegate noted that France had been studying the matter of space debris together with its European Space Agency partners. He felt the time would soon come when the matter could be considered in specific terms, but as yet it required more in-depth study and until that was completed the time was not ripe for the Legal Subcommittee to take up the matter.[29]

The Legal Subcommittee's 1991 session also provided participants with a tangible illustration of the danger that an out-of-control space object with NPS on board might create. In response to several requests, the Soviet delegate gave information on the final stages of the abortive flight of Salyut-7/Cosmos-1686 orbital combination, the debris of which had fallen on Argentina close to the Chilean border on February 7, 1991, without causing casualties or damage to property. He noted that all the necessary information about the flight had been transmitted to all countries concerned which were assured that the issue of compensation for any damage would be settled in the context of bilateral talks in accordance with the 1972 Liability Convention.[30] However, the Argentine delegation noted that since the search operation for fragments of the station and the work of assessing the damage had not yet been completed, Argentina had been placed in a very peculiar position, for it had neither the experience nor the information needed to take appropriate preventive measures. Therefore, it was necessary for States on whose territory the space object might fall to have access to information prior to the fact of the accident and not after the fact, as in the case of Argentina.[31]

(b) Concerns in the Committee

During the 1991 meeting of the Committee, the chairman voiced his assessment of the debris situation noting that there appeared to be general agreement that sooner or later the question of space debris had to be addressed at the international level and that the General Assembly had already recognized this by noting in its resolution 45/72 that space debris could be an appropiate subject for discussion by the Committee in the future.[32]

A more elaborate statement in favor of the prompt consideration of the space debris issue as an agenda item came from the delegate of Czechoslovakia. He noted that the problem of space debris and its possible harmful interference with active space objects, including manned missions, had been recognized by the scientific community and was not a new subject for the Committee. Several studies on the danger of space debris were among the lists of documents of past sessions. Then he went on to state that:

–   the danger of a collision and consequent damage to or
    loss of a satellite is inexorably rising and that we can

envisage an avalanche of collisions of debris and the growth of unstable regions;

- no time should be lost in embarking on the discussion of the issue and preventive measures should be considered inasmuch as these are much more effective and less expensive than remedial measures attempting to remove the debris already in orbit;

- the time has already come to discuss in the Committee and its subcommittees the subject of space debris. There were already hundreds of scientific papers, and at least a thousand studies before the Committee and yet the Committee was still waiting for more papers, more studies and more debris in addition to the close to seven thousand trackable space objects and some fifty thousand nontrackable pieces of debris;

- the proposal of Czechoslovakia to have the space debris issue put on the agenda of the Committee and its subcommittees was put forward in the spirit of international cooperation and the only motivation behind it was the strength of scientific facts.[33]

While all delegations shared the concerns about space debris, and many stressed the need for international cooperation, with some raising the issue of its possible effect on the accessibility of outer space,[34] as in prior years, there were a number of delegations which favored prompt inclusion of the space debris issue on the agenda of the Committee or its subcommittees.[35] Other delegations reiterated that more knowledge was needed concerning space debris,[36] and that a better scientific and technical understanding of the problem was required, not only in the major spacefaring nations but in the international community as a whole.[37] Some countries stressed the need to minimize the risks created by space debris,[38] while others emphasized that it was imperative to improve the technology for monitoring space debris, to disseminate in a timely manner the data on it and adopt measures to reduce its generation.[39] Still others were content with urging the Committee to come to an agreement on how to deal with the issue in the future and stressed that international cooperation was required if future generations were to reap the benefit of planet Earth.[40] Belgium also emphasized that the time had come to think seriously about the elaboration of appropriate rules, which would constitute the legal framework to tackle this specific form of environmental pollution and assure the safe use of satellites.[41]

Among the spacefaring nations, both the United States and France, while recognizing the importance of the subject, continued to take the position that consideration of the space debris issue as an agenda item was premature until further national research on it had been completed.[42] While there appeared no specific indication from the Soviet Union pro or con of such consideration, China only indicated its willingness to hold discussions as to whether the issue should be included on the agenda of the Scientific and Technical Subcommittee.[43]

At the end of the Committee's session, there was agreement on a need for the continuation of research on space debris, for the development of improved technology for the monitoring of space debris and for the compilation and dissemination of data on space debris. With reference to the work of the Scientific and Technical Subcommittee, the Committee agreed that information on national research on space debris should, to the extent possible, be provided to the Subcommittee, in order to allow the Subcommittee to follow this area more closely. In this connection, the Committee requested the Secretariat to invite Member States to provide such information on their national research to the Subcommittee.[44] The gist of the preceding language expressing this consensus was then incorporated in General Assembly resolution 46/45 of December 9, 1991, which also reaffirmed its prior year recommendations.[45]

LATEST DEVELOPMENTS

In taking note of resolution 46/45 and also of the information that had been submitted by Member States in response to that action,[46] the Scientific and Technical Subcommittee during its 1992 session reaffirmed the importance of reducing the generation of space debris. It also underlined the need for further research concerning space debris, for the development of improved technology for monitoring space debris and for the compilation and dissemination of data on space debris while stressing the importance of international cooperation in addressing these issues.[47] Additionally, the Subcommittee emphasized that further studies should be conducted on the problem of collision of NPS with space debris and that it should be kept informed of the results of such studies.[48]

As to the all important issue of considering space debris as an agenda item, there has been no noticeable change in earlier positions.[49] However, it may be of interest to record that Czechoslovakia submitted a working paper, which was taken note of by the Subcommittee, concerning a definition of space debris and a possible format for relevant discussions in the Committee and its subcommittees.[50]

The 1992 session of the Legal Subcommittee reportedly left the space debris issue pretty much as it was but during the most recent COPUOS meeting both China and the Russian Federation indicated their willingness to have the space debris issue discussed in the Scientific and Technical Subcommittee as an agenda item and also France no longer appeared to raise any objection to such a move.[51] Thus, for all practical purposes, the United States remained the only country officially still set against having the issue placed on the agenda although perhaps not with exactly the same intensity as in the past.

ASSESSMENT AND PROSPECTS FOR A TREATY ON SPACE DEBRIS

In order to make a proper assessment of U.N. initiatives in the field of the protection and preservation of the near space environment for future generations, with a focus on the space debris issue, it is essential to bear in mind the potentials and limitations inherent in the relevant institutional setup.

The Committee on the Peaceful Uses of Outer Space was established by the United Nations shortly after the beginning of the Space Age. It was instrumental in drafting a number of major international treaties and U.N. resolutions pertaining to the peaceful utilization of outer space which were subsequently approved by the General Assembly. Consisting of representatives of fifty-three U.N. Member States and assisted by observers of a number of international institutions, the Committee followed, almost without exception, the time tested principle of decision making by consensus throughout its existence. Its two subcommittees, the Scientific and Technical and the Legal Subcommittee, consisting of the same members as their parent body, have provided much needed assistance to it during their regular annual meetings.

While the remarkable track record of the Committee in building a new body of international law to govern space activities has given it an aura of prestige and a sense of confidence potential, the impediments inherent in a multifaceted decision making process have virtually eliminated the possibility of quick and easy accomplishments. Seen in such a light, it is not surprising to find a relatively slow pace of relevant U.N. initiatives as revealed by the preceding review.

In light of the needed consensus in the Committee and its subcommittees, realistically, the General Assembly could only make very limited progress in its resolutions dealing with the near space environment, in general, and the space debris issue, in particular. This is evident from the overall tone and substance of the respective three resolutions. The first called on Member States that they pay more attention to all aspects related to the protection and preservation of the outer space environment, especially those potentially affecting the Earth's environment, and also to the problem of collisions with space debris and other aspects of space debris, and that Member States continue national research on that question. The second resolution added a rather cautious statement that space debris could be an appropriate subject for discussion by the Committee in the future. Even in the third year, the General Assembly could only go so far as to include a call for the continuation of national research on space debris, for the development of improved technology for the monitoring of space debris and for the compilation and dissemination of data on space debris, with the request that, to the extent possible, information thereon should be provided to the Scientific and Technical Subcommittee in order to allow it to follow this area more closely.

The limited action of the General Assembly and its inability to make the space debris issue an agenda item, is clearly traceable to the lack of consensus on this issue in the Committee and its subcommittees. From among the spacefaring nations which bear the responsibility for the debris creation and would also have to incur the expenses of remedial measures, the United States and, until recently, also France opposed such a move as premature. Prior to the 1992 COPUOS session the Soviet Union and China had equally exhibited no specific interest in it though they were repeatedly prodded by a large number of mostly developing nations which were concerned about the detrimental effect of space debris on future space activities.

Notwithstanding this impasse, the Scientific and Technical Subcommittee which, for some years, had on its agenda the use of NPS in outer space, could legitimately discuss the collision of NPS with space debris. However, interestingly enough, this fact alone did not result in more than an agreement that further studies should be conducted on the matter and that the Subcommittee should be informed about the results of such studies. This is revealing because it indicates that, even if space debris could have been discussed as a part of the agenda in the Committee, quite conceivably a consensus could not have been reached on more than what was incorporated in the General Assembly resolutions.

Be it as it may, it appears that no substantial steps will be taken on space debris, beyond the continuation of studies and dissemination of information on it, until it becomes an agenda item and, based on prior experience, probably only many years thereafter. Thus one may wonder, when can it be expected that the U.S. will agree to the space debris discussion as an agenda item and a follow-up action? At least one notable commentator suggests that 1995 is the likely date.[52]

All this does not imply that the time spent on research, collection and dissemination of data on space debris is necessarily lost time. While many scientists believe that we already have enough data, this writer has repeatedly stressed the need not only for more comprehensive information, but also for a continuous monitoring and updating of the space debris situation.[53] Because of the vital importance of detecting space debris, the call for the development of improved technology is an important and necessary intermediate step for any subsequent international action.

The question may be asked whether, during the current holding pattern, any other action apart from those incorporated in the U.N. resolutions might be useful in the long run and, at the same time, have a chance of acceptability to the spacefaring nations. It would appear that, beyond the collection and dissemination of data on space debris, there should be a call for information about action already taken by U.N. Member States to reduce space debris or prevent its further proliferation. Such a call may not run into much opposition, inasmuch as several countries, including the U.S., are known to have already undertaken various steps and the availability of this information might provide an indication of the type of measures that may have a chance of acceptability if, and when, internationally recommended standards and practices would come up for eventual consideration.

Should the space debris issue become an agenda item, the question may arise regarding the procedure and the substance of what can be expected to be accomplished. As to the former, clearly the space debris issue has both scientific and legal aspects, and there is no reason why the Czechoslovak suggestion to have the scientific aspects dealt with in the Scientific and Technical Subcommittee and the legal aspects in the Legal Subcommittee, should not be followed.[54]

As to substance, in addition to a norm stipulating the scope and content of article IX of the 1967 Outer Space Treaty, the need for the development of universally acceptable principles, rules and regulations aimed at protecting the space environment against pollution by debris, have been suggested as necessary steps. The norm would serve as a basis for the establishment of effective machinery for prior consultation in cases in which the outer space activities of one State threatened to interfere with the exercise by other States of their right to conduct such activities.[55]

In this connection, it may be noted that the representative of the International Astronautical Federation's International Institute of Space Law, in presenting the outline of a Draft for a Convention on Manned Space Flight during the 1991 session of the Legal Subcommittee, drew attention to its Article V which, *inter alia,* dealt with some desirable steps to be taken to avoid harmful space debris, pollution and contamination. Among them were the obligation to study the feasibility of appropriate measures, the dissemination of the respective information through the U.N. Secretary-General, and compulsory international consultation upon request, if a State Party had reason to believe that the activities of another State or its nationals might interfere with the manned space flight of the State Party.[56]

Insofar as prospects for the conclusion of a treaty on space debris are concerned, in the short run, there appears little likelihood of such development. It is much more likely that if guidelines and operational practices can be agreed upon, they would initially take the form of a set of principles and recommendations.[57]

In the long run, attitudes of the spacefaring nations which would have to bear the cost of remedial measures may change, if for no other reason than enlightened self-interest, should the space debris situation deteriorate to such an extent as to become an intolerable impediment to their own space activities. Let us hope that an effective space debris treaty will materialize before the eleventh hour for the benefit of both present and future generations.[58]

ENDNOTES

+   Unless otherwise indicated, the views expressed in this article are those of the author and do not necessarily represent the views of any organization with which he is connected.

1. UN doc. A/AC.105/409, at 30 (1988).

2. *Id.*

3. UN doc. A/AC/105/L.179 (1989).

4. Two such international symposia may be singled out for a brief reference. For instance, the International Institute of Space Law devoted

sessions during the preceding 30th Colloquium on the Law of Outer Space to discussions on the legal aspects of outer space environmental problems. *See* 30 Proc. Colloq. L. Outer Space 121–190 (1988). Also about the same time, another international colloquium on space environmental issues took place in Cologne, Federal Republic of Germany. For details, see Environmental Aspects of Activities in Outer Space (K.-H. Böckstiegel ed. 1988).

5. As a matter of concrete example, it was pointed out that the consequences of a spacecraft with NPS on board colliding with space debris could be studied using a variety of collision models, particularly, in the altitude range of 800 to 100 kilometres, where most nuclear sources, together with a large number of other satellites and the maximum density of space debris, were to be found. It was emphasized that since different types of objects could collide methods had to be developed for calculating the effects of elastic and nonelastic impacts, and of spacecraft destruction with the scattering of secondary fragments. *See* UN doc. A/AC.105/456, at 33 (1990).

6. UN doc. A/AC.105/456, at 21 (1990).

7. UN doc. A/AC.105/C.2/SR.530, at 4 (1990). Support for a discussion of the subject of pollution of the outer space environmenmt, especially by space debris, was also forthcoming from other delegations. *See* UN docs. A/AC.105/457, at 14 (1990), A/AC.105/C.2/L.171 (1990).

8. UN General Assembly, Official Records (hereinafter GAOR), doc. A/45/20, at 30–31 (1990).

9. UN docs. A/AC.105/PV.337, at 6–7 (1990) and OS/1478 (1990), at 4 (1990). For information regarding the studies on the question of space debris recently carried out by the European Space Agency in 1988, the Interagency Group on Space for the National Security Council of the United States in 1989 and the International Astronautical Federation, see UN doc. A/AC.105/420 (1990).

10. UN docs. A/AC.105/PV.339, at 42 (1990), OS/1478, at 3 (1990).

11. They included Czechoslovakia, the Federal Republic of Germany, India, Japan, Mexico, Nigeria and Pakistan (as a specific item). *See* UN docs. A/AC.105/PV.337, at 14 (1990); A/AC.105/PV.340, at 6 (1990); A/AC.105/PV.343, at 44–45 (1990); A/AC.105/PV.344, at 13 (1990), A/AC.105/PV.346, at 33, 37 (1990); OS/1478, at 3 (1990).

12. *E.g.* Federal Republic of Germany, *see* UN doc. A/AC.105/PV. 343, at 44–45 (1990).

13. *E.g.* Colombia and the German Democratic Republic, with India mentioning the possibility of international cooperative projects on data collection and monitoring of the environment. *See* UN docs. A/AC.105/PV.340, at 8 (1990); A/AC.105/PV.343, at 28 (1990); A/AC.105/PV.344, at 8 (1990), OS/1478, at 2 (1990).

14. UN doc. A/AC.105/PV.344, at 24–25 (1990).

15. UN doc. A/AC.105/PV.343, at 32–33 (1990).

16. UN doc. OS/1478, at 3 (1990).

17. GAOR, doc. A/45/20, at 16 (1990).

18. UN doc. A/AC 105/483, at 22 (1991).

19. *Id.* at 14, 15 (1991).

20. *Id.* at 22 (1991).

21. *Id.* at 14, 15 (1991). The Subcommittee agreed that information on national research on space debris might be provided to the

Subcommittee, as had been done in the past. This could be done through provision of working papers, interventions in the general debate and technical presentations by specialists from Member States, as well as COSPAR and IAF. Papers on space debris were circulated by the Indian Space Research Organization and a working paper was also submitted by Germany. *Id.* at 4, 22, and 27 (1991).

22. *Id.* at 22 (1991); UN doc. A/AC.105/PV. 358, at 38 (1991).

23. *E.g.* Argentina, Brazil, Czechoslovakia, Mexico, Nigeria and Pakistan. *See* UN docs. A/AC.105/C.2/SR.540, at 3, 8 and 11 (1991), A/AC.105/C.2/SR.542, at 5, 7 (1991), A/AC.105/C.2/SR.545, at 2–3 (1991).

24. *See* statements by Argentina and Pakistan, UN docs. A/AC.105/C.2/SR.540, at 8 (1991); A/AC.105/C.2/SR.542, at 7 (1991).

25. Attention to this was drawn by Nigeria and Pakistan. *See* UN docs. A/AC.105/C.2/SR.546, at 4 (1991), A/AC.105/C.2/SR.542, at 7 (1991).

26. UN doc. A/AC.105/C.2/SR.540, at 8 (1991).

27. The Colombian delegate supported the proposal put forward by the Argentine delegation for the establishment of rules and regulations on space debris, which it deemed necessary in view of the danger of congestion of the geostationary orbit and of the dangers posed by the presence of such debris in outer space and by its re-entry. *See* UN doc. A/AC.105/C.2/SR.541, at 4 (1991).

The Ukranian delegation noted that international law relating to the enviroment already included well-established norms that could be used. *See* UN docs. A/AC.105/C.2/SR.545, at 3 (1991) and OS/1510, at 3 (1991).

28. UN docs. A/AC.105/C.2/SR.541, at 5 (1991), OS/1506, at 3 (1991).

29. UN docs. A/AC.105/C.2/SR.541, at 6 (1991) and OS/1506, at 3 (1991).

30. UN doc. A/AC.105/C.2/SR.543, at 3–4 (1991).

31. UN doc. A/AC.105/C.2/SR.548, at 8 (1991).

32. GAOR doc. A/46/20, at 41 (1991).

33. UN docs. A/AC.105/PV.351, at 68–70 (1991), A/AC.105/PV.356, at 43–45 (1991).

34. *E.g.* Sweden. *See* UN doc. A/AC.105/PV.358, at 38–41 (1991).

35. Including for instance, Czechoslovakia, India, Nigeria, Pakistan, Philippines and Sweden. *See* UN docs. A/AC.105/PV.353, at 9–10 (1991); A/AC. l05/PV. 354, at 28, 42, 56 (1991); A/AC.105/PV.358, at 38–41 (1991).

36. *E.g.* Austria, Belgium and Brazil. *See* UN docs. A/AC.105/ PV.352, at 23 (1991), A/AC.105/PV.356, at 41 (1991); A/AC.105/ PV.358, at 14–16 (1991). Italy stressed the need for further research, with Brazil desiring especially to focus on the problem of collision of NPS-carrying spacecraft with space debris. *See* UN doc. A/AC.105/PV.358, at 14–16, 32 (1991).

37. *E.g.* Sweden. *See* UN doc. A/AC.105/PV.358, at 38–41 (1991). In this connection Poland proposed the elaboration of a report on the protection and preservation of the outer space environment, especially on the activities potentially affecting the Earth's environment. The main goal of the proposed report would be to identify space activities which cause hazardous phenomena to the space environment and to point out the consequences for the further development of space activities. The aim of the report, which would not be directed against any nation, body, or space activities, would be to highlight the unwanted and dangerous by-products of space activities. The report should consider the subject of space debris, nuclear explosions in space and their influence on space objects, the atmosphere and the Earth's environment, the influence of rocket propulsion systems, the use of NPS and solar-power plants, and finally the rapid increase of satellite communications which lead to increased intensities of electromagnetic waves in outer space and the atmosphere. *See* UN doc. A/AC.105/PV.355, at 16–18 (1991).

38. *E.g.* Italy. *See* UN doc. A/AC.l05/PV.358, at 32 (1991).

39. *E.g.* Austria, China, Nigeria and the Philippines. *See* UN docs. A/AC.l05/PV.352, at 23 (1991); A/AC.105/PV.354, at 42, 56 (1991); A./AC.l05/PV.358, at 26 (1991).

40. *E.g.* the Netherlands, *see* UN doc.A/AC.105/PV.353, at 23 (1991). Australia supported the role of the United Nations and called for the strengthening of that role. *See* UN doc. A/AC.105/PV. 361, at 34 (1991).

41. UN doc. A/AC.105/PV. 356, at 41 (1991). Belgium also expressed concern about radioactive contaminations that might endanger human life in space as well as on Earth, especially when they involve space objects containing radioactive material. In the long run, space debris might create a threat to the natural environment on Earth. *See* UN doc. A/AC.105/PV.356, at 41 (1991).

42. GAOR doc. A/46/20, at 17 (1991).

43. UN doc. A/AC.105/PV.358, at 26 (1991)

44. The Committee also noted the importance of international cooperation in addressing those issues and endorsed the recommendation of the Scientific and Technical Subcommittee that further studies should be conducted on the problem of collision of objects carrying NPS on board with space debris and that the Subcommittee should be kept informed of the result of the studies. GAOR doc. A/46/20, at 12, 17 (1991).

45. GA res. 46/45 (1991). In a slight change of its earlier recommendation, the General Assembly considered that space debris could be an appropriate subject for "in-depth" discussion by the Committee.

46. This was contained in document A/AC.105/510 and Add. 1 to 3. The Subcommittee also heard a special presentation on the subject by an expert from India. *See* UN doc. A/AC.105/513, at 21 (1992).

47. UN doc. A/AC.105/513, at 21 (1992). As in 1991, the Subcommittee agreed again that information on national research on space debris should continue to be provided to the Subcommittee and that this could be done through provision of working papers, interventions in the general debate and technical presentations by specialists from Member States and international organizations, in particular, by COSPAR and IAF. As an added element, the Subcommittee agreed that other matters related to the protection and preservation of the outer space environment should also be a subject of further research. In this connection, it also took note of a working paper submitted by Poland (UN doc. A/AC.105/C.1/L.181) calling for the preparation of a report on the preservation and protection of the outer space environment, especially activities potentially affecting the earth environment, and proposing the establishment of a working group to prepare the report. *See* UN doc. A/AC.l05/513, at 21 (1992).

48. The Subcommittee took note of a working paper submitted by the Russian Federation (UN doc. A/AC.105/C.1/L.180) containing preliminary results of studies on the question. *See* UN doc. A/AC. 105/513, at 14 (1992).

49. UN doc. A/AC.105/513, at 21 (1992).

50. UN docs. A/AC.105/C.1/L.184 (1992); A/AC.105/513, at 22 (1992).

51 *See* Vladimir Kopal, *The 31st Session of the Legal Subcommittee of the United Nations Committee on the Peaceful Uses of Outer Space, 23 March – 10 April 1992,* 20 J. SPACE L.46, at 49 (1992). *See also* the statements made by the delegations of China and the Soviet Federation on June 19, 1992. *Cf.* UN doc. OS/1568, at 3, 6 (1992).

52. Edward R. Finch, Jr., *Outer Space Environmental Pollution: An International Law-Science Update and Comment on "Heavenly Junk,"* 20 J. SPACE L. 65, at 67 (1992).

53. *See* Stephen Gorove, *Man-Made Space Debris: Data Needed for Rational Decision,* reproduced in STEPHEN GOROVE, DEVELOPMENTS IN SPACE LAW: ISSUES AND POLICIES 157–161 (1991).

54. UN doc. A/AC.105/C.2/SR.530, at 4 (1990).

55. UN doc. A/AC.105/C.2/SR.540, at 8 (1991).

56. UN doc. A/AC/105/SR.544, at 11–12 (1991). The provision of this Draft requires studies on the feasibility of appropriate measures to avoid harmful space debris which is far different from just calling for national research on space debris and aspects of space debris. For a text of the Draft which was prepared under the auspices of three leading institutions in Germany, the U.S. and the U.S.S.R., see 18 J. SPACE L. 209 (1990).

57. It may be noted that the observer of the International Law Association during the Committee's 1991 meeting drew attention to the work of the ILA Space Law Committee on a draft of an international instrument on the environmental aspects of space activities, and on space debris, in particular. *See* UN doc. A/AC.105/ PV.353, at 34 (1991).

58. In the course of space debris discussions, there were occasional references by delegates to the interests of future generations. *See,* for instance, statement by the Ukranian delegation, UN doc. A/AC.105/C.2/SR.545, at 3 (1991).

59. Press Release USUN 94-(93). June 10, 1993.

Postscript

As this publication goes to press, a postscript of a hopeful sign should be added. During the 1993 meeting of the Committee, the United States dropped its opposition to having the space debris issue placed on the agenda of the Scientific and Technical Subcommittee on the condition that the agenda item was appropriately focused on research being done on the debris environment.[59] Then the door is now open in the United Nations for meaningful deliberations on the debris issue which hopefully will lead to eventual international action.

25: A Legal Regime for Orbital Debris: Elements of a Multilateral Treaty

Pamela L. Meredith

President, Space Conform, Washington D.C., Adjunct Professor, American University, Co-Chair, AIAA Orbital Debris Committee

INTRODUCTION

After 35 years of space exploration and exploitation with little regard for the space environment or the consequences for future space operations, the potential hazards posed by accumulating orbital debris are increasingly being recognized as a problem. Government agencies and International organizations involved in space exploration are studying remedial measures. Several spacecraft operators have begun instituting or are planning voluntary practices aimed at orbital debris mitigation (Part I.). While studies and voluntary measures are laudable, they are not sufficient. Current international law and national regulations are not adequate to ensure consistent implementation of technically feasible and economically sound mitigation measures, on a pervasive basis, in a manner which is necessary to protect the space environment (Part II.).

Several options are available for strengthening the legal regime applicable to orbital debris, ranging from the amendment of existing treaties to the adoption of a new international legal instrument (Part III.). The concept of a "framework convention" with procedures for subsequent protocols is a particularly attractive model which has been used successfully in the regulation of other man-made environmental problems, such as trans-boundary air pollution and ozone depletion, and which now is being proposed for controlling global warming (Part III.C.2.c). Part IV. of this paper outlines the issues which might be addressed in a legal instrument dealing with orbital debris, whether it is a framework convention, a traditional international agreement, a resolution, or any other statement of principles.

I. ORBITAL DEBRIS AND VOLUNTARY MITIGATION MEASURES

A. Classification of orbital debris

The U.S. Space Command presently maintains a catalog of more than 7,000 man-made objects in space.[1] Only objects that are 10 cm in diameter or larger and mainly in the low-Earth orbit[2] environment are being monitored due to limitations in tracking capabilities.[3] Catalogued space objects represent only a very small fraction of a percentage of the total orbital population, which is estimated at many millions.[4]

Of all the catalogued objects in space, only about 5% constitute operational spacecraft. The rest is orbital debris in one form or another.[5] Inactive payloads, e.g., satellites that have reached the end of their operational lives, account for about 20% of the total debris population.[6] Spent, intact rocket stages that are left in space following the launching of a payload account for approximately 14%.[7] About 12% is operational debris, which refers to objects intentionally discarded during the normal course of space operations, and might include payload shrouds, packing devices, lens caps, and more.[8] Fragmentation debris makes up the remaining 49%.[9] This type of debris may result either from an explosion [10] or from a collision with another object[11] in space.

B. Voluntary mitigation measures

Despite minimal legal regulation or guidance (see Part II., below), some spacecraft operators currently perform maneuvers that serve to mitigate debris generation. They do so voluntarily, presumably out of consideration for the condition of the environment in which they wish to continue to operate safely in the future. For example, operators of geostationary[12] satellites sometimes boost their satellites into higher, so-called "graveyard" orbits at the end of the satellite's useful life.[13] Some current and prospective operators of low-Earth orbit spacecraft either have or plan to execute controlled reentry maneuvers.[14] A mitigation technique now used by several launch vehicle operators involves the expulsion of excess propellants and pressure contained in spent upper stages.[15] This maneuver is significant because such upper stages, or rocket bodies, often remain in space for many years, and without venting of residual fuels they are prone to explosion, which creates fragmentation debris. Some spacecraft operators take deliberate measures to reduce operational debris released during the course of their normal space operations; for example, lanyards are used to attach releasable items. [16]

C. The inadequacy of voluntary measures

While voluntary debris mitigation is commendable, it is not

adequate to ensure that the orbital environment remains safe and conducive to future space missions. This is true particularly in the context of manned space operations, where safety standards need to be extremely high. It is also true, however, in view of the projected increase in low-Earth orbit non-manned space operations over the next years. For example, satellite constellations of the kind proposed by Motorola Satellite Communications,[17] NASA's planned Earth Observing System,[18] and the Strategic Defense Initiative Organization's Brilliant Pebbles [19] are at the same time both debris generators and potential victims of debris hazards.

Voluntary debris mitigation measures also suffer from a lack of consistency in implementation and coordination among spacecraft operators. For example, geostationary satellite operators have selected different disposal orbits, or none at all, with no apparent coordination. With the advent of low-Earth orbit satellite constellations proposed at various altitudes between 760 km to 10,360 km,[20] random and uncoordinated – albeit controlled – spacecraft reentries into the atmosphere may not be acceptable disposal methods. Coordination and management would seem to be necessary to ensure that reentries are conducted in an orderly fashion without jeopardizing functional satellites in lower-altitude constellations.

II. CURRENT LAW AND POLICY APPLICABLE TO ORBITAL DEBRIS

Very little exists in the way of legal regulation of orbital debris. A handful of United Nations-promulgated space treaties contains little more than vague admonitions to would-be space polluters. Various international conventions and declarations for protection of the environment are either not applicable to space debris, or their provisions are too vague and general to be relied upon to safeguard the orbital environment. Policies adopted in the United States are very general in purpose and expression, and while they are directed at all space sectors, including military, civil, and commercial operators, as well as regulators, they only require that debris mitigation be one among other important spacecraft design and operational considerations. Moreover, implementation of the policy has been somewhat lacking to date.

A. Public international law

Public international law with possible application to space debris includes international space law, international environmental law, and customary international law. These three bodies of law are examined here to determine the extent to which their provisions apply to orbital debris, whether they impose a duty to mitigate or notify of the formation of orbital debris, and whether liability may attach for damage caused by orbital debris.

1. International space law

International space law refers to a series of United Nations promulgated treaties and resolutions intended to govern activi-

ties in space, including the Treaty on Principles Governing the Activities of States in the Exploration and Use of Outer Space, Including the Moon and Other Celestial Bodies (Outer Space Treaty),[21] the Agreement on the Rescue of Astronauts, the Return of Astronauts, and the Return of Objects Launched Into Outer Space (Rescue and Return Agreement),[22] the Convention on International Liability for Damage Caused by Space Objects (Liability Convention),[23] the Convention on Registration of Objects Launched Into Outer Space (Registration Convention),[24] the Agreement Governing the Activities of States on the Moon and Other Celestial Bodies (Moon Treaty),[25] Principles Governing the Use by States of Artificial Earth Satellites for International Direct Television Broadcasting,[26] and Principles Relating to Remote Sensing of the Earth From Space.[27] All of these legal instruments were formulated by the United Nations Committee on the Peaceful Uses of Outer Space (COPUOS), the main international space law-making body.[28] The United States is not a party to the Moon Treaty.

a) *Extremely limited duty to mitigate.* None of the treaties or resolutions listed above mentions orbital debris, although a few provisions make tangential references. Of all the treaties to which the U.S. is a party, the Outer Space Treaty contains the most significant provisions with respect to debris mitigation. Its Article IX requires that States parties to the treaty conduct their space activities "with due regard to the corresponding interests of all States,"[29] and requires them to undertake consultations with other States before proceeding with space activities that potentially could cause "harmful interference"[30] with the activities of such other States. Any duty imposed by these vague and very general provisions is extremely limited and probably cannot be relied upon to infer negligence in the event of a breach, let alone to protect the space environment from debris accumulation, as evidenced by the present situation.

b) *Extremely limited duty to notify.* The requirement under Article IX of the Outer Space Treaty to consult with other States "before proceeding with [an] activity or experiment" which "potentially" could cause "harmful interference" with the activities of such other States[31] certainly is not a general obligation to notify or consult with other States on the creation of debris. The duty applies only if another State's space operations potentially will be interfered with, and only with respect to that State. Moreover, the duty only applies to an activity or experiment, *i.e.* a deliberate act (from which debris will result).

The utility of this provision is diminished not only by the vagueness of its formulation, which detracts from the legal duty, but also by the fact that it may be difficult or impossible to identify the States whose operations potentially could be interfered with. Furthermore, the provision probably does not apply to the creation of operational debris, spent upper stages, or inactive payloads, *i.e.* the normal by-products of a space mission, or to an accidental fragmentation, because the creation of these kinds of debris is not an "activity" which triggers the provision. On the other hand, the requirement probably would apply to the intentional creation of fragmentation debris, that is, an induced explosion or deliberate collision typically as part of a military target acquisition experiment, since such acts would be considered activities or experiments.

Also the Registration Convention imposes no general duty to register information on the creation of orbital debris. The convention requires parties to register objects launched into outer space in a national register,[32] as well as to provide information on the orbital parameters and purpose of such objects to the United Nations Secretary General who maintains an international register.[33] The requirements to register and submit information are triggered by the launching of a "space object" into Earth orbit or beyond.[34] A space object is defined to include "component parts of a space object as well as its launch vehicle and parts thereof." [35] In practice, payloads (e.g., a satellite) and their launch vehicle (upper stage) are registered, while operational or fragmentation debris is not. There is no requirement to notify the Secretary General when a previously registered satellite reaches the end of its useful life or otherwise becomes disabled, although States must notify the Secretary General "to the greatest extent feasible" when a registered space object no longer is in Earth orbit,[36] e.g., if it reentered the atmosphere.

c) *Liability.* The Outer Space Treaty, Article VII, and the Liability Convention impose liability on States for damage caused by their space activities. Under the Outer Space Treaty, the State which launched an object into outer space or which procured the launch, or which authorized a launch or its procurement,[37] or from whose territory or facility the launch took place, may be held liable for "damage" caused to "another State Party" by such an object, if the former state was at fault.[38] Under the Liability Convention, the same state may be held liable 1) regardless of fault if "damage is caused by [a] space object on the surface of the Earth or to an aircraft in flight"[39] and 2) if at fault, when "damage is caused to a space object elsewhere than on the surface of the Earth by a space object . . . ."[40]

It is clear that damage caused by orbital debris to the space environment in general is not actionable under any of these treaties.[41] The issue then is to what extent damage caused by orbital debris to another State, including its property and its citizens and their property, is recoverable. Recovery for damage occurring in space hinges upon whether the debris causing the damage may be considered an "object [launched] into outer space"[42] or a "space object"[43] and whether fault can be inferred.

It is not clear from the provisions of the Outer Space Treaty or the Liability Convention, nor is it evident from the official records of their negotiating histories, whether or to what extent space debris is covered by the terms "space object" and "object [launched] into outer space." The definition of "space object" as including "component parts of a space object as well as its launch vehicle and parts thereof,"[44] is somewhat helpful, but not conclusive. Scholarly opinion differs on the interpretation of "space object."[45] Some scholars have attempted to differentiate among the different classes of debris,[46] holding for example that operational debris is covered, while other kinds of debris are not. So far, no incident has occurred which has forced an interpretation.

Even assuming that all debris, as such, is covered by the Outer Space Treaty and the Liability Convention, recovery, nevertheless, may elude the victim because it may be impossible to identify the tort-feasor, establish a causal link, or proximate causation. Even more problematic may be the burden of proving fault. Neither the Outer Space Treaty nor the Liability Convention offers a definition or standard of fault. Assuming that intentional and negligent conduct is covered, negligence will be difficult to prove since currently the duties to mitigate and notify of debris are so limited and legally weak that a breach may be impossible to establish.

2. International environmental law

A number of conventions and declarations for protection of the environment have been adopted over the last two decades, including, e.g., the United Nations Conference on the Human Environment (Stockholm Declaration),[47] Convention on the Prohibition of Military or Any Other Hostile Use of Environmental Modification Techniques (Environmental Modification Convention),[48] Geneva Convention on Long Range Trans-Boundary Air Pollution of 1983 (Geneva Convention),[49] Convention on the Prevention of Marine Pollution by Dumping of Wastes and Other Matter,[50] and the Vienna Convention for the Protection of the Ozone Layer of 1986 (Vienna Convention).[51] A signing ceremony was scheduled at Rio de Janeiro, Brazil, in June 1992 for the United Nations Framework Convention on Climate Change (Rio Convention).[52] President George Bush attended the ceremony for this controversial convention on reducing the likelihood or severity of a future global warming of the Earth's atmosphere.[53] Of these legal instruments, only the Stockholm Declaration and the Environmental Modification Convention apply to space activities.

a) *Extremely limited duty to mitigate, no duty to notify, and no liability.* While most of the provisions of the Stockholm Declaration apply explicitly to the Earth environment, Principle 21 appears to apply to space, as well. It provides that:

States have, in accordance with the Charter of the United Nations and the principles of international law, the sovereign right to exploit their own resources pursuant to their own environmental policies, and the responsibility to ensure that activities within their jurisdiction and control do not cause damage to the environment . . . of areas beyond the limits of national jurisdiction.

Assuming that "areas beyond the limits of national jurisdiction" encompass outer space, States would be responsible, pursuant to the Stockholm Declaration, for ensuring that activities involving spacecraft over which they exercise jurisdiction and control do not cause damage to the space environment. Under Article VIII of the Outer Space Treaty, a State retains jurisdiction and control over an object launched into space for which it is the State of Registry pursuant to the Registration Convention. However, the Stockholm Declaration probably is too general in its formulation to impose a real legal obligation, and moreover, is not legally binding.

Under Article I of the Environmental Modification Convention, to which the U.S. is a party, States parties "undertake not to engage in military or any other hostile use of environmental modification techniques having widespread, long-lasting or severe effects as the means of destruction, damage or injury to any other State Party." Article II defines environmental modification techniques as "any technique for changing – through the deliberate manipulation of natural processes – the dynamics, composition or structure of the Earth, including its biota, lithosphere, hydrosphere and atmosphere, or of outer space." However, the convention does not prohibit environmental modification techniques for "peaceful purposes," a term which has been interpreted in the space law context to mean non-aggressive, as opposed to non-military, purposes.[54]

The Geneva Convention, the Vienna Convention, and the Rio Convention are mentioned here not for their direct applicability to orbital debris, since they do not apply, but rather because they may serve as useful models for the regulation of debris. They embody the concept of a "framework convention" containing very general provisions imposing a minimum of legal obligations while also establishing procedures for subsequent protocols that might impose control requirements. The concept is ideal for regulating man-made environmental problems surrounded by uncertainty as to their cause and effect, such as ozone depletion, global warming, and orbital debris. It is ideal because it allows the parties to agree to general statements of concern which carry virtually no commitment while at the same time providing a vehicle for development of control requirements in step with the evolution of scientific research and knowledge. See Part IV.C.2.c. for further discussion of the framework convention approach.

3. Customary international law
Two elements are required for a rule of customary International law to form: 1) consistent State practice, and 2) *opinio juris,* i.e., the belief by States that a particular practice is legal or legally required.[55] In other words, unless a particular conduct or practice has been adhered to consistently over a period of time in the belief by States that the conduct or practice is legal, or required, a rule of customary international law

does not exist. When a rule of customary international law has been established, it is legally binding on all States.

a) *No duty to mitigate.* Since the launch of Sputnik in October of 1957, when space exploration first began, nations and their authorized private entities have launched a multitude of spacecraft with a variety of configurations and mission objectives. Within the broad limits of the Outer Space Treaty, States have considered themselves entirely free to select spacecraft designs, operations, and missions, as well as methods, if any, for end-of-life disposal. They have done so hitherto largely without consideration for the space environment *per se,* or for subsequent space operations. Although, as mentioned above, some spacecraft operators (including governments, international organizations, and private companies) over the last few years have taken certain limited measures to mitigate the creation of orbital debris, these practices are by no means consistent. Because orbital debris mitigation measures to date have been performed randomly and without the perception of an underlying legal duty, there is currently no rule of customary international law requiring that such measures be taken.

b) *No duty to notify.* There is no duty under customary international law to notify other States or international organizations of the creation or presence of orbital debris as a result of a space mission. Nations with objects in space have not made it a practice to notify other States of debris generation, nor do they perceive that they have a general duty to do so. For example, in December 1991, the Soviet Union performed a typical deliberate fragmentation of a reconnaissance satellite,[56] which the U.S. was able to track, although it was not notified. Consequently, the elements necessary to create a rule of customary international law are not present.

c) *Liability.* Under customary international law supported by judicial precedent, a State may be held liable for damage it has proximately caused if it was at fault.[57] Consequently, a State whose space debris has caused damage may be held liable if it acted intentionally or negligently. As discussed in II.A.3.a., above, negligence cannot be established if there is no duty to mitigate or notify of the creation of orbital debris since there is no duty to breach.

B. U.S. law and policy
United States national policy calls upon the military, civil, and commercial space sectors to minimize and mitigate the creation of orbital debris. Only to a very limited extent has this policy been implemented formally by space operating and regulatory agencies. Nonetheless the policy is evidence of an emerging national consensus that space operators should not be permitted to conduct their space missions with total disregard for the orbital environment and the consequences of debris they might be generating.

1. The policy

In a November 1989 Presidential Directive on National Space Policy, the Bush Administration stated that

> all space sectors will seek to minimize the creation of space debris. Designs and operations of space tests, experiments and systems will strive to minimize or reduce accumulation of space debris consistent with mission requirements and cost effectiveness. The United States government will encourage other space-faring nations to adopt polices and practices aimed at debris minimization.[58]

While this policy statement, as well as others like it,[59] urges the military, civil, and commercial space sectors to make orbital debris minimization and mitigation a spacecraft design and operational consideration, it also recognizes that mission objectives and cost effectiveness should not be sacrificed. Such references to mission objectives and cost effectiveness serve to weaken the policy with respect to orbital debris. Moreover, by its very nature, the policy is general and non-specific, and can accomplish nothing unless it is implemented by the space operating and regulatory agencies.

2. Implementation

So far, only the Department of Defense (DOD) U.S. Space Command has responded to the National Space Policy with regulations adopted on June 6, 1991. Entitled "Minimization and Mitigation of Space Debris,"[60] the regulations do not impose specific spacecraft design or operational requirements for debris minimization or mitigation, but refer to the National Space Policy of 1989.

NASA currently is developing a space safety policy that encompasses debris. The agency so far has dealt with debris on an ad hoc basis, and has stated repeatedly that more research is needed on debris flux, hazard characterization, and mitigation before a comprehensive agency policy can be adopted.

None of the regulatory agencies, including the Department of Transportation (DOT), which regulates launch vehicles[61], the Federal Communications Commission (FCC), which regulates telecommunications satellites, and the Department of Commerce[62] (DOC), which regulates[63] remote-sensing space systems, has adopted regulations or even begun the rulemaking procedures that might lead to the adoption of such regulations. The DOT's Office of Commercial Space Transportation (OCST) does, however, to some extent consider space debris consequences during its safety review in the context of launch license applications, and is actively engaged in orbital debris research. An Interagency Working Group on Orbital Debris, which includes representatives from all of the agencies listed above, is currently reviewing and evaluating agency actions.

III. REGULATING ORBITAL DEBRIS: OPTIONS AND APPROACHES

A. Voluntary measures, technical standards, or legal action

Several approaches are available for dealing with the orbital debris problem. One such approach is to encourage the continuation of voluntary mitigation practices. This approach, alone, is not sufficient, however, to ensure the future safety of the space environment for reasons discussed in Part I.C., above. Another approach is to rely on the technical community[64] in the U.S. and internationally to develop standards aimed at debris mitigation. While this approach has an obvious appeal, it is, in itself, not adequate to ensure consistent and coordinated implementation of debris mitigation measures on a pervasive basis. This objective can only be achieved through a third approach, i.e., the use of legal mechanisms, the advantage of which is that any standards that might be adopted are backed by legal force.[65]

The optimal solution, in fact, lies in a combination of these approaches. Voluntary debris mitigation measures should be continued while technical experts in government and industry strive to reach a consensus on standards for debris flux modeling and hazard characterization, debris terminology, and spacecraft design and operation. Law makers, government regulators, and international civil servants should remain prepared to assume responsibility for translation of technical standards into debris management solutions, and ultimately debris mitigation requirements.

B. International implementation: all spacefaring nations

To be truly effective, any debris mitigation practice, technical standard, or legal requirement must be implemented internationally. The debris problem is inherently global, and any solution that is purely national in scope cannot achieve the objective of ensuring that space remains safe and accessible for all future manned and unmanned missions. This is not to say that States should not begin at the national level first. However, the ultimate solution must encompass all spacefaring nations, through one multilateral or several bilateral legal instruments.

C. Options for regulatory action

Options for regulatory action with respect to orbital debris exist at the national and at the international levels. In the U.S., Federal agencies operating or regulating spacecraft could take such actions pursuant to existing statutory authority. Two international bodies already regulating space activities or, alternatively, a conference of nations could take action internationally.

1. At the U.S. national level

Spacecraft operating agencies such as NASA, DOD, and DOC's National Oceanic and Atmospheric Administration (NOAA) can adopt debris mitigation guidelines and regulations for their own programs, which also will affect their supply contractors, typically through the spacecraft procurement process. As mentioned in Part II.B., above, these agencies are in varying stages of talking such actions.

Regulatory agencies such as DOT's Office of Commercial Space Transportation (OCST), the FCC, and NOAA[66] each have sufficient statutory authority to impose debris mitigation requirements[67] on the private spacecraft operators it regulates, pursuant to a proper rulemaking. Section 7 of the Commercial Space Launch Act of 1984[68] charges DOT (OCST) with the licensing of launch vehicles "consistent with the public health and safety, safety of property, and the national security interests and foreign policy interests of the United States."[69] DOT also has the power to prevent the launch of certain payloads[70] if such safety, national security, or foreign policy interests would be jeopardized by the launch.[71] OCST's mandate to consider the safety aspects of a launch clearly would encompass the imposition of debris assessment, mitigation, and notification requirements.

The Communications Act of 1934 created the FCC to regulate "communications... by radio [which includes a satellite] so as to make available, so far as possible, to all people of the United States a rapid, efficient, nationwide and worldwide... radio communication service."[72] The premise of FCC's regulatory action would be the need to eliminate the risk of debris impact, which might otherwise disable a satellite and impede communications. Finally, the Land Remote-Sensing Commercialization Act of 1934 empowers DOC (NOAA) to impose requirements with respect to end-of-life disposal of a remote-sensing satellite.[73]

2. At the international level
Several options are available for regulation of orbital debris at the international level. Rules or principles could be formulated in existing fora charged with regulating space activities, or certain aspects thereof, or in a new forum created for the purpose, such as a conference of nations. Such a conference could be confined to spacefaring nations, or it could encompass the community of nations.

A wide range of types of legal instruments or actions are possible within each of these fora, such as amendments to existing treaties or the development of new international agreements (including treaties and conventions), protocols, memoranda of understanding, declarations, resolutions, and other statements of principles. A legal instrument may be more or less ambitious in scope and purpose, and may impose legally binding requirements (i.e., international agreements, amendments, and protocols), or merely moral obligations.

a) *Action within the UN COPUOS.* The United Nations Committee on the Peaceful Uses of Outer Space (COPUOS) was set up as a permanent body of the U.N. General Assembly in 1959 to regulate the conduct of space activities by States.[74] As stated earlier, COPUOS already has adopted

five treaties and two resolutions (see Part II.A.I.), and is currently in the process of formulating a set of principles governing the use of nuclear power sources in space.[75] The U.N. General Assembly has suggested that space debris might be an appropriate item for future inclusion on COPUOS agenda.[76]

If orbital debris is adopted as a COPUOS agenda item, while deliberations probably would begin in the Scientific and Technical Subcommittee, the end result might be a set of principles emanating from the Legal Subcommittee. Protracted negotiations held over several years would likely precede the formulation of such principles relating to orbital debris, especially if these principles were to be modeled on those elaborated for nuclear power sources in space. The principles governing nuclear power sources are relatively ambitious in scope and purpose, and address *inter alia* the following issues: 1) notification to the United Nations Secretary General of the presence of a nuclear power source; 2) guidelines and criteria for safe use; 3) safety assessment; 4) notification and consultation in the event of reentry; 5) assistance to States; 6) State responsibility for national activities; 7) liability and compensation for damage; and B) settlement of disputes.[77]

It is conceivable, of course, that COPUOS would adopt a different approach to regulating orbital debris, i.e., one that establishes broad and very general principles of cooperation and information exchange without substantial legal commitment, as opposed to a set of principles that seek to implement the ultimate control measures immediately. See Part III.C.2.c., below. All of the treaties and resolutions formulated by COPUOS subsequent to the Outer Space Treaty, which did contain broad and general principles to be elaborated in subsequent treaties, have tended to seek the ultimate solution.

b) *Action within the ITU.* The International Telecommunication Union (ITU) is a specialized agency within the United Nations system[78] charged with managing the international radio frequency spectrum, among other responsibilities.[79] Jurisdiction over space communications was assumed by the ITU in 1959, when the organization for the first time allocated frequencies for space communication services.[80] The ITU's jurisdiction is limited by the International Telecommunication Convention (ITU Convention),[81] and refers primarily to the communications aspect of space activities.[82] Based upon the same premise as that upon which the FCC may assert authority over orbital debris (III.C.1), the ITU has at least ancillary jurisdiction over debris, i.e., to the extent it might affect communications.[83]

On June 12, 1991, the ITU's International Radio Consultative Committee (CCIR), which is charged with studying the "technical and operating questions relating specifically to radio communications,"[84] formulated a draft recommendation urging that "as little debris as possible should be released into the geostationary orbit during [satellite deployment]," that "every reasonable effort should be made to shorten the lifetime of debris in a transfer orbit," and that geostationary satellites should be transferred to "supersynchronous graveyard" orbits at end-of-life.[85] The CCIR does not have "law-making"[86] power, however; the Plenipotentiary Conference and the Administrative Radio Conference do, subject to ratification by member States.[87]

If the ITU were to make orbital debris mitigation an agenda item at a future World Administrative Radio Conference, it could adopt a resolution based on the CCIR recommendation, or it could revise the Radio Regulations to reflect the CCIR recommended mitigation measures. While an ITU resolution is non-binding, the Radio Regulations have treaty status, and revisions are binding upon member States, subject to ratification.

c) *Action by a Conference of nations: environmental framework conventions.* A Conference of nations could be convened, in several sessions. It might be expedient to confine the Conference to spacefaring nations. "Spacefaring nations" could be defined broadly, to encompass every country and intergovernmental organization that operates a spacecraft, or more narrowly, to include only nations or intergovernmental organizations with launching capability or substantial involvement in space operations. The mandate of such a Conference could be broad or narrow, depending upon the issues the participants are willing to address and the commitments they are willing to make. In accordance with the mandate of the Conference, one or more legal instruments could be adopted.

For example, the Conference could adopt a very general declaration of concern for the space environment, akin to the Stockholm Declaration of June 1972.[88] The thrust of that declaration was to convey a sense of concern for the deterioration of the human environment; impute responsibility to nations for the future of the Earth's environment; and to urge all nations to take measures to prevent pollution, depletion of resources, and destruction of wildlife. However, no specific legal obligations were imposed upon States.

In a somewhat more ambitious approach, the Conference could model its efforts on those which have led to several United Nations-sponsored environmental "framework conventions," such as the Geneva Convention on Long Range Trans-Boundary Air Pollution of 1983, the Vienna Convention for the Protection of the Ozone Layer of 1986, and the United Nations Framework Convention on Climate Change signed at Rio de Janeiro in Brazil in June 1992 (see Part II.A.2). They are called framework conventions because they establish certain broad principles, i.e., a legal framework, aimed at environmental control, without imposing strong legal commitments on the States parties, while at the same time providing for a procedure for the future negotiation and formulation of protocols that might contain control measures.

For example, the Vienna Convention creates a framework for international cooperation on research, monitoring, and information exchange concerning modifications of the ozone layer and the health, environmental, and climatic effects of such modifications.[89] It also provides procedures for negotiating and formulating control measures by way of protocols to the convention, but does not require the adoption of such protocols.[90] The Protocol on Substances that Deplete the Ozone Layer (the Montreal Protocol) was adopted in 1987.[91] It imposes specific obligations on States parties to limit and reduce their use of chlorofluorocarbons (CFCs)[92] and potentially other substances, as well.[93] Basically, the Montreal Protocol requires a phased reduction in the production and consumption of certain CFCs by 50% of current levels by 1999. (Complete phase-out by year 2000 is required by the "London amendment" to the protocol, which is not yet in force[94]). Special provision is made for developing countries.[95] Other items covered by the Montreal Protocol are methods for calculating acceptable levels of CFCs,[96] restrictions on imports of CFCs,[97] research and exchange of information,[98] technical assistance to facilitate the implementation of the protocol,[99] and reporting requirements.[100]

The environmental framework conventions, exemplified by the Vienna Convention and the Montreal Protocol, provide a suitable model for dealing with orbital debris because the considerations germane to the environmental problems they address apply equally to orbital debris. As with other environmental problems created by man, such as trans-boundary air pollution, depletion of the ozone layer, and global warming, where much uncertainty still surrounds issues of cause and effect, many questions are still unanswered with regard to the formation of, and potential hazards presented by, orbital debris. Consequently, States are unwilling, and prudently so, to commit to solutions that would require severe control measures, and which, in retrospect, may turn out to have been unwise. The concept of a framework convention is attractive because it allows for the evolution of the law in step with advances made in scientific research and analysis. It also is attractive because it allows nations to express their concern in the form of very general statements which give the perception of action which States are eager to convey. Finally, and most importantly, it establishes a consultative mechanism that can be used to respond to a worsening situation in a more timely and effective manner.

IV. REGULATING ORBITAL DEBRIS: ELEMENTS OF A MULTILATERAL LEGAL INSTRUMENT

Regardless of whether the traditional COPUOS approach to regulating space activities, *i.e.* attempting to achieve consensus on the ultimate solution in one legal instrument, be it a treaty or a resolution; or whether the concept of a framework convention with subsequent protocols is adopted, certain issues will have to be addressed, as set forth in provisions A. through Q., below. If a framework convention is contemplated, items A. through I. and Q. could be included in the initial convention, while Items J. through Q. could be addressed in a protocol to the convention. Any other legal instrument might attempt to address all or some of these Items.

A. Preamble

A preamble establishes the context in which the legal instrument is being adopted, and usually refers to existing related international agreements and resolutions. The intent or motivation behind, or other observations that might shed light on the goals and objectives of the legal instrument may be stated, as well. Usually the preamble will list a set of considerations that appear to justify the operative provisions which follow.

For example, the preamble of a legal instrument dealing with orbital debris should "recall" the Outer Space Treaty and other pertinent space treaties and resolutions and possibly certain environmental conventions and declarations, depending on the tenor of the particular agreement, resolution, etc., which is being adopted. Moreover. the preamble might "recognize" the importance of space and space operations to the future economic and social well-being of mankind, and express "concern" for the potential consequences of unmitigated and uncontrolled accumulation of orbital debris and crowding of the space environment for present and future generations.

B. Definitions

Definitions are intended to clarify the meaning of certain key terms and to allow for shorter references in the operative provisions of a legal instrument, and often are stated at the outset. They are of utmost importance to the application of the agreement, resolution, etc., and must be carefully crafted to ensure that the definition does not give rise to more questions than it answers. An example of a definition that has caused much uncertainty as to the meaning of the word it attempts to explain, is that provided for "space object" – in Article I of the Liability Convention. See Part II.A.1.c., above.

The most important term requiring a definition in this context is, of course, "orbital debris," or "space debris," whichever term is deemed to be appropriate. While orbital debris would refer to debris within the orbital environment, space debris would encompass debris in other areas of space, as well. In any event, the definition should distinguish between man-made and natural debris, and include only the former.

The definitional touchstone for orbital/space debris is crucial, not only because of the mitigation and control requirements that might be associated with debris, but also because of the implications of the categorization of an object as debris for purposes of salvage rights.[101] Touchstones, or criteria, that might be considered are those that refer to whether the object in question is non-functional, or non-operational, with no reasonable expectation of assuming or resuming its intended mission or any other lawful mission for which the object is, or can be expected to be, authorized. Fragments and component parts of such objects should be included explicitly. A definition of debris in terms of "usefulness," or rather "non-usefulness" is not advisable because that standard necessarily requires a sub-

jective judgment. It also may be necessary to define different classes or categories of orbital/space debris, such as "operational debris" and "inactive payloads."

If debris mitigation measures or performance levels are required, these, too, will have to be defined. For example, "controlled reentry," as an end-of-life disposal requirement for certain low-Earth inactive payloads, could be defined in terms of the altitude of the perigee of the elliptical orbit into which the payload would have to be placed to achieve rapid reentry.

C. Scope and applicability

To be effective, a legal instrument dealing with orbital debris must be made applicable to military, civil, and commercial space activities, alike. With the exception of the Principles Relating to Remote Sensing of the Earth from Space, which explicitly except military reconnaissance and surveillance, all the U.N. space treaties and resolutions adopted to date have been applicable to all space sectors.

To the extent objects considered orbital debris have nuclear generators or other nuclear power sources on board, reference should be made to Principles Relevant to the Use of Nuclear Power Sources in Outer Space,[102] if and when such principles are adopted. See Part III.C.2.a.

D. Purpose clause

A clear statement of the purpose of the legal instrument is important. It is important not only because it elucidates the intent of the drafters and explains what the agreement or resolution aims to achieve, but also because it sheds an interpretative light on other provisions, as well. Pursuant to the Vienna Convention on the Law of Treaties of 1969,[103] to which the U.S. is a party, treaty provisions shall be interpreted in accordance with the ordinary meaning given to the terms[104] in light of the object and purpose of the treaty.[105] If a framework convention is contemplated (see Part III.C.2.c.), the purpose clause may state the ultimate goal or objective of the convention and subsequent protocols.[106]

The purpose of any legal instrument dealing with orbital debris, should be to minimize, mitigate, and control debris and the hazards posed by it. Total elimination of debris is unrealistic, at least in the foreseeable future, as debris will exist in supersynchronous "graveyard" orbits and in various stages of planned disposal, *i.e.*, in elliptical orbits with low perigees that will cause reentry.

E. Encouragement of continued research

A provision should be included which encourages continued research and investigation on debris formation, flux, cloud dispersal, cratering, and hazard characterization, as well as on mitigation or control measures.[107]

F. Cooperation and exchange of information

Provisions should be made for the cooperation on orbital debris research and tracking, and for the exchange of information on the debris environment and hazards presented by it, as well as on mitigation measures.[108] Provisions for transfer of technology (technical data), consistent with national laws and policies,[109] should be included here.

G. General obligation to minimize debris

A general, vaguely phrased obligation to minimize and mitigate debris and to protect the space environment, consistent with national security interests and industrial competitiveness concerns is important.[110] Such an obligation should apply to all parties, while taking into account the stage of development of their respective aerospace industries and the mitigation methods and technologies at their disposal. Again, provisions for transfer of technology, consistent with national laws and policies could be included.

H. Special provisions for developing countries

Special account should be taken of the special needs of the developing countries, particularly their needs for information and technical data. Their situation also should be considered in the imposition of debris mitigation requirements. See Provision J., below.

I. If a framework convention, procedures for protocols

If a framework convention with subsequent protocols is contemplated, mitigation requirements and other control measures (see provision J., below) should not be contained in the initial convention. Rather, a procedure should be established for the adoption of protocols or other legal instruments that might require the parties to implement mitigation and control measures. To carry out these procedures, a Conference of the Parties (to the framework convention) should be established as the supreme body of the convention, which would meet at regular intervals to establish, as appropriate, or revise debris mitigation and control requirements.[111] A Secretariat to handle administrative matters may also have to be established.[112] See Part III.C.2.c.

J. Mitigation and control measures

In the event a framework convention is adopted, orbital mitigation and control requirements would be contained in subsequent protocols to the initial convention (see Provision I.). Any other legal instrument might attempt to include them immediately.

In any event, mitigation and control requirements should be performance rather than design oriented, thus affording spacecraft operators and their manufacturers maximum flexibility in achieving the desired levels of debris control. Performance requirements tend to spur innovation and efficiency, while design requirements are apt to stifle technology. Performance standards are the preferred method also with respect to U.S. regulatory action,[113] and are often used in government spacecraft procurements .

Mitigation requirements may refer to inactive payloads,

spent rocket upper stages, or operational debris, as well as to the formation of fragmentation debris through deliberate acts, such as anti-satellite satellite or ballistic missile defense target acquisition testing. Geostationary satellites might be required to retire to graveyard orbits,[114] and controlled reentry could be required for certain low-Earth orbit satellites.[115] Venting of spent rocket upper stages and minimization of operational debris to acceptable levels also might be mandated. Mitigation procedures and acceptable performance levels would have to be specified for each of these categories. "Escape" clauses, such as "to the extent such practices would be consistent with the national security interests and mission objectives," should be avoided. See also Provision H. concerning the special situation of the developing countries. Provision should be made for review of the control and mitigation measures.

K. Assistance and technology transfer

Provisions would have to be made for assistance and technology transfer, particularly to developing countries, consistent with national laws and policies and national security interests.

L. Restrictions on import and use of non-conforming space hardware

The parties should agree to restrict the import and use of spacecraft, including launch vehicles, or component parts, that do not meet the minimum performance levels specified in Provision J., above. Again, special considerations might have to be made for developing countries in order to allow them to develop their respective space industries.

M. Notification of debris formation

Elaborating on rudimentary notification and consultation requirements contained in the Outer Space Treaty and the Registration Convention (Part II.A.1.b.), States could be required to notify the United Nations Secretary General, to the extent possible, of the formation of debris, or certain kinds of debris, including orbital parameters, sizes, densities, etc., as well as of potential hazards posed to the operations of other nations.

N. Verification

Although a few countries, such as the U.S., have sophisticated tracking and monitoring capabilities that would allow them to determine at least in certain cases whether a required debris mitigation procedure had been carried out, such measures, alone, cannot be relied upon to verify compliance with debris mitigation and control requirements.[116] Compliance reports to the United Nations Secretary General or some other neutral international body could serve as a supplementary means of verification. Due to the sensitivities of States and their tendency to guard their space technologies, especially if the technology has military applications, an agreement to permit inspection of space hardware and manufacturing facilities may be very difficult to attain.

O. Non-compliance and sanctions

Measures that might be taken in the event of a violation or deliberate non-compliance should be considered. An aerospace products and services trade embargo directed at the non-complying party is an extreme measure, and States are not likely to agree to it.

P. International liability

As mentioned earlier, international liability does not lie under current law for damage caused to the space environment, itself, and States probably would not want to commit to it. The notion of "environmental liability" in space may be premature.

A provision establishing international liability for damage caused to other States, including their property and citizens and the property of such citizens may not be necessary. As discussed in Part II.A., above, States may be held liable for damage under the Outer Space Treaty, the Liability Convention, and customary international law, provided certain conditions are met. It may be sufficient to refer to these legal sources, with one important clarification: It is not clear today if, or to what extent, the two treaties would apply to damage caused by orbital debris because they speak of damage caused by "objects [launched] into outer space" and "space objects." A statement would be useful to the effect that "the parties to this [legal instrument] agree that orbital debris [or certain classes of orbital debris] constitutes an object or a space object within the meaning of the Outer Space Treaty, Article VII, and the Liability Convention."

Another defect often attributed to the Outer Space Treaty and the Liability Convention is the lack of a definition or a standard of fault in instances where fault is required to bring them into play (see Part II.A.1.c.). The problem is not so much the lack of a definition or standard, however, as it is the current lack of a duty to act, e.g., to mitigate or notify. Without a duty there can be no breach and therefore no fault. A legal instrument as outlined here may create the requisite duty, depending on its formulation. Breach of a duty to act then will create a presumption of fault, thus shifting the burden of proof to the State whose debris caused the damage, *i.e.,* if that State can be identified.

Q. Standard treaty provisions

If the legal instrument adopted is an international agreement, certain standard provisions must be included for dealing with amendments, withdrawal, signature and accession, entry into force, and authentic texts.

CONCLUSION

In conclusion, voluntary debris mitigation practices should be encouraged. Efforts should continue within the technical community, including representatives from government, industry, and professional organizations, towards developing guidelines and standards for debris mitigation. Policy coordination

between and among U.S. government agencies with space operating and regulatory responsibilities, as is currently taking place within the interagency Working Group on Orbital Debris, is important. As more experience is gained with voluntary debris mitigation practices, as consensus builds in the technical community, and as policy issues are being clarified, hopefully within no more than a couple of years, U.S. regulatory agencies should begin the rulemaking process which might lead to the adoption of debris mitigation requirements. In the meantime, preparations should be made for an international solution. A framework convention could be adopted during the agency rulemaking proceedings without much risk of interference with those proceedings or the competitiveness of the United Sates space industry, due to the general nature of the obligations undertaken. The framework convention would contain procedures for protocols which, in turn, could adopt international mitigation obligations reflecting U.S. regulatory requirements.

Acknowledgements of research assistance are made to Alfred Malena and Michael Gross, American University, Washington College of Law.

ENDNOTES

1. *Report on Orbital Debris.* Prepared by the Interagency Group (Space) for United States National Security Council (Feb. 1989) [hereinafter cited as NSC Report], at 3.

2. No authoritative definition exists of "low-Earth orbit". Objects at altitudes below 6,000 km generally are considered to be in low-Earth orbit. Sometimes, objects up to 12,000 km are said to be in low-Earth orbit, as well.

3. NSC Report, *supra* note 1, at 3.

4. *Id.*

5. *Id.,* at 6.

6. *Id.,* at 7.

7. *Id.*

8. *Id.*

9. *Id.*

10. Rocket upper stages that are left in space containing residual fuels are particularly prone to explosion.

11. To date, known collisions generally have been deliberate, *i.e.,* caused in connection with anti-satellite satellite or ballistic missile defense testing.

12. Geostationary satellites are located in the geostationary orbit, which is a circular orbit in the plane of the Earth's equator, about 36,000 km from the surface of the Earth. In this orbit, a satellite revolves around the Earth at a velocity which matches the Earth's rotation and therefore appears to be stationary relative to the Earth.

13. *Orbital Debris Mitigation Techniques: Technical, Legal, and Economic Aspects.* Special Project Report, Prepared under the Auspices of the American Institute of Aeronautics and Astronautics (AIAA), 1992, at 24.

14. *Id.*

15. *Id.*, at 23.

16. *Id.*, at 26.

17. Motorola Satellite Communications, Inc. of Chandler, Arizona is proposing a constellation of 77 satellites ("Iridium"), at an orbital altitude of about 700–800 km, to be used for purposes of radiodetermination and global mobile voice communications services to and from cordless, handheld units.

18. NASA's Earth Observing System currently contemplates six low-Earth orbit satellites for gathering of environmental data to assist in monitoring of global climatic changes. J.R. Asker, *NASA Reveals Scaled-Back Plan for Six EOS Spacecraft.* A.W.& S.T., Mar. 2, 1992.

19. The Strategic Defense Initiative Organization contemplates space-based smart projectiles (brilliant pebbles) to destroy on impact enemy warheads that are still attached to their thrusting boosters, as well as in the coast phase.

20. Several companies have applications pending before the United States Federal Communications Commission for permission to launch satellite systems between these altitudes, including Motorola Satellite Communications of Chandler, Arizona, *supra* note 17; Ellipsat Corporation of Washington, D.C.; Constellation Communications, Inc. of Herndon, VA; Loral Qualcomm Satellite Services, Inc. of Palo Alto, CA; and TRW of Redondo Beach, CA, proposing satellite systems for radiodetermination and mobile voice communications services to cordless, handheld units in the L-band, and Orbital Communications Corporation of Fairfax, VA; Starsys, Inc. of Washington, D.C.; LEOSAT Corporation also of Washington, D.C.; and Volunteers in Technical Assistance of Arlington, VA, proposing radiodetermination and mobile data communications services primarily in the UHF and VHF bands.

21. *Done* Jan. 27, 1967, *entered into force* Oct. 10, 1967. 18 U.S.T. 2410, T.I.A.S. 6347.

22. *Done* Apr. 2, 1968, *entered into force* Dec. 3, 1968. 19 U.S.T. 7570, T.I.A.S. 6599.

23. *Done* Mar. 29, 1972, *entered into force* Sept. 1, 1972. 24 U.S.T. 2389, T.I.A.S. 7762.

24. *Done* Jan. 14, 1975, *entered into force* Sept. 15, 1976. 28 U.S.T. 695, T.I.A.S. 8480.

25. U.N.G.A. Res. 34/68 (1979), *entered into force* Jul. 12, 1984.

26. U.N.G.A. Res. 37/92 (1982).

27. U.N.G.A. Res. 41/65 (1986).

28. COPUOS became a permanent committee of the United Nations General Assembly in 1959. U.N.G.A. Res. 1472 (XIV), Dec. 12, 1959. COPUOS has 53 members, and has established a Legal Subcommittee and a Technical Subcommittee. Since 1962, COPUOS' practice when formulating treaties or resolutions has been to attempt to reach agreement by consensus (a compromise between unanimity and majority vote). Once COPUOS reaches consensus on the text of an international agreement (treaty or convention) or resolution, the matter is referred to the First Committee of the United Nations General Assembly, and then to the United Nations General Assembly, itself, for a vote. If an international agreement is approved by the General Assembly, it is opened for signature by member States. Signatory States are then free to ratify the agreement in accordance with their own national procedures. When a certain number of States have ratified (usually the number needed is stipulated in the agreement), the international agreement will enter into force. It is then binding on all Signatories which have ratified it, *i.e..* Parties. Resolutions are not binding.

29. Outer Space Treaty, art. IX.

30. *Id.*

31. *Id.*

32. Registration Convention, art. II.

33. *Id.,* art. IV, *cf.* art. III.

34. *Id.,* art. II.

35. *Id.*, art. I(b).

36. *Id.*, art. IV.3.

37. Outer Space Treaty, art. VII, *cf.* art VI. States are responsible for their national activities in outer space, whether these are carried out by governmental agencies or non-governmental entities, and are required to authorize and continually supervise such activities, *id..* art. VI.

38. *Id.,* art. VII.

39. Liability Convention, art. II.

40. *Id.,* art. III.

41. *See* H.A. Baker, *Liability for Damage Caused in Outer Space by Space Refuse.* XIII ANNALS OF AIR & SPACE L. 183, 204 (1988) (providing that damage by the space environment, *per se* not covered by the Liability Convention). *See also* H.A. Baker, *Space Debris: Law and Policy in the United States.* 60 U. Colo. L. Rev. 55 (1989), and SPACE DEBRIS, LEGAL AND POLICY IMPLICATIONS (1989).

42. Outer Space Treaty, art. VII.

43. Liability Convention, art. III.

44. *Id.,* art. I(d).

45. *See* Baker, *supra* note 41 (discussing the term "space object" in the context of orbital debris). *See also* C.Q. Christol, THE MODERN INTERNATIONAL LAW OF OUTER SPACE (1984), at 83 (discussing the definition of "space object"). *See further* W.F. Foster, *The Convention on International Liability for Damage Caused by Space Objects.* 10 CANADIAN Y.B. OF INT'L L. 137 (1972), B. Cheng, *International Liability for Damage Caused by Space Objects.* 1 MANUAL ON SPACE L. 83 (1979), N.M. Matte, AEROSPACE LAW (1977), and S. Gorove, STUDIES IN SPACE LAW: ITS CHALLENGES AND PROSPECTS (1977) (discussing liability for damage caused by space objects).

46. *See* Baker, Liability for Damage Caused by Space Refuse, *supra* note 41 (distinguishing among different categories of orbital debris).

47. *Report of the United Nations Conference on the Human Environment.* held at Stockholm, Sweden, June 5–16, 1972, U.N. Doc. A/CONF. 43/14 and Corr. 1 (June 16, 1972).

48. T.I.A.S. 9614, *done* May 18, 1977, *entered into force* Jan. 17, 1980. *See* P.F. Uhlir and W.P. Bishop, *Wilderness and Space.* BEYOND SPACESHIP EARTH: Environmental Ethics and the Solar System (Ed. E.C. Hargrove, 1986) (discussing this and other treaties and their application to outer space).

49. 18 I.L.M. 1445 (1979).

50. 11 I.L.M. 1291 (1972).

51. 26 I.L.M. 1516 (1987).

52. *See* U.N. Doc. A/AC.237/L.14 (May 8, 1992) (containing the most recent draft of the convention as of this writing).

53. N.Y. Times, May 13, 1992, at 4.

54. *See* Gorove, SPACE LAW, *supra* note 45, at 85–94; S.H. Lay H. Taubenfeld, THE LAW RELATING TO ACTIVITIES OF MAN IN SPACE (1970), at 25; and D. Goedhuis, THE CHANGING LEGAL REGIME OF AIR AND OUTER SPACE (1978), at 27 (supporting the position that "peaceful" in the context of Article IV of the Outer Space Treaty means non-aggressive).

55. *See e.g.,* M.N. Shaw, INTERNATIONAL LAW (1986), at 59–81 (discussing customary international law as a source of international law).

56. 5 ORBITAL DEBRIS MONITOR 1 (Jan. 1, 1992).

57. *See* Shaw, *supra* note 55, and the *Corfu Channel Case.* 1949 I.C.J. 23 (concerning Albania's duty to make reparations for failing to prevent a disaster resulting from mine fields laid with Albania's actual or presumed knowledge) and the *Chorzow Factory Case.* 1928 P.C.I.J., ser. A, No. 17.

58. Presidential Directive on National Security Policy (Nov. 1989).

59. *See* recommendations of the Interagency Group (Space), Report on Orbital Debris, *supra* note 1, at 53–54, which provide that

   minimizing orbital debris should be a design consideration for all future commercial, civil and military launch vehicles, upper stages, satellites, space tests and missions;

   Each agency with operational or regulatory responsibilities for spacecraft should develop and distribute internal policy guidance consistent with National Space Policy regarding debris minimization;

   Current agency operational practices for debris mitigation during launch and space operations should be continued and where feasible and cost effective, improved.

   Furthermore, the National Aeronautics and Space Administration Authorization Act F.Y. 1991, Pub. L. 101–611, sec. 118 (1990) provides that

   It is the sense of Congress that the United States policy should be that

   (1) the space related activities of the United States should be conducted in a manner that does not increase the amount of space debris; and

   (2) the United States should engage other space-faring Nations to develop an agreement on the conduct of space activities that ensures that the amount of orbital debris is not increased.

60. U.S. Space Command Reg. 57–2 (June 6, 1991).

61. *See* the Commercial Space Launch Act of 1984, as amended, 49 U.S.C. secs. 2601–2623 (1988) and its implementing regulations, 14 C.F.R. Parts 400–499 (1990) (providing the regulatory regime for launch vehicles).

62. *See* the Communications Act of 1934, 47 U.S.C. secs. 151–613 (1988), the Communications Satellite Act of 1962, 47 U.S.C. secs. 701–757 (1988) and implementing regulations, 47 C.F.R. Parts 25 and 100 (1990) (providing the regulatory regime for telecommunications satellites).

63. *See* the Land Remote-Sensing Commercialization Act of 1984, 15 U.S.C. secs. 4201–4292 (1988) and its implementing regulations,

15 C.F.R. Part 960 (1990) (providing the regulatory regime for remote-sensing space systems).

64. The technical community refers to technical organizations, or committees within organizations, such as the American Institute of Aeronautics and Astronautics (AIAA); the Committee on Space Research (COSPAR) of the International Council of Scientific Unions; the International Academy of Astronautics (IAA); the International Astronautical Federation (IAF); and the International Organization for Standardization.

65. Generally, international agreements, treaties, and conventions are legally binding, while resolutions and declarations are not, although the latter carry considerable moral force.

66. NOAA is at the same time an operator of weather satellites and a regulator of private remote-sensing space systems pursuant to the Land Remote Sensing Commercialization Act, *supra* note 63.

67. For example, subject to a proper rulemaking, these agencies could require private spacecraft operators to assess the debris impact of their missions; take certain specified measures or achieve certain levels of performance of debris minimization, including end-of-life disposal maneuvers; and notify of the creation of certain kinds of debris, including orbital parameters and other important information.

68. 49 U.S.C. sec. 2606 (1988).

69. *Id.*

70. The Commercial Space Launch Act distinguishes between payloads requiring a "license, authorization, or permit . . . by any Federal law," and those that do not, 49 U.S.C. sec. 2606(b)(2), *cf.* (b)(1) (1988). The former group includes telecommunications satellites, which require a license from the Federal Communications Commission, *supra* note 62, and remote-sensing space systems, which require a license from the DOC, *supra* note 63. With regard to these payloads, the DOT must "ascertain" whether the license has been obtained. *Id.* sec. 2606(b)(1). The latter group includes foreign payloads and U.S. payloads that do not require authorization under Federal law. The DOT may examine these payloads from a safety perspective, as well as for consistency with U.S. national security and foreign policy interests. *Id.* sec. 2606(b)(2).

71. *Id.*. 2606(b)(2).

72. 47 U.S.C. sec. 151 (1988).

73. 15 U.S.C. sec. 4242(b)(2) (1988). The disposal measure is subject to the approval by the President of the United States.

74. *See supra* note 28 (providing the cite for the resolution which constituted COPUOS, and discussing its mode of operation).

75. The agenda item is called "The Elaboration of Draft Principles Relevant to the Use of Nuclear Power Sources in Outer Space." *Report of the Legal Sub-Committee of the Committee on the Peaceful Uses of Outer Space on the Work of its Twenty-Eight Session (20 March – 7 April. 1989).* U.N. Doc. A/AC.105/430 (Apr. 20, 1989), at 16–28 (discussing the then current status of the negotiations of such principles). Subsequent COPUOS Legal Subcommittee reports discuss only a few of these principles, as additional agreements were achieved.

76. U.N.G.A. Res. 45/72 (1991). *See Report of the Committee on the Peaceful Uses of Outer Space to the U.N.G.A..* U.N. Doc. Supp. No. 30 A/46/20 (1991), at 17.

77. *See* The Elaboration of Principles Relevant to the Use of Nuclear Power Sources in Space, *supra* note 75.

78. Charter of the United Nations, 3 Bevans 1153 (1945), art. 57.

79. *See* International Telecommunication Convention, T.I.A.S. 8572, *done at* Nairobi, Kenya, Nov. 6, 1982, *entered into force* Jan. 4, 1984 (ITU Convention), art. 4.

80. The frequency allocation was for a space research service.

81. ITU Convention, *supra* note 79.

82. *Id.*, art. 4.

83. For example, the ITU exercises certain jurisdiction over the geostationary orbit. *See e.g.* ITU Radio Regulations, *Final Acts.* World Administrative Radio Conference, Geneva 1979, as revised, art. 11, Section II, (concerning frequency coordination of geostationary satellites). Orbital locations and spacing between satellites in the geostationary orbit are important factors in this coordination.

84. ITU Convention, art. 11.1.

85. *Draft Recommendation on Environmental Protection of the Geostationary Orbit.* ITU Doc. 4A/TEMP/50-E (CCIR Study Group June 12, 1991).

86. Quotation marks are added because, with very few exceptions, no international organization has law-making power in the domestic sense of the word, where the term is associated with the power of a government to bind legally its subjects. International organizations generally cannot bind legally member States unless the members agree to be bound, *i.e.*. they sign and ratify, a particular obligation. Moreover, only certain organs within an organization have the authority to formulate rules which may become legally binding, subject to signature and ratification by a member State.

87. *See* ITU Convention, arts. 6 and 7.

88. *See supra* note 47 (providing a cite for the Stockholm Declaration).

89. Vienna Convention, *supra* note 51, arts. 2–5.

90. *Id.*, art. 9.

91. 26 I.L.M. 1541 (1987).

92. *Id.*, art. 2.

93. *Id.*

94. *Report of the Second Meeting of the Parties to the Montreal Protocol on Substances that Deplete the Ozone Layer.* CITE

95. Montreal Protocol, art. V.

96. *Id.*, art. 3.

97. *Id.*, art. 4.

98. *Id.*, art. 9.

99. *Id.*, art. 10.

100. *Id.*, art. 7.

101. *See* Hall, *Comments on Salvage and Removal of Man-made Objects from Outer Space.* COM. BASED ON VOLS. I-XV OF THE COLLOQ. ON THE L. OF OUTER SPACE, 153 (Ed. M.D. Schwartz, 1976); H. DeSaussure, *The Application of Maritime Salvage to the Law of Outer Space.* XXIV IISL 128 (1985); R. M. Jarvis, *The Space Shuttle Challenger and the Future Law of Outer Space Rescues.* 20 THE INT'L LAW. J. 591 (1986); and R.M. Jarvis, *Space Salvage: A Proposed Treaty Amendment to the Agreement on the Rescue of Astronauts. the Return of Astronauts. and the Return of Objects launched into Space.* 25 VA. J. INT'L L. 965 (1986) (discussing salvage of space objects). 102] *Supra* note 75.

103. 8 I.L.M. 579 (1969).

104. The ordinary meaning of a word must be assessed in the "context" of the treaty in which it appears, and is subject to "good faith" interpretation. Vienna Convention on the Law of Treaties, *supra* note 103, art. 31.1.

105. *Id.*

106. *Compare* Rio Convention, *supra* note 52, art 2.

107. *Compare* Vienna Convention, *supra* note 51, art 2; and Rio Convention, art. 5.

108. *Compare* Vienna Convention, art. 4.

109. *Compare id.*, art. 4.2.

110. *Compare* Vienna Convention, art. 2

111. *Compare* Vienna Convention, art. 6; and Rio Convention, art. 7

112. *Compare* Vienna Convention, art. 7; and Rio Convention, art. 8

113. Regulatory Program of the United States Government, Executive Office of the President, Office of Management and Budget, Mar. 31, 1987.

114. Suggested graveyard altitudes are 300–500 km above the geostationary orbit.

115. Suggested perigees for ensured reentry are between 180–200 km above the surface of the Earth.

116. Other treaties have relied on the parties' own national means of verification. *See e.g.* Interim Agreement Between the United States of America and Union of Soviet Socialist Republics on Certain Measures With Respect to the Limitation of Strategic Offensive Arms, *signed* May 26, 1972, *entered into force* Oct. 3, 1972, 944 U.N.T.S. 0, art. V.1 (providing that "[f]or the purpose of providing assurance of compliance with the provisions of this Interim Agreement, each Party shall use national technical means of verification at its disposal . . ."). The reference to "national technical means" is a reference to reconnaissance and surveillance from space.

VII. Panel Discussions

26: Panel Discussion

Diane P. Wood, chair
University of Chicago

1. QUESTIONS FOLLOWING THE PRESENTATION OF IRVING WEBSTER

B. Bloom: What part of the second stage is not aluminum?

I. Webster: A good bit of it. The second stage tanks are 410 stainless steel, the bottles are titanium, and the second stage engine injector is titanium.

D. Rex: Do you have any knowledge to what extent the rocket body burns up during re-entry if it comes down?

I. Webster: I do not have first-hand knowledge on that. We believe that it essentially all burns up. We have never found any fragments anywhere, but we do not know.

D. Rex: This is closely linked to your point that you are not willing to use it as a de-orbiting thrust.

I. Webster: One could play the numbers game. As I said, we have launched 211 of them. At this time, we have never recovered fragments from a re-entered stage that I am aware of, and I've been on the program 30 some years. You might ask how many Delta second stages are still up there with propellant aboard? There can't be more than three; I believe that only two vehicles, which were launched in the third quarter of 1980, could still be in orbit, and probably are.

2. QUESTIONS FOLLOWING THE PRESENTATION OF DANIEL JACOBS

H. Gursky: We heard a lot of discussion this morning about the debris as a problem. Are there any standards that you're working to? It sounded as if at this point there is just a general desire to reduce the debris and the potential for debris.

D. Jacobs: There are two parts to the answer to your question. The first is that these efforts so far have not addressed potential standards or policies. They have been trying to reduce our uncertainty and increase our understanding of the debris envi-

ronment and its implications and effects. That, in my opinion, has to a large extent been achieved. A great amount of uncertainty has been reduced through the Haystack data that Don [Kessler] mentioned, and other activities. We are getting to the point that we really understand what the environment is and what its implications are.

Some things are still missing: to date no one has yet done cost effectiveness studies to determine the economic impacts of the various remedies you could apply, and determine which ones are going to be the best from a commercial point of view. Those still need to be done, and hopefully will be done soon. However, with that said, a few things have started happening. The first is the CCIR recommendation that I mentioned. Even though all of the CCIR recommendations are voluntary, most members do follow them. There is a standard in the area of end-of-life disposal of [objects in] geostationary orbit. We now have to apply that approach in other areas, such as required venting of upper stages. I would like to have a treaty (or something like one) on that problem, as well as the issues of disposal of upper stages and lower orbit systems. But those are the next steps.

3. DISCUSSION FOLLOWING PRESENTATION FROM NATIONAL PROGRAMS

H. Gursky: This morning a lot of the concern related to the manned program. I am wondering what would happen if there is no longer a manned program. What would the impact be on the debris problem?

D. Kessler: Without a manned program, you would obviously have a higher threshold of allowable debris. You would lose a lot of the sensitivity that we now have to the problem, but that could be very negative in the long term, because we would then allow ourselves to reach this critical density and way surpass it. Let it go for very much longer and you get a fast runaway, and even unmanned spacecraft then eventually lose their ability to operate in low earth orbit.

R. Reynolds: It is also worth noting that the impact on science programs is going to get greater if you do not control the environment. That would be an additional reason for not letting the thing slide, regardless of (for example) what happens to the space station.

H. Gursky: I wonder, if the risks are reduced to the point where they're acceptable to the commercial users of space, whether there would be any interest at all in dealing with any possible science impact. I have heard nothing at all from any of the national reports regarding the radioactive material in space, as an example.

J. Primack: Since you brought up the issue of science, it is easy enough to multiply the numbers together. If you just look at the trackable debris, the figures that Kessler or Professor Rex showed tell you that the probability of a catastrophic failure of something like the Hubble Space Telescope is perhaps a tenth of a percent in 10 years. But, that is only trackable debris. If you look at the extrapolations down to particles of a few millimeters in size – which are more or less confirmed by various kinds of observations – you get the probability of a catastrophic accident which, over the lifetime of the Hubble Space Telescope (which I'll take to be ten years) gets up to several percent, maybe more. So that is the current environment, and that is large. If we had a runaway, so that the amount of debris in the centimeter size class starts to grow by an order of magnitude, then it would be very hard to do science in lower earth orbit, leaving aside commercial activities from GEO.

D. Jacobs: In a GEO, the commercial user is probably the primary one. In the long run, the impacts on the commercial sector may be more harmful than any others, so those need to be considered too.

J. Loftus: The discussion tends to focus on the manned facilities because they are the largest and they have the longest committed residence times. But there is no reason to believe that we may not be operating a Hubble Space Telescope for 30 years rather than 10, and so the risks tend to increase.

It is important to know that the orbits which are most at risk and which have generated the problem are precisely those low Earth orbits used for weather and for earth observations. It turns out that the orbits in which manned space craft operate are relatively debris free, except for that which rains down from above, as Nicholas Johnson pointed out in his initial presentation. So I think the answer is that commercially, for things like earth observation satellites, it is an issue for large spacecraft now, and it will become an increasing issue unless we manage the environment.

D. Wood: Bob Warren mentioned that companies are already taking mitigation measures in their own best interests, because they do not want to see their investment endangered. When do we reach a point that the company will take mitigation measures that are not good enough for the world as a whole?

R. Warren: I guess it's a bit like having the mice decide that they should bell the cat. They all agree this is a super idea, and then they don't have many volunteers to go up and put the bell on. It is difficult at best in these days of recession to get a space program approved. I have been working for 13 years just to get Radarsat going, and most of it was in getting approval for the cost of having Radarsat. So it's not as easy as you might think, even though we are all enthusiastic, to do good things when it simply does add to the cost.

If I have to add 200 kilograms of fuel, it means I have to have a launch vehicle that will carry the additional 200 kilograms. There isn't any easy solution here. I am encouraged to hear about de-orbiting, and I would like to learn more. But my question now, relating to re-entering a vehicle, is whether anyone has done research to determine under what conditions a satellite will burn up and not hurt anybody? Can I go to my designers and say if this is designed in a certain way it will burn up on the way back? If I can convince my boss at least that it won't hurt anybody on the ground when it comes back in, I might be able to talk him into the cost of the additional fuel to re-enter.

D. Rex: The question of whether a body does or does not burn up during re-entry is indeed a very difficult one. We looked into that problem for some time, but we stopped the work because so many influencing parameters were involved. On the one side, it is influenced by the orbital aspects of the angle of incidence. On the other, it is influenced by the material and the construction and the design of the object. The angle of incidence does not really completely describe the amount of energy put into the object, because with a steep angle of incidence, you have heat flux that is very high momentarily, but it may not last long enough to burn up everything in the object. Thus, it depends very much on how the object is constructed: is it compact, does it consist of several shells, etc. To give a general guideline on how to construct a vehicle and how to design re-entry is nearly impossible; it must be done for every case. Generally speaking, there is a tendency that materials with a higher melting point do survive the re-entry. That is especially the case for all rocket bodies that re-enter, so the tank may burn up, but the rocket motor, which was designed to withstand high temperatures from the material side, may survive. Also, we have identified certain materials that never burn up. For example, if you have uranium oxide used as a fuel for nuclear reactors, it will never burn up, because during re-entry the temperature reached never exceeds the melting point.

D. Jacobs: I would like to go back to the question how to cause commercial firms to adopt more practices that they would be willing to do, or that we might deem sufficient. This is why we need international agreements, but everyone has said we are not ready to do this yet. However, there are some things that individual space agencies can do to help move that process along. In their procurement activities for their own space hardware and systems, they can require certain debris mitigating hardware or practices to be built in. By doing so, they are helping their own industries develop the technologies and the abilities to put these into the systems that they can then offer later commercially, and that is a major transition step we could help them with. It is a difficult sell to make in times of shrinking budgets, but it is one that each agency could do on its own if it wished.

27: Panel Discussion

Paul F. Uhlir, Session Chair

National Research Council, U.S.A.

P. Uhlir: We have a lot of good questions here. I've tried to arrange them in order beginning with the threshold questions of definition and technical issues, leaving some of the more complex issues dealing with process and agreements for the end. If you don't hear your favorite questions coming up right away, don't worry, we'll get to them. One of the reasons I am doing it this way is that we only have about an hour of so of discussion time, and therefore we probably won't get through all of these questions tonight. John Simpson has suggested that we address some of the remaining questions in the morning. Therefore, I am saving the kind of questions that address the issues that we were going to address tomorrow in any case for that time. What I'd like to do is read the questions and then if I have an idea of who might be able to answer or provide a comment initially in response we'll start off with that individual. After that anyone else who wants to say anything is welcome.

DEFINITION OF TECHNICAL AND LEGAL TERMS

The first question – and actually there were several posed on this issue – is one of definitions. There is a question of defining space or orbital debris. This, of course, was raised by Pamela Meredith in her talk. The question is as follows: The definition of debris appears essential for moving forward legally. How may this term be defined to the satisfaction of the international community?

I would like to say a few words in response to that. There is the International Organization for Standardization (ISO), which has a Technical Committee-20, on aeronautics and space vehicles. Working Group 6, under TC-20, is being established to address orbital debris issues and one of the first topics on its agenda, as I understand, will be the standardization of terminology related to orbital debris. It's quite obvious that in order to make progress – as a threshold condition for making progress on any international agreements – the parties that are negotiating an agreement need to be talking the same language and understanding what the terms they are using mean very specifically. Some of these terms, such as "space object," are quite slippery, especially from the legal standpoint. Certainly

from a technical standpoint many of the terms are either obscure or difficult to grasp or understand. So I think a lot of work needs to be done on defining the various terms related to these problems and that will then lead to a better understanding of what the issues are and what the nuances and subtleties might be in preparation for any meaningful agreements to deal with the problems. I don't know if anyone here is aware of any formal process other than the ISO one that is underway. Are the space agencies addressing this issue, or any of the professional societies?

S. Gorove: The International Law Association (ILA) has an International Study Group dealing with space law. This group has been considering the drafting of a convention dealing with space environmental issues, particularly with space debris. We had a conference most recently in Cairo. There, a second draft of a convention was discussed and I think the final one is expected to come out at the next ILA meeting, which is going to be held, I believe, in Argentina in 1994. This is a convention which is prepared by the Reporter after receiving comments from space experts in different countries all over the world.

P. Uhlir: The two countries you mentioned aren't exactly renowned space powers. I was wondering to what extent the launching states are represented and what is the membership of the group?

S. Gorove: Well, of course there are no countries as such involved. There are individuals who are on the ILA Space Law Committee from different countries, which includes the U.S. and most of the spacefaring nations. That is one example. Another one is within the International Astronautical Federation, which has established a committee to consider the issue of space debris mostly from a technical point of view. At the same time, the scientific and legal liaison committee in the International Aeronautical Federation's (IAF) legal arm, the International Institute of Space Law, is also considering together with the IAF group drafting something that might appear acceptable or amenable for acceptance or consideration in this work. So those two are in addition to the one you referenced.

P. Uhlir: The one I mentioned I think will focus primarily on technical terms. But this raises the issue that as the problem is addressed in more fora, both formally and informally, that close coordination among the various efforts needs to be maintained so that they're not working at cross purposes, so that the work is maximized. Does anyone else know of such activities in the U.S. government or at one of the agencies? Okay, well, terminology obviously is important and perhaps since the groups mentioned are non-governmental that might be an appropriate starting place, but the official government players in this ought to be concerned about beginning this process as well in preparation for any multilateral types of talks.

There is another question on defining the meaning of "pollution" or "contamination" in space. I would think that is really covered by the comments made already, unless anyone has a specific comment to make on those terms.

S. Gorove: The reference was made in connection with Article 9 of the 1967 Outer Space Treaty. Once you deal with the space debris issue, in addition to trying to identify or give a definition of space debris, you also will have to discuss the meaning of contamination, pollution, or the special term used in Article 9, which is "harmful contamination." One needs to ask whether that is the same as contamination or pollution, and whether those are terms that can be used interchangeably. In any event, at the time when you draft a convention dealing with space debris, you will also touch upon Article 9 and will have to determine what you should do in terms of the requirements, standards, and consultation procedure. So you will need clarification of these issues as well.

P. Uhlir: Can we define the specific steps needed for monitoring of particles smaller than 1 centimeter on a continuing basis? What ranges of particle sizes are most needed for monitoring?

H. Baker: Paul, could I respond to the contamination question before we move on? I would suggest that a good area to look is international environmental law, because there are different views and different scholars have defined it differently. For example, Springer in his book *The International Law of Pollution* has analyzed the meaning of "pollution" and how it is used in various international environmental instruments. I think that might be a good place to start.

J. Primack: Well, just on this issue of contamination, pollution, etc., I'm not sure that the right way to deal with the difficulties of charged particle emission and the creation of an artificial radiation belt or gamma ray emission from operating reactors is best dealt with by definitions. On the other hand, since law is greatly concerned with these things, maybe that is something with which one should be concerned. People have made the suggestion that one can, for legal purposes, consider this radiation to be interference and I guess one also could consider

it to be a form of pollution. I'm not sure which of either of these two or maybe some other definition is the most appropriate one to fit this problem under existing international law. I just want to remind you that you might want to consider that as well as the space debris issue that we've mostly been considering.

DEBRIS MONITORING

P. Uhlir: The other question dealt with the sizes of debris that are most in need for monitoring, with the fragments smaller than 1 centimeter being specifically mentioned.

D. Kessler: In the past we wanted to monitor the one millimeter to one centimeter particles mainly because that was the size range that affected most spacecraft. However, I think we've gotten a lot smarter in the last few years. There are things going on in space that we don't understand and some of those things are in much smaller size regimes. One of the things that has come up recently, for example, is that we need criteria, some sort of measurement, to verify that we're going over this critical density concept. That may well mean going into different size regimes like the microparticles, very small particles that could only be generated as products of collisions. I think the right answer now is that we would like to see the entire size spectrum, from the centimeter size and smaller, monitored continuously and we've made great advances in how you can do that. The technology is here. It is just a matter of the will to do it.

D. Rex: When we want to have data on the millimeter size range of orbital debris it seems to me that radar measurements from the ground in the long run will not be sufficient. What we need is optical detection from Earth orbit. So we need a mission with a powerful optical sensor that can detect millimeter sized particles at distances of some 100 to 1000 kilometers. Also we need to not just know how many particles there are, but on which orbits they are. I think it was said yesterday, that there is some indication that some fraction of these particles are on highly elliptical orbits, especially if you go in the low millimeter size range, and in order to assess their origin it is essential to know about their eccentricity. This is also important for the design of shielding. The optical device that is needed is somewhat complicated, for it takes money and the problem is again which nation would spend the money for that purpose. But I would very much urge that we have such a mission. Studies for that have already been done in many places, but we should come to a situation where it really is performed.

P. Uhlir: Perhaps cost-sharing through international cooperation might be an appropriate topic for discussion.

A. Reinhardt: I would like to follow up on that. Currently, as was well documented yesterday, the space surveillance network can see objects down to approximately 10 centimeters in size. With the technologies that we are working on it will be possible to see optically perhaps down to right around the 1 centimeter level, with the goal of determining orbital element sets for every piece. If you look at the models, how large is that problem? Professor Rex, I believe I saw 80,000 particles in that size regime and the Haystack data are suggesting on the order of 150,000 objects. The cataloguing problem you're faced with is quite immense. Right now the net count is 6,800 objects. There are plans by 1995 or '96 with the SPADOC upgrades to be able to handle 16,000 objects from the United States. I think that looking ahead a little bit, what might be food for thought is if the other nations wanted to establish their own sensors and then communicate on a world-wide net to help catalogue objects. Clearly a world-wide net of sensors is warranted, if you want to stay on top of the debris. If you're going to try to catalogue debris you will need several world-wide sensors. The U.S. Space Command is downsizing along with the rest of the American military systems. So we'll be losing sites in the very near future. I would just say that perhaps this is food for thought that can be taken up in the scientific committee. We're going to be pushing for in the short term a very modest, low-level, institutionalized continuous measurement program, so we can keep updating the catalogue numbers on the small debris. The debris measurement is crucial.

VERIFICATION ISSUES

P. Uhlir: The next question is actually related to debris monitoring and partially answered already. It says a major problem in establishing a formal international agreement is adequate verification of compliance by all parties. What technologies are or soon will be available for effective monitoring or verification, and what is their approximate cost? The same technologies for defining the problem would be also applicable for monitoring and verifying, subsequent to an agreement. It sounds like the same techniques that have been already mentioned would be applicable.

D. Jacobs: I'm not sure that I agree totally with that Paul. It would be applicable if you're looking to see if people are continuing to have things blow up on them or fragment. If you want to do validation on the ground, you may want to go look at the spacecraft and the objects before they're launched. It's totally a different regime. Then you start to get into things like arms control, and that is not cheap or easy to do.

D. Wood: I wanted to add to that along similar lines. I think the verification depends on which problem you're talking about – whether its venting, etc. There is a different problem on the ground. If we're looking at designs or other kinds of construction criteria for spacecraft, you will get in the commercial area quickly into trade secrets and proprietary information, and a great reluctance to let somebody wander in and look at all of your blueprints.

P. Uhlir: There are obviously a lot of difficulties other than technical ones in terms of verifying and monitoring, but I guess there is still the question of the cost. It was mentioned that some of these methods were quite expensive, but no figures were given. Are there any figures attached to any of them?

D. Kessler: We're spending about 15 million dollars total to use the ground radar to monitor the centimeter and larger population for about a five-year period. Most of that has to do with construction. Once it is constructed the operational cost will be much lower than that. The same could be said roughly for about down to 1 millimeter using ground radar, maybe a little higher because you need a little bit more complicated radar. When you go into space, you're talking about spending hundreds of millions of dollars, and that is for construction costs.

RE-ENTRY MANEUVERS

P. Uhlir: The possibility of unsafe re-entry has been used as an excuse to reject the deorbiting solution. Could the detonation of a satellite in very low orbit be used as a safe re-entry method? This would permit complete burn up during re-entry, thus eliminating the need to worry about the possibility of it falling on a populated area. This obviously raises an important question of public safety and liability under the Liability Convention, which assigns absolute liability for any damage caused to the surface of the earth. If there is any mandatory requirement for deorbiting then that is a very important consideration. There are obviously objects of certain mass or density are not going to burn up completely. At the same time, the risk involved of hitting anything actually is very remote indeed. But it is the kind of thing that can cause public hysteria if it happens to fall in the wrong place. It is an important issue and the question is fairly straight forward.

D. Rex: May I give some comments on that. The idea that the deorbiting of a space object would create some additional risk on the ground was already mentioned here yesterday. I think it is to the contrary. If we deorbit an object that means only that we bring it down now and we do not wait until it will come down naturally in 10 years, or 50 years, or whenever. So eventually it will make a re-entry and then there will be some risk to the ground, but if we make an active deorbit maneuver, we have also the opportunity for this maneuver to be performed in a way to determine the point of impact on the ocean. We could guarantee that we dump it into, say, the Pacific Ocean – where there is enough space for such purposes – and also there is no contamination of the ocean by this material, there is no poison in it or whatever. The deorbiting is really a safe maneuver and there is no extra risk to the ground associated with it.

I. Webster: What I said yesterday was that without adequate fuel in a vehicle and without knowing that you have adequate fuel in a vehicle you cannot control where you are going to re-enter or when you're going to re-enter. You may not have enough fuel aboard to have a controlled re-entry. The way to solve that problem is to design the vehicle from the beginning to set aside the amount of fuel that is necessary, or to not fly missions that you cannot fly without having re-entry capability. The point that I want to make is not that you can't re-enter safely, because you can, and, yes, there is much of the area that is ocean and one can re-enter in an ocean area by and large. The problem comes when you don't have fuel enough to do it in an orderly and planned manner, you run out of fuel during the re-entry maneuver and you're in trouble. Now you've done something, you've tried to do something, and you've failed to do it properly. The lawyers will have a field day.

A. Reinhardt: Well, the question also addressed explosions in lower earth orbit. This is very similar to an intercept test and we are evolving procedures now to make sure that those are conducted as safely as possible. First, you'd have to carry an explosive charge, which is a weight penalty, so it means less payload on orbit. The second consideration is you'll now be imparting energy so there will by definition be particles – probably only a few, but particles – that will achieve enough velocity to go into a skate or go into orbit. So now you have to run the COMBO (Computation of Miss Between Orbits) codes to make sure that as this is coming down – it's a real-time problem you're trying to deal with so there will be a lot of variables involved – that you're not going to be in the vicinity where this evolving cloud could take out an operational satellite because then the lawyers would really get excited. The third thing is you also have to consider the Federal Aviation Administration's rules. We block off huge areas of air space in order to conduct intercept tests so you have to deal with our pilot friends flying international airplanes and if you explode you might actually spread the cloud into regions where there is air traffic. So can you do it? Sure. But there are a lot more considerations that might not be apparent at first that would have to be worked out. I'm not quite so sure that in a real-time problem of decay that the timing factors would be such that you could safely do it over a broad ocean area and have enough air space blocked off and not be worried about a cloud running into an operational satellite. There are a lot more factors in there than just meets the eye.

D. Rex: First, I would support what has just been said about the explosion device. I think the safety issues involved there are tremendous and I would rather say that we should not follow such a procedure. Just one remark to this: even if we do not have enough propellant to make a safe re-entry, that is to say to make a predetermined re-entry, if you just go into an elliptical orbit where the perigee is 150 kilometers you stay for another couple of weeks on orbit and then come down. Even then the risk to the ground does not become larger, but it is the same risk which you would have had before that retarding

thrust maneuver. I can't see that we increase the risk to the ground by any such life shortening maneuver.

P. Uhlir: Couldn't you get around the uncertainty of the amount of fuel left by design?

I. Webster: What we do is like what I presented in my paper for an elliptical orbit: we start in the perigee, that is in the 80 to 100 nautical miles, and we add the incremental velocity normal to velocity factors so that we don't graze the orbit, or that is the intent. Doing exactly what Professor Rex just discussed.

R. Reynolds: We seem to be agreeing on too many things; I like the idea of blowing things up to reduce the ground footprint. Presumably you'd do it in an intelligent way, so you wouldn't get the large delta V's which might lead to a problem at orbital altitudes. You can probably do parametric studies to determine whether or not you're going to have a ground footprint a lot easier with a breakup than if you have intact structures. I think it's something we should look into, because we are going to have to deal with the re-entry problem and, if there are ground footprints, that is going to put constraints on what your programs can do.

P. Anz-Meador: Also there might be cases where the onboard destruct systems could be utilized to more or less just open up the satellite, thus exposing more of it to the atmosphere on re-entry.

P. Uhlir: There is also one technique that has already been used by the former Soviet Union for detaching radioisotope thermoelectric generators, and that is another example that may be relevant. The next question is: Do we have any consensus on actions beyond those already taken by the main space agencies to reduce debris? In particular, do we agree that it is desirable that starting with spacecraft launched in the year 2000, all large objects to be placed in the approximately 800 to 1200 kilometer zone and the 1500 to 2000 kilometer zone be designated to de-orbit within say approximately 5 years after the end of their useful life? If people object to this, what alternative approach would they suggest to avoid the debris cascade problem? Or do they not agree that this is a problem?

A. Reinhardt: The position that we're in, from the Department of Defense viewpoint, is that we're still in the Phase I research position of understanding the problem. With regard to the cascading effect we can agree that it's possible, but I think we can disagree on where we're at right now in the environment until more measurements are taken. That is the position that we're working towards. I think we are working with NASA and have initiated some efforts – I don't think that's been discussed here yet – on developing a debris minimization or mitigation handbook from the DOD viewpoint that will include cost-effective or cheap measures to implement. As I explained in my paper today, it seems a bit stringent to start to shoot for a date when all payloads, especially in the higher altitudes like 1600 kilometers, are going to be forced to de-orbit because that would impose an extremely large cost on the development of those systems now. Furthermore, I would contend that the DOD has been able, by measuring carefully the amount of fuel onboard in station keeping, to extend the life of those systems for much longer than they were originally planned. If you have a deboost, you're going to shorten their lifespans and have to then have replacement launches, which, I think, could conceivably add to the debris problem more than just leaving them there for the short-term because otherwise you would be adding more flights and bringing more things up and it's during the boost phase that you have the highest potential for a problem. I don't think that analysis has been done well enough yet to look at the tradeoffs. That is why we go back to the point that we're still in Phase I. Hopefully, we'll have a position by the end of 1993 on when we might need to start bringing in those types of stringent measures.

D. Rex: I would support the ideas put down in that question to de-orbit such objects from those altitudes, except for the timeframe which is stated. We should be very cautious about that because as was mentioned just now, this imposes quite a penalty on these missions and we really have to design these objects a long time before their launch. I want to mention one other aspect. The altitude region of 800 to 1200 kilometers has been stated. The same would apply at 1500 kilometers. The situation is somewhat different for the higher orbits. We have the GPS system, for instance, and other such constellations may emerge at about a 20,000-kilometer altitude. So far, the density of objects there is so low we could afford to bring some more into that space without the de-orbiting requirement. Indeed, for that altitude, the de-orbiting requirement is imposing a very, very high penalty on the satellites there.

P. Uhlir: I'm aware that two years ago the NASA Earth Observing System was being designed to be de-orbited, when it was still a large spacecraft configuration. Does anyone know whether that is still in the design of the descoped version? It's out? That is not surprising.

D. Kessler: One comment about de-orbiting. There is a conflict of policy here. To de-orbit right now, the easiest way to do that is use an energy storage device on orbit, i.e., store the fuel. One of the things we don't want to do is store fuel. Is it possible we could actually create more breakups as a result of such energy storage devices. Tethers also have similar problems, when you start stringing things out. There are some technology issues to be resolved before we can re-enter things or bring things back.

R. Reynolds: Since we've been discussing this issue (deorbit), I would like to mention that NASA is circulating a manage-

ment instruction for its program managers that would require them, as a part of their assessments during program development, to address debris issues that may be introduced during program operations, either as a result of normal operations or through anomalies such as explosions. They will also (be asked to) address their potential contribution via collisions and to plan for disposal at the end of mission. The primary issue for long-term control of the orbital debris problem is the removal of structures at the end of their mission life from regions in space that affect other users of space. For low earth orbit and highly eccentric transfer orbits this probably implies reentry of these structures.

We are developing a handbook to support the program managers in performing the assessments, a project that is just getting started. The handbook and the associated instructions cover all orbit regimes, but the emphasis will be on low earth orbit (LEO), geostationary earth orbit (GEO), and the transfer orbits. We are planning to complete the handbook by October, 1993. We have models that let us assess the "debris consequences" of various debris deposition events, such as explosions or abandoning structures in orbit, which contribute to the orbital debris problem. We are currently most limited in our ability to assess cost impacts and cost benefits, and to understand the potential ground footprint for reentering systems. Cost vs. benefits will be needed to justify requirements having high cost impact, and some ability to assess the ground footprint on reentry will be required if we are going to explicitly require systems to remove themselves from orbit. NASA recognizes the need for technical input to help determine requirements and to assess cost/benefits of debris mitigation options, and we will be reviewing those.

DE-ORBIT FROM GEOSTATIONARY EARTH ORBIT

D. Jacobs: I just wanted to make one comment. Even though I personally agree with the gist of the question, in that I think there are at least one or maybe a couple of things that you could do very soon, I want to give an example of how things can backfire on you, if you don't know exactly what you're doing when you do decide to do it. The example deals with the use of graveyard orbits in GEO. That has been advocated for many years and many people around the world have actually used the practice. However, NASA advocated for some time that we didn't know enough about it and that we were probably creating more of a hazard than we were alleviating. It turns out that further studies have shown that to be true and now the CCIR (the International Telecommunications Union's Consultative Committee on International Radio) adopted a recommendation of a minimum of 300 kilometers above GEO, correcting the 40, 50 and 80 kilometer range that some spacecraft operators were using earlier. So that is an example of something that on paper sounds like it is a good thing to do. People who are non-technical say that is great, we ought to go ahead and do it right away because there is no reason to wait. But no one quite understood all the technical details, and it ended up being a worse thing than if you left it alone.

P. Uhlir: That is a good entry to the next question. In geostationary orbit there are currently no minimum requirements for performance standards for spacecraft at the end of their useful lives. GEO spacecraft operators either leave them where they are or de-orbit them. However, those that de-orbit spacecraft use different techniques with various levels of success in what may be best described as a free for all. In light of the potential harm to the GEO from ineffective or poorly designed de-orbit techniques, why does the U.S. oppose establishing a mandatory common approach to dealing with GEO spacecraft at their end of life?

D. Jacobs: If I could just follow up, that is what we just did. In fact, the CCIR recommendation was drafted by the U.S.

P. Uhlir: Okay, but there are different operators and manufacturers that have already designed different techniques for effecting that and some have been less than successful. It seems to me that there has to be some minimum performance or minimum design standards to effectuate the 300 kilometer distance, for instance, knowing how much fuel you have so that you don't run out in the middle of a maneuver, that kind of thing. Is that also being addressed in a CCIR forum?

D. Jacobs: I don't believe CCIR recommendations go into those kind of technical requirements, but if this recommendation is adopted and the nations then use it, as the Federal Communications Commission would probably use it in granting licenses in the U.S., that would then become a procurement requirement. If someone buys a satellite with that procurement requirement and it doesn't operate that way, then the manufacturer won't sell too many satellites.

P. Uhlir: The collateral question is whether there is any research being done on the most effective techniques for doing that kind of maneuver. I don't know if that is really necessary or if enough knowledge of that is already in hand. It does seem that the maneuvers haven't all been uniformly effective, including those of some U.S. spacecraft.

A. Reinhardt: There is still a lot of work left to be done in the GEO, and there is still a lot of analysis similar to the LEO analysis that needs to be done. We'll update each other on where we're at in July, at least the DOD and NASA will.

D. Jacobs: One more follow-up comment is that these questions are dealing with satellites at the end of life and, as Joe Loftus mentioned earlier, the upper stage and apogee kick motors are probably a bigger problem. This recommendation

included another recommendation that the appropriate group, which is I think Study Group 7 in CCIR, take up that issue because they're the only ones who have jurisdiction over it within CCIR. So hopefully they'll begin to address that in the near future.

SURVEILLANCE OF RE-ENTRY

P. Uhlir: The next question is on surveillance of re-entry. In the case of re-entry, free access to unclassified data shall be granted to any countries that need it for their own assessments of the safety of their population. This was not the case in the past. Therefore, what is the procedure to get this implemented? I'm not entirely clear on the question, but does anyone have a comment on that?

J. Marcé: I already addressed this yesterday. The question is extremely simple. For the re-entry of Salyut 7, as I presented yesterday, we had satellite re-entry surveillance on behalf of the French government, but as you know, in France we did not have our own means to do the surveillance. So we were relying on the data coming from the U.S. Two days or one day – I am not sure – before nominal re-entry, we were short of any raw data. This meant we just had the prediction, but we are not able to make our own assessment of the prediction, which is normally what we wanted to do to compare the data with the forecast. I think Asia was in the same position as we were. In this respect I already discussed that with Joe Loftus when he visited us recently. It's not an acceptable position, because I think any country should have access to the raw data in order to make its own assessment. This is why I would like a response to this question because it is a problem of safety of people.

P. Uhlir: I understand now. I wasn't here for your talk yesterday.

A. Reinhardt: I'm not familiar with the background of the Salyut mission.

J. Marcé: It was because of a decision by the U.S. government, I think, which prevented us from getting the data. I think it was not a decision of NASA itself, but it was a decision by the State Department, or something like that. In this case, I think something should be done to allow data to be freely distributed to anybody who needs it or wants it.

DEPARTMENT OF TRANSPORTATION ACTIVITIES

P. Uhlir: We only have time for one or two more questions this afternoon. Most of the rest of the questions that we have here are related to the legal and international agreement issues. Those will be appropriate to begin the session in the morning. Let me just address this one here. What is the Department of

Transportation doing with respect to debris? Do we have anyone from DOT here?

E. Rosenberg: The Department of Transportation licenses launches of commercial launch vehicles to ensure that there is no jeopardy to public health and safety, safety of property, or U.S. national interests. In that respect, the DOT does examine issues of orbital debris risk in re-entry hazards. In licensing the launch of a vehicle, it requires that the launch operator demonstrate some understanding of those hazards and risks.

SPACE TEST RANGE POLICIES

P. Uhlir: Of course, DOT is part of the Interagency Working Group on Orbital Debris. Last question here: the Department of Defense has adopted a responsible debris policy within the framework of the space test range. Could the military representative discuss the current status and policies of the space test range?

A. Reinhardt: As I understand the policies of the space test range – this is now Reinhardt's interpretation – they are to try to minimize orbital debris and also on each test range. This is the same for NASA, which has its own flight safety requirements. It depends on what part of the problem you are talking about. With regard to orbital debris, the idea is to model what the cloud is going to be and then define the critical radius where the chances of a 10 m^2 satellite flying through that has a one in a million chance of getting hit by debris one centimeter or larger. That defines a point in time and then they put an error bar around that, an ellipsoid in space if you will, and then run that against the operational catalogue. If there is nothing that comes within close proximity of that ellipsoid, they say they have a window in effect. That is what is happening now by committee, that is what we want to try to simulate on our work station. With regard to the range, there are various requirements –and I'm not familiar with all of them – but most of the ranges try to contain all their tests on range by definition. There are some places, like White Sands, where they can extend the limits a little bit through negotiations with the surrounding areas and then they have a minimum of approximately 58 foot pounds coming down. I'm not sure of the exact numbers here.

DEFINING ORBITAL DEBRIS

P. Uhlir: The questions we have remaining are perhaps the more difficult ones to answer, and maybe won't be answered satisfactorily at all. They certainly will bring up some issues that need to be considered and discussed. The first one is: What legal issues are of most interest to scientists? The other side is: What legal issues are of most interest to industry? Perhaps we could get a few representative ideas from some of the engineers and scientists here in reflecting on the legal presentations of yesterday and otherwise.

B. Bloom: I'd like to ask the lawyers to what extent they think the definition of orbital debris has to be tied to the ability to measure it, and to what extent it can be decoupled, particularly with regard to the issue of our detection and tracking problems with the submillimeter size material?

P. Uhlir: I don't know how important it would be for definitional purposes as it is for actually implementing some agreement.

H. Baker: Bernie, I would answer your question by suggesting that when you're formulating a working definition, that you don't want to eliminate any possibilities for debris in your definition. Therefore, you make it as broad as possible and you don't link it with any requirements to measure. In other words, you include all the different possibilities that you can come up with. The link shouldn't be with measuring the amounts that you can find. Rather, the link should be with some of the other issues that I was raising yesterday, namely, what kind of control somebody has over an object. I didn't talk about it yesterday, but in my paper I mentioned that there is a permanent link between the definition of space debris and jurisdiction, control, and ownership issues. In order to successfully deal with space debris, especially when it comes to retrieval, you have to have some kind of test that can sever that jurisdictional and ownership link, and what I have proposed is the idea that there is no more control at all. Some people have suggested that there be a lack or no more permanent control over that vehicle. However, this test would not account for satellites which are launched and stored in orbit for future use. Satellites in these storage orbits are not controlled for a period of time, but their operators intend to reestablish control in the future. Therefore, the satellite only becomes space debris when there is absolutely no control over the satellite. And a definition of space debris with an "absolutely no control" test would break the link and would permit removal of the debris.

W. Lang: Very briefly on legal techniques concerning definition of main items of a treaty. We have had similar problems in various environmental treaties. There is certainly a risk in giving any definition at all, because it would never fit. The most relevant example is the Convention on Biological and Bacteriological Weapons. What is a biological weapon? What kind of substances are to be considered? So if ever we come to draft a model treaty or something like that I would be very hesitant to give, once and for all, a definition of orbital debris. I would rather suggest that one give a description whether for a certain period of time, there is an index and then that index could be reviewed every 2, 3, or 5 years by the main organ existing under the convention. I think this would be the way out in all those extended technical areas where developments are very fast.

P. Uhlir: I would guess that would hold true for all definitions.

J. Primack: I was just going to raise again the question of interference. The example that I've mentioned – and it's come up a couple of times – is the emission of charged particles by operating nuclear reactors or, for that matter, also the emission of gamma rays which interferes with essentially any satellite with sufficiently sensitive instruments that happens to be in the same region of space. The question I'm asking is under what legal doctrines should this be approached? The simplest solution would be to ban all nuclear reactors in Earth orbit or in a certain range of orbits. But to the extent that one wants to get at the problem rather than at a proposed solution, one could try to use some legal doctrine of non-interference, non-pollution, something of that sort. I think that is an interesting legal question, how to do that?

P. Uhlir: It also goes to the question of how you define debris since that is really not "harmful emission."

J. Primack: It's not clear that it is reasonable to consider that debris, but there is some relationship.

P. Uhlir: There is a definite relationship, but one can see a debris agreement not covering it, because it is a different set of issues.

VERIFICATION THROUGH PRE-LAUNCH INSPECTIONS

J. Primack: Another interesting set of legal questions that we've mentioned a couple of times, but have skirted around, has to do with pre-launch inspections. It's far easier to verify that a spacecraft meets certain performance requirements or design requirements if some sort of cursory inspection is allowed before a launch. That whole area is hard to get examined from a legal point of view.

P. Uhlir: That tends to be more of a political problem in terms of nations willing to have intrusive inspections of their facilities.

J. Primack: But the old attitude ten years ago was that, of course, the USSR and the US intelligence community would never allow such and such. In fact, intrusive inspections are becoming the norm.

P. Uhlir: It's not necessarily the national security angle as it is perhaps the proprietary issue.

J. Primack: Sure. But the International Atomic Agency does increasingly intrusive inspections and protects proprietary rights. One could easily imagine that some sort of international organization be chartered to do that. The U.S. government of course does all kinds of inspections involving proprietary

rights and strives to protect these rights. So it certainly isn't out of the question.

P. Uhlir: It isn't and it's certainly an option in terms of verification and enforcement. It's certainly a valid option, but I think a politically difficult one. For an initial kind of agreement, that [option] would be further down the road to agree on something like that. All the ramifications of it should be discussed. All the objections that States might have ought to be identified up front and analyzed as to how they might be resolved, because otherwise it won't go anywhere.

W. Lang: Just a complement of information on this issue of verification. Earlier this week, the draft of the Chemical Weapons Convention was practically finished with a very intrusive system of on-site inspections. So we will see how quickly this convention enters into force and how soon this verification machinery will really work. If it works then we'll see if it could be adapted to the needs of space objects.

P. Uhlir: The other issue here is even if inspections are allowed they're not always effective. For instance, countries tend to try to hide any transgressions in the area of nuclear facilities. I think the inspections have been of mixed effectiveness.

J. Primack: The simple way to deal with that is if, for example, there is an agreement that spacecraft are to be designed to do certain things – de-orbit, vent, or burn fuel – the burden of proof that the spacecraft design meets the desired objectives could be on the launching agency or on the spacecraft owner. They will simply have to present adequate proof. By putting the burden on the owner you avoid many of these problems. These are not areas where there is a great incentive to hide anything anyway.

P. Uhlir: It's also a matter of costs associated with implementing these provisions. That is not trivial.

D. Jacobs: I want to make one more comment back to the original question of what legal issues are of most interest to the scientists and go back to Howard Baker's point on the definition. It seems to be the definition or description that is important, because whatever you include or don't include will drive the design. If you are establishing a treaty, or whatever it is, that determines liability and responsibility for debris objects. If you don't include some objects, then the designers will have little incentive to try to control those kinds of objects, whether it be paint flakes or debris from a fragmentation or whatever. If you have liability for what is in that definition, they will obviously try to design to control that. I think that would be one of the most important issues, even though we'd like not to define things exactly or not to have complete descriptive lists. Whatever you don't include will probably not be designed to protect or to prevent.

LIABILITY ISSUES

P. Uhlir: That is a good point and I'd just like to add that the liability issue, in general, from anyone's perspective, whether you are a scientist, engineer, or lawyer is going to be the thorniest issue. There is a Liability Convention now in place, but it's not clear that it's either desirable to make it stronger in the outer space context or that it's possible to do that. It will always be something that is open to a great deal of interpretation. If you look at, for example, the Cosmos-954 incident in the late 70s as an obvious example of liability, even there, it wasn't really admitted or very satisfactorily resolved, and that was damage on the ground. If you're talking about damage in space it's really assigning blame and owning up to responsibilities. It is going to be a very difficult task no matter how many rules are in place.

J. Primack: In the case of Cosmos-954, the Russians said they'd do the cleanup. The Canadians said no, we'll do the cleanup and you pay for it. The Russians said we'll just pay a little bit, as much as it would have cost if we'd cleaned it up. It wasn't that they didn't admit liability, it was a dispute over costs.

P. Uhlir: I don't know the exact details, but I know the payment was quite small in compensation and I know that negotiations were protracted and it wasn't a clearcut decision, even though it seems like that was a fairly clearcut case under the Liability Convention.

J. Primack: The Liability Convention never made it clear because there isn't any dispute resolution mechanism you see. So, it was never clear what exactly the costs are that the country at fault has to pay.

P. Uhlir: This raises another issue. An effective dispute resolution mechanism is essential. It certainly is something that needs to be considered in any of these discussions.

D. Kessler: As far as liability in space, to me LDEF (Long Duration Exposure Facility) may make a good example. Here we've had a hundred scientists looking at LDEF for over a year now and we know the debris impacts but have no idea where they have come from. There is one particular surface on the rear that is really hard to get to, it takes a special orbit. We've narrowed it down to either European Space Agency or the United States. That is about as far as you can get. But for the rest of LDEF – or where most of the impacts are – we have no idea who is responsible for those.

P. Uhlir: It's a problem of sufficient evidence in a legal context.

S. Gorove: What happened in the Cosmos-954 accident was that the Soviet Union paid 3 million Canadian dollars. The Canadian claim was a little bit over 12 million Canadian dollars. The Soviet Union never admitted liability on the basis of the Liability Convention. They said they would settle the matter diplomatically, which they did. Now, actually the Convention also recommends first of all that you try to settle the matter on the basis of negotiations. So in a way, they followed that procedure, even though they denied liability in connection with the Convention.

RIGHT OF SELF DEFENSE

P. Uhlir: What is the scope of permissible action against threatening space debris? Someone raised a question with me earlier about who has right of way? Is there a duty to move out of the way of another object, or how does that work?

B. Bloom: Self interest is going to dictate.

P. Uhlir: Right. But regarding the scope of permissible action, well I guess if it were a military spacecraft and it had the capability of destroying the spacecraft that appeared to be heading on a collision course that this would be a self defense issue.

S. Gorove: I think you're absolutely right. It would be I think determined on the basis of the rules of self defense. So if it's a life threatening type of situation, it certainly seems to me that the party that is exposed to the threat should be able to take action against the threatening object.

P. Uhlir: Moving out of the way no doubt is the most preferable in a non-combat situation.

S. Gorove: Yes, but action is permissible even to the extent of destroying the threatening object. The only problem that you get into is the problem with Article 8 of the Outer Space Treaty, which says that the State party which registered the object would have jurisdiction and control and declares also that ownership of the object in space would not be affected. So ownership remains with the party that put up or registered the object unless otherwise decided or agreed upon. Therefore you would have these two principles conflicting. But I think that the self defense rule should prevail in that type of conflicting situation.

P. Anz-Meador: I'd like to point out, however, that destroying the spacecraft is probably not what you would want to do. You'll just wind up with a new debris cloud whose center of mass is still on a collision course with yourself.

H. Baker: It depends what kind of action you want to take. Professor Gorove was talking about a possible life threatening situation and there is the option he suggested. I believe I suggested yesterday that there may be enough evidence on the table to invoke sentence 3 of Article 9 of the Outer Space

Treaty which says that if a State believes that its space activity is going to cause potentially harmful interference with the space activities of another State, it should initiate consultations. There are a lot of weaknesses.

P. Uhlir: There's no time to consult.

H. Baker: That is something that has to be worked out. The more important one I would suggest is sentence 4, which says if a State has a reason to believe and can demonstrate – I say can demonstrate because I don't think they can show a reasonable belief any other way – that another State party to the Treaty has a space object that is causing potentially harmful interference then the first state can request consultations. That is one way to take action. I have no idea what the success rate of those actions will be, but as I suggested yesterday based on my analysis of that language, the consultation procedures are pretty weak, but this may be, in fact, an option. Whether it will be taken is another question.

P. Meredith: I don't think we can say there are clear legal rules for right of way today. I guess it hasn't really become a problem. But I'm envisioning a situation where it could become a problem, with multiple layers of low-earth orbit satellite systems, if satellites from a higher system are re-entering through the lower systems, or if satellites for a higher system are launched through the lower systems. In that event you'll have to address the right-of-way questions. We haven't gotten there yet from the legal viewpoint.

P. Uhlir: There are collision avoidance capabilities.

P. Meredith: Well that is true. But that is really not addressing it from the legal viewpoint.

P. Uhlir: It still is a self-interest angle.

P. Meredith: Sure. That I agree on.

B. Bloom: Maybe some technical information should be put on the table for the lawyers. If it's a large object, we're going to try to avoid it. If it's smaller than 10 centimeters, the entire decision time of the whole engagement scenario is on the order of 30 seconds. So for all these discussions about Article 9 and the interest of States and so forth, you need to consider that we're talking seconds.

H. Baker: I just want to clarify something about Article 9. Of course if it's 30 seconds away you're not going to be able to invoke any Treaty provisions. But what I'm suggesting is that Article 9 talks about potentially harmful interference, and I would underline potentially, so you don't have to wait until something gets to the point where you have to call together the parties to the Treaty. There is enough evidence now to put

forward an argument that there already is potentially harmful interference, based on some of Don Kessler's work on the cascading effect as well as on the critical density problem, assuming that the numbers are correct and that some of the plans for the Iridium constellation and other kinds of multiple satellite systems come to fruition.

W. Lang: I want simply to remind you of what has been said already yesterday and might have been forgotten. The launching State or the State under which authority the launching takes place is under legal obligation to prevent disturbances. Irrespective of the Outer Space Treaty we do have certain applicable customary rules of international law. So if the launching State is not able to prevent damage, then certainly the corresponding situation is that of destruction by the State that is threatened. But what happens if the State that feels threatened does not have the means of destroying the object by which it feels threatened?

D. Kessler: I have a comment on the potential for damage. If you use a space station as an example, there are several hundreds of objects at any one time that have the potential to create a collision avoidance situation. You don't know which ones they are, but basically anything that is at space station altitude, which will vary as a function of time. What you would have to do under the scenarios just described is tell everybody who crosses space station altitude that they have the potential of causing a problem and should be eliminated.

LIABILITY ISSUES

D. Jacobs: I would like to make another point with regard to liability. Everyone as far as I know has been running their collision avoidance on launch and on orbit calculations using the U.S. Space Command catalogue. This makes the U.S. Space Command the *de facto* supporter of the future insurance business and this type of future treaty. I don't know if they want to be in that business, but even if they are and everyone's using their catalogue, what if they miss something and someone runs into it. Are they then liable? Is someone going to sue the U.S. Government? At some point you would have to have some independent means of trying to perform this function.

J. Loftus: I find the discussion sort of interesting because it is static. It seems to me that one of the issues that lies in this question of liability is one of temporal precedence. If you go in harm's way, it seems to me it is inappropriate to hold the other party liable. That is implicit in much of this discussion. Those who are there first have established some right of precedence and if you then come along and operate in such a manner that they are a hazard to you, that is a situation that you have elected. Somehow or other this time sequence aspect needs to be dealt with because we're talking about systems in place which will have physical properties literally for millions of years, particularly in the geosynchronous region. That aspect has not been addressed at least in the discussions I've heard to date.

S. Chekalin (through translator): The most attention right now is being focused on the legal aspects of this problem. But from the technical point of view you can easily protect your object in space from debris less than 1 centimeter in size. For objects larger than 10 centimeters you can avoid a collision, but the most dangerous objects are those 1–10 centimeters in size. We should make careful estimation of how many such objects there are in space and avoid future contamination by debris of this size.

R. Reynolds: It seems to me that we're talking about two issues – debris control and traffic management – where the considerations of liability are quite different. If we were concerned about one intact object striking another, where liability can be determined, this would be a space traffic management symposium. However, we are meeting today because of fragmentations. The most likely source of loss of an operating spacecraft will be damaging impacts from fragmentation events – explosions now, collisions in the future if steps are not taken to control the environment. The originating structure for the fragment causing this type of loss will be known at best as a probability, and there could be a number of breakup events that might have comparable probabilities. Under the circumstances I don't see how you could assign liability for any given event.

P. Uhlir: Well certainly this is just another aspect of assigning liability, the problem associated with identifying the sources of debris.

R. Reynolds: If we're trying to work out some kind of international agreement, this is what we should be trying to work out.

P. Uhlir: Sure. At this point we're just discussing issues. We'll be addressing the terms of an agreement or the specific elements later.

D. Rex: In my view the whole discussion drifts a little bit too much into the area of liability for damage and compensation for damage. These are typical legal approaches for such things. The whole question of space debris, however, is only very loosely connected to such liability questions. We are trying to deal with the prevention of debris in the orbital environment. As has been said before by other colleagues, most of the damage that can be caused by space debris is from particles of indeterminate origin. So the liability question really is not relevant here. Also the whole question of damage to the Earth's surface is only loosely connected to the space debris issue. I would rather advocate that we concentrate in this discussion on the prevention aspects, rather than on the liability and compensation aspects.

P. Uhlir: I agree that would be a better focus. However, I think once we start having collisions that the injured parties will be seeking to find who is responsible and trying to assign blame. The issue of liability, therefore, will become an important aspect even if it will be very difficult to legally establish.

P. Meredith: I just have one follow-up to the comments by Joe Loftus. I don't know if I'm putting words in your mouth now, but did I understand you correctly when you said that a rule of customary international law already may have formed regarding some sort of time priority; in other words, any late-comer should move out of the way for a system already in an orbital path? Are you suggesting that should be the rule?

J. Loftus: No, I think in principle that what I'm saying is that issue at least must be addressed. If there is a set of activities going on in a particular orbit location and you subsequently enter that region you can hardly hold the person who is there liable for the risks that you have chosen to accept. I agree with the concern of Professor Rex that there is a tendency to confuse the liability and the environmental management issue. I think we can define courses of action that will protect the orbital environment. I think the nature of the activity when we fail to protect is such that we will never be able to identify a liable party in a reliable manner. So I think what you're going to wind up with is something equivalent to no-fault insurance. It's such an uncontrollable process from a liability point of view, that I think it is the wrong issue on which to focus. The primary issue is to control the environment.

P. Meredith: I agree with that.

TECHNIQUES FOR REMOVAL OF DEBRIS

J. Primack: A while ago we were discussing what the alternatives are for dealing with a space object that could strike your space object. We've skirted around this question several times during the last two-and-a-half days. I thought it might be useful to bring it up as an explicit question to be addressed by those people, if there are any here, who have actually looked into it. For large objects it's fairly clear that a possible solution – expensive, but at least possible – would be to use manned or robotic spacecraft to achieve the same orbit and capture such objects, de-orbit them, or something of that sort. The question I'd like to ask is are there any other alternatives, especially for the small objects, which are of course the most numerous and the most dangerous, are there any realistic alternatives to deal with them besides prevention? I've read a number of articles in the popular press in the last year or two about one or another gadget that some space engineer is proposing or has supposedly invented. None of them has made any sense to me. I'm curious if people who have looked into it a little more could comment on that.

D. Kessler: I've done some calculations as to how you might get rid of the small stuff. If you can eliminate all small objects that pass within about 10 kilometers of something, if you can make them magically disappear, you can sweep space fairly clean within a 10-year period or so. The engineering trick is to come up with some gadget that will do that without generating more debris. I'll leave that to the engineers. Things like a foam disk have been proposed, it's passive and it will slow a particle down as it passes through it so that it will drop out of orbit, but it will take out big objects as well as the small stuff.

J. Primack: You mean like satellites?

D. Kessler: Yes, satellites. The amount of mass you have to put in space, even something very thin, would take about 100 shuttle launches to do it. That is not really practical. A more practical system might well be something that is more active and can manoeuver to get whatever you want and sweep it out. That I think you could do with a fairly small instrument. You need something that can sense an object and get in the way. It's all futuristic and there is a lot of engineering to do, but conceptually it is possible.

J. Loftus: As Don mentioned there are a lot of schemes that have been looked at for the low earth orbit for such cleansing mechanisms. When we price them, they cost many times what the preventive measures would cost. There is another case we've looked at which is in the geosynchronous arc, where a large number of satellites and rocket bodies have been left derelict. We've done back-of-the-envelope studies that show if you were to send into the geosynchronous orbit a special-purpose device to take each of those derelict objects and put them in a graveyard orbit more than 300 kilometers above the geosynchronous arc, the cost would be essentially $30 million per object removed. This is quite prohibitive.

A. Reinhardt: I have a couple of quick points. First, back to the original question that started this discussion on whether it is feasible to knock out something on its approach. The time constraint is overwhelming. It wasn't until last year or the year before that we actually had an on-orbit manoeuver and that was by the shuttle. There were two of them recently and those were for large pieces. By my count, there have been roughly 130 breakup events (111 actual breakups through December 1992) that have occurred, with the uncertainties that are in the catalogue and the uncertainties about the breakups, models, and what is going on in general. Even if there is something that appears it might come close, we still are not going to move an operational unmanned system. We just hope it won't get hit. That is the current operational strategy. Another item is regarding liability. There have been many satellites that have broken up, but the Cosmos 1275 breakup is the only one that is suspected of having been the result of a collision. There have been a couple of others that we suspect may have been, but

there is not enough evidence to say what happened. The engineers will go through the evidence and ask whether the battery or solar panel failed. They will go through a detailed failure analysis and still be left with an unresolved circumstance. Back to your question, Dr. Primack, about clean-up systems, I've got a whole drawer full of things that people submit to me, and some of them are ground-based systems. Various technologies: lasers, kinetic devices, soft-catch mechanisms, some launched from the ground. With technology being developed, some of them might become economically feasible. Now you get into the arena that you still have a Congressionally mandated ban in this country on testing any potential ASAT (anti-satellite) weapons. Even if you are trying to remove space debris, you're going to have to convince Congress that it's not an ASAT weapon or has no ASAT applications, which of course would be viewed rightly that it does. There is a lot on the drawing boards out there, but until the mandate is lifted we're not going to test it. Maybe this group could work to that.

P. Uhlir: We've gotten through two questions in half an hour. I've got a whole stack left. Most of the rest of the questions are dealing with the issue of agreements and how best to arrive at solutions to the problems.

ORBITAL DEBRIS AND INTERNATIONAL ORGANIZATIONS

As acknowledged by various speakers, the U.S. is one of the nations reluctant to sign any international treaty dealing with "the common heritage of mankind," such as the Moon Treaty. Therefore, what is the chance of getting some kind of international treaty dealing with mitigation of orbital debris?

I would say that there is a difference in terms of the more idealistic and imposed resource sharing legal regimes, such as the Moon Treaty or the Law of the Sea Treaty, than in the case of debris in which there is a great degree of self-interest involved in terms of trying to arrive at solutions. I don't think that the U.S. reluctance to sign the Moon Treaty is at all indicative of whether we'd be more inclined to sign an agreement limiting orbital debris. But there is still the issue of when the U.S. will be willing to formally address the issue in a multilateral agreement. That is something we've been discussing throughout the morning. The writer also said that he strongly supports the idea of creating a consultative committee on orbital debris in order to prepare the ground on this delicate topic. I think certainly that is the kind of minimum action that should be taken on a multilateral level and not simply on a bilateral one in order to arrive at common understandings, rather than on separate understandings. Are there any comments on that particular issue?

J. Loftus: I don't know that it's a comment so much as a question and may reflect my ignorance more than anything else. But it seems to me there is a presumptive question as to what body if any, has the appropriate charter and authority to address such an issue. The International Telecommunications Union (ITU), for example, has some interest in positional location of spacecraft, but it is derivative, not primary. The ITU controls position primarily to dictate radiofrequency interference conditions and not for the sake of maintaining positional control. I can't identify, at least in what I've seen, anybody or any body that has an agreed upon authority to approach such issues.

P. Uhlir: Orbital debris issues generally?

J. Loftus: Well, essentially if you take that and extend it, what you're really talking about is an organization that would have a counterpart authority in space to those authorities exercised by the International Air Transport Association and the International Civil Aeronautics Organization wherein we do establish positional criteria for how aircraft shall be managed in air space nationally and internationally. So there is a set of rules by which we do those things. There is no counterpart that I've been able to identify that has an existing charter [for regulating activities in space] and so one of the questions is, does that mean there has to be a *de novo* organization?

P. Uhlir: There may in fact need to be, but the United Nations Committee on the Peaceful Uses of Outer Space (UN COPUOS) certainly has subject matter jurisdiction to discuss these issues. The question is whether it has enough authority to establish another organization that could actually perform such functions. I don't believe there are any other international organizations that have the established subject matter, jurisdictional authority as a forum to comprehensively address these issues.

P. Meredith: No. I agree with you there probably isn't. But the question is, is it necessary, do we need a standing international regulatory body? Probably not. A better solution might be just an international convention that is implemented in each of the nations through their regulatory agencies, of which we in the United States already have several, governing or regulating space activities. I would suggest that approach rather than a standing international regulatory agency such as the ITU for space debris anyway.

D. Jacobs: I just wanted to sort of repeat what Pam said. I asked the same question of our attorneys at NASA and they had recently gone through this question because of the last World Administrative Radio Conference (WARC). There was a lot of discussion regarding geostationary slot allocation that was brought up by the developing world. The U.S. position on that was that no one has authority concerning physical allocation; they only have frequency management authority. There is a vacuum in that area and therefore the ITU could not assign physical spots as the third world was asking for. You could

create conventions or agreements, however, that would do this outside of a standing organization.

P. Uhlir: I would like to add that assigning physical spots is really counter to existing space law because space is the "province of mankind" and there are no ownership rights to any part of it. To assign it is to assume that there is the ability to have exclusive permanent right to ownership of a part of something that is not subject to ownership.

S. Gorove: The UN General Assembly has recommended that the space debris issue is an appropriate subject for discussion in the Committee on Peaceful Uses of Outer Space. Now this would mean that if U.S. opposition blocks placing the item on the agenda of the full Committee or one of its Subcommittees, then this question of whether there should be some form of organization established outside of the UN COPUOS would be discussed there. Apart from the public international organizations, of course, you have private international organizations which have been involved a great deal in the discussion of these issues, one of these being the International Astronautical Federation (IAF) and the other the Committee on Space Research (COSPAR). Both of those organizations have been discussing matters pertaining to space debris and what to do about it. These discussions have been mostly on the technical level, but not entirely because the IAF is also discussing the legal issues.

P. Uhlir: There are some other professional organizations, international and national, that are discussing this issue as well, but I think those tend to be more information dissemination activities, addressing some specific technical issues, rather than a political process that has the ability to impose solutions that all parties would agree to. But that is an important part of the whole process in any case and certainly should be encouraged.

W. Lang: On this issue of international organizations, there are organizations that are now in charge of the issue as just explained, but nothing prevents the spacefaring nations from getting together to agree among themselves and to draft a treaty. What I believe is that even the spacefaring nations among themselves would need some kind of international organization simply to monitor compliance with the treaty and in order to further develop the instrument. So I do not believe that they could simply get along without some permanent body in which they themselves are represented. With regard to the reference to private organizations, I wish simply to record the example of marine pollution, in which you have the official level, the so-called MARPOL Convention of 1973, and then you have a number of private agreements among the owners of the tankers and among the ship building companies, through which funds are provided by the private interests that cover whatever accidents may happen. So I think we have enough models, especially in the field of marine pollution, on which we could draw when drafting an instrument for orbital debris.

LIABILITY ISSUES

P. Uhlir: The following question addresses the limitations of the models. That is, are there any technical, economic, or legal issues that make orbital debris a unique environmental issue such that the lessons learned from other environmental regulatory contexts do not apply? For instance, what are the differences, rather than the similarities in terms of applying maritime, or ozone, or Antarctic, or whatever legal regime one can think of as analogous? What are the limitations of grafting such an approach to this particular context? I'm sure we're not going to answer that immediately here, but it is an important issue that needs to be analyzed in detail. It is perhaps the kind of topic that we might want to consider as a follow on to this Conference, to start looking at the details of possible solutions and both the similarities and differences of other contexts.

H. Baker: Paul, I'd just like to respond to the concerns that have been expressed about where our priorities lie. I would agree that we don't want to make liability the focus and that we do want to make environmental protection the focus. But one of curious things about international law is that if you're going to have a treaty on environmental protection, you're going to need some kind of general principle about responsibility. States should endeavor to make their best efforts to mitigate or reduce the amount of space debris, but without a specific regime to enforce that, you're not going to get anywhere. In order to carry out the responsibilities somewhere along the line there has to be consideration given to liability, if not in the Liability Convention, then perhaps as part of an environmental agreement. In other words, if you don't carry out your responsibility, if you don't remove your object, what is the sanction?

D. Jacobs: I'd like to respond to that by disagreeing with it to some extent. I think that if you do have to approach it from an environmental point of view, you do have to have responsibilities to meet certain environmental standards in terms of preventing things from blowing up and not spreading around operational debris and that sort of thing. You can do that and hopefully come up with sanctions, although I'm doubtful about sanctions. But I don't think the liability question will be helpful at all based on the points that Al Reinhardt made. From a legal proof point of view, I don't think you could ever prove legally that someone's fragment was responsible for a breakup. They never see the breakups happen. With the radars, they see a cloud where there used to be an object and they assume that this happened. From a legal point of view, I doubt that you could ever prove that. So I'm not sure the liability issue would be of much help. What would help is the responsibility to meet certain environmental standards.

H. Baker: Let me respond very briefly by pointing out that there are at least two kinds of liability. One is negligence which is based on fault, in which case you have to prove as you indicated and it's going to be impossible when you can't even identify the part let alone tell who it belongs to. The other liability is strict liability. Strict liability requires damage, a cause of damage and a connection between the damage and the cause.

P. Uhlir: Then there is collective or distributed liability, which is the concept behind no-fault insurance. That is where everyone assumes their own liability. There is a question here on liability issues regarding damage to environmental commons. In order to have standing to bring a claim in law you normally have to demonstrate personal harm. In the case of damage to commons it's generally ruled that no one has standing to bring a claim because that harm hasn't happened to them specifically. So that would be the short answer to this question.

W. Lang: The only case that comes close to a space treaty on this subject matter would be the Treaty on Nuclear Accidents because there are fewer States with nuclear industries that may possibly cause some hazards than States that have no nuclear industries and can only be victims of nuclear hazards. There the problem has been settled, and it's a question of numbers, balance, and justice.

ISSUES IN NEGOTIATING A DEBRIS AGREEMENT

P. Uhlir: There is a follow-up question that has been posed. Should the space debris problem be left only to the major space faring nations to solve, or can the other nations make a contribution? In the case of the Treaty on Nuclear Accidents, was that a broad multilateral, or was it just the nuclear countries?

W. Lang: The two conventions on early notification and on assistance in case of a nuclear accident are worldwide treaties to which all States belong. There are three categories of States: the nuclear weapons States, which have a special regime, the nuclear industry States, and those that are simply ordinary citizens of the world without any nuclear industry. With some deficiencies, these treaties have been concluded and they may be a model for space debris.

P. Uhlir: For negotiating an orbital debris treaty, you would say it would be best to involve just the spacefaring nations, but then that agreement should be open to signing by all countries?

W. Lang: Yes. Let us take a specific case, the Convention on Chemical Weapons. It was drafted by 39 States, but it shall be signed and ratified by 100 States. So I could imagine that this whole group of States, mainly spacefaring, could take the lead and that the other States sooner or later would jump on the wagon. I think this is possible and as I explained yesterday you can settle the problem between spacefaring and non-spacefaring nations by having two different categories of States and two different categories of rights and duties. Those non-spacefaring nations will have very few duties and practically no rights, whereas the others will have more duties and more rights.

S. Gorove: I would like to come back to the question in relation to the commons. Under the Liability Convention there is no action you could take because the commons are not included. So far as general international law is concerned, I think it is fair to say that if you had substantial damage to the commons then action would lie in relation to any party that may have an interest in making use of the commons.

P. Uhlir: There is a significant difference of opinion both in the U.S. and internationally about the timing of bringing the orbital debris problem to established multilateral fora for discussion and action. At the same time, practically all experts in this area agree that a formal multilateral approach to effectively addressing the orbital debris problem will be necessary. The question to the technical community therefore is what specific requirements need to be satisfied before the topic is considered ripe for formal multilateral discussion?

D. Rex: Well if we really go along the lines that have been proposed by Mr. Lang and Ms. Meredith to choose a framework treaty approach, which from time to time can be supplemented by additional protocols, then there is no such problem really because as we all know there are some steps that are clearly visible, clearly defined from the technical point of view, and I think there is not much divergence of views on that. Then there are other steps where certainly more research has to be done and we will not get consensus any time soon. The first step would be to make that framework treaty that was proposed to us yesterday. I don't see that there is any need for delay there because the contents of that would be pretty general, and would reflect no more than what has been discussed here on a consensus basis. I don't think there is anything that is controversial in such a text. Only when we come to the protocol texts there may be some divergence of views. By using that approach we have the opportunity to start now with the drafting of a text to a treaty.

J. Loftus: I think that one of the questions has been a definitional question. Many people have wanted to address orbital debris in various legal and political contexts without having very much of a technical definition of what the issues are. I think many of us who are in the technical community have been reluctant to see the subject introduced in those contexts, particularly by those who have expressed what I would characterize as a moralistic posture, that any contamination is unac-

ceptable. We have wanted to see a consensus in the technical community as to what the issue really is and what is effective as both a preventive measure and what is necessary to achieve some adequate control. This is important so that there is some corpus of technical material to be addressed so that things are grounded in some pragmatic issues and not dealt with in a purely political sense.

P. Uhlir: Unfortunately, you weren't here yesterday. We had a fairly extensive discussion of definitions and the need to define terms and issues. In order to begin arriving at a consensus or international agreement on the issues, they have to be discussed on a technical level in a multilateral way rather than in bilateral types of processes. Even within this framework type of approach, one can make it a technical approach, but it would have two benefits to it. One is it would be multilateral so that all the interested parties would be talking at the same time and defining the issues in the same way. Otherwise you will not have a common definition if you don't have all the parties discussing it at the same time. Secondly, if you formalize the process you will actually start making some progress in a multilateral way on the technical issues as well. So I think there are arguments of favor in beginning a multilateral process. It will speed up actions that everyone already can agree to and institute those internationally instead of simply doing it on an ad hoc bilateral basis, which is less effective.

J. Loftus: The problem I have with that is that in point of fact the U.S. space surveillance network is the primary source of information. There is a comparable network that has been operated by the former Soviet Union, which has some sensors that can provide this information. Apart from those two networks there are occasional facilities that can provide certain point observations. So it is not an activity to which very many people really have technical access and who can get anything to truly understand the issue. I also would not characterize the discussions technically as being ad hoc or merely bilateral. So I think that we really have two different discussions in parallel here. One is what can be done in terms of principles of law and international procedure and another one about what can be done with regard to the physics of the problem.

D. Jacobs: I would like to follow-up on that. Let me speak personally for a moment, not from NASA's position. Actually you [Uhlir] weren't here the first day during my presentation in which I talked about our bilateral relationships, but then stated that I believe we will have a multilateral organization formed within six months to a year on the technical level to do exactly what you're talking about. I think that is in the works and it will start soon, hopefully. Regarding the framework type of agreement, I have to agree with Professor Rex that I personally see no reason why a general framework of intentions to work on environmental issues couldn't be done now. There are nontechnical issues involved, so that you could state the basic principles and conditions, and then work out the protocols as they come about. On the U.S. side that would require several agencies to agree on this matter and hopefully all those agencies or most of them are represented at this Conference. Hopefully, when we return to Washington we can begin discussions about that. I don't know how soon the protocols would follow that. We've only identified one or two things that everyone agrees we could do technically. The others are still somewhat down the road.

V. Utkin (through translator): The last 35 years human beings have been dealing with space. Right now, we can get results from our research in space and make some decision about the need to protect from space contamination, not only to remove what exists in orbit. Most of all we need to do this for future space shuttle operations. We need to make better designs and make some convention in order to control it. It's very complicated at this time now to remove this debris from the space, but from my point of view, the most important problem is that more countries are going into space. The former Soviet Union made a lot of mistakes which were done before we knew better. That is why the most important thing is to protect developing countries from making these mistakes, to give them some suggestions about this, and to develop some criteria about contamination and debris. If you start to work on a new rocket it takes 5–7 years to develop. During this time we can improve the situation if we develop new criteria now. It will be easier to make decisions from the point of view of law and liability, because the technical side of this problem is the most important to resolve initially. We can make any decision for liability and so on, but if we will continue to develop contaminated rockets any such decision will be hopeless. We can either go another round of spacecraft development and make the same mistakes as before, or we can learn from our experience and give the benefits of this experience to developing countries.

Suggested Further Reading on Orbital Debris

1. *Orbital Debris*, D. J. Kessler, and Shin-Yi Su., (eds.), NASA Conference Publication #2360, Proceedings of a workshop sponsored by NASA-JSC, Houston, Texas, July, 1982.

2. *Space Debris, Asteroids and Satellite Orbits*, D. J. Kessler, E. Grün, and L. Sehnal (eds.), COSPAR Advances in Space Research. Vol. 5, no. 2. (Elmsford, NY: Pergamon Press, 1985).

3. *Artificial Space Debris*, Nicholas L. Johnson and Darren S. McKnight, (Malabar, FL: Orbit Book Company, 1987).

4. *Cosmic Dust and Space Debris,* J. A. McDonnell, M.S. Hanner, and D. J. Kessler, (eds.), COSPAR Advances in Space Research. Vol. 6, no. 7. (Elmsford, NY: Pergamon Press, 1986).

5. *Space Debris – A Report from the ESA Space Debris Working Group.* ESA SP-1109, November, 1988.

6. *Space Debris: Legal and Policy Implications*, Howard A. Baker, Martinus Nijhoff Publishers, Boston, 1989.

7. *Orbital Debris from Upper-Stage Breakup*, Joseph P. Loftus, Jr., (ed.) Progress in Astronautics and Aeronautics, Vol. 121 (Washington, DC: AIAA) 1989.

8. *Report on Orbital Debris*, Interagency Group (SPACE) for National Security Council, Washington, D.C., February 1989.

9. *Smaller Solar System Bodies and Orbits*, S. H. Runcorn, M. H. Carr, D. Mohlmann, H. Stiller, D. L. Matson, B. A. C. Ambrosius, and D. J. Kessler, (eds.), COSPAR Advances in Space Research. vol. 10, nos. 3, 4, Elmsford, NY: Pergamon Press, 1989.

10. *Environmental Aspects of Activities in Outer Space: State of the Law and Measures of Protection*, Karl-Heinz Bocksteigel, (ed.), IBSN 3-452-321356-0, Bonn: Carl Heymanns Verlag KG, 1990.

11. *Space Program: Space Debris, a Potential Threat to Space Station and Space Shuttle*, General Accounting Office, GAO/IMTEC-90-18, Washington, DC: US GPO, 1990.

12. *Orbital Debris: Technical Issues and Future Directions*, Andrew Potter (ed.) (Proceedings of a conference sponsored by AIAA, NASA and DOD in Baltimore, MD, April 16–19, 1990) NASA Conference Publication (CP), 10077, 1992.

13. *Orbital Debris: A Space Environmental Problem – Background Paper*, Office of Technology Assessment, OTA-BP-ISC-72, Washington, DC: US GPO, September 1990.

14. *Orbital Debris Mitigation Techniques: Technical, Economic, and Legal Aspects*, AIAA SP-016-1992.

15. *Orbiting Space Debris: Dangers, Measurement, and Mitigation*, R. T. McNutt, Report #26, Program in Science and Technology for International Security, Massachusetts Institute of Technology, Cambridge, Mass., December 1992.

16. *Space Dust and Debris*, D. J. Kessler, J. C. Zarnecki, and D. L. Matson (eds.), COSPAR Advances in Space Research, Vol. 11, No. 12 (Elmsford, N.Y.: Pergamon Press, 1991).

17. *Proceedings of the First European Conference on Space Debris*, (W. Flury, ed.), Darmstadt, Germany, 5–7 April 1993 (ESA-SD-01).